DESIGNPREIS
2007

Designpreis der Bundesrepublik Deutschland 2007

Design Award of the Federal Republic of Germany 2007

Herausgeber Editor Rat für Formgebung — German Design Council

Verlag Publisher Birkhäuser Verlag AG — Basel, Boston, Berlin

DESIGNPREIS
2007

Designpreis der Bundesrepublik Deutschland 2007

Design Award of the
Federal Republic of Germany 2007

Bundesministerium
für Wirtschaft
und Technologie

TECHNISCHE INNOVATION UND GUTES DESIGN – das sind die entscheidenden Instrumente, um Produkte und Kommunikationsleistungen aus dem Massenangebot hervorzuheben und ihre Marktchancen zu steigern. Gutes Design trägt zu Akzeptanz bei den Kunden bei, fördert die Produkt-Qualität, hilft dem Unternehmer, bessere Preise zu erzielen und transportiert das Image eines Unternehmens. Kurz gefasst: Design ist ein unverzichtbares Marketing-Instrument! ——— Der *Designpreis der Bundesrepublik Deutschland* wird jährlich für hervorragend gestaltete Produkte und Kommunikationsleistungen an Unternehmen und Designer sowie für besondere Verdienste für die Vermittlung von Design in die Wirtschaft an Persönlichkeiten oder Institutionen als Ehrenpreis vergeben. Für Produkte und Kommunikationsleistungen können bis zu 5 Preise in Gold und 20 Preise in Silber an Unternehmen und Designer verliehen werden. Für besondere Verdienste um die Vermittlung von Design kann darüber hinaus noch eine Persönlichkeit oder Institution den *Designpreis der Bundesrepublik Deutschland* erhalten. ——— Dem *Designpreis der Bundesrepublik Deutschland* kommt als Preis der Preise eine herausragende Bedeutung für innovative Gestaltungsleistungen in Deutschland zu. Der *Rat für Formgebung/ German Design Council,*

TECHNICAL INNOVATION AND GOOD DESIGN – these are the deciding instruments that set products and communication services apart from the masses available, thereby increasing their market chances. Good design plays a role in how customers accept products, promotes quality, helps companies achieve better prices and transports a corporate image. In short: design is an indispensable marketing instrument! ——— The *Design Award of the Federal Republic of Germany* is an honourary award presented each year to companies and designers for outstandingly well designed products and communication services, as well as to persons or institutions for special services in creating an awareness of design in industry. For products and communication services there are up to 5 Gold Prizes and 20 Silver Prizes that can be awarded to companies and designers. In addition, for special services in creating an awareness of design in industry, a person or institution can receive the *Design Award of the Federal Republic of Germany.* ——— As the "prize of prizes" the *Design Award of the Federal Republic of Germany* is out of outstanding importance with regard to innovative design in Germany. The *German Design Council* stages the competition on behalf of the *Federal Ministry of Economics and Technology.* Only products and

Grußwort Word of Greeting

Michael Glos Bundesminister für Wirtschaft und Technologie
Federal Minister of Economics and Technology

richtet den Wettbewerb im Auftrag des *Bundesministeriums für Wirtschaft und Technologie* aus. Am Wettbewerb zum Preis können ausschließlich Produkte und Kommunikationsleistungen teilnehmen, die in den vorangegangenen fünf Jahren bereits eine regionale, nationale oder internationale Designauszeichnung erhalten haben, aus dem europäischen Wirtschaftsraum stammen und auf dem deutschen Markt angeboten werden. Die Nominierung erfolgt überwiegend über die Wirtschaftsminister, -senatoren der Bundesländer und im Auftrag des *Bundesministeriums für Wirtschaft und Technologie* durch das Präsidium des *Rat für Formgebung.* ———— Über 900 Nominierungen wurden für den Wettbewerb 2007 eingereicht. Diese rege Beteiligung ist ein erfreulicher Beleg dafür, dass immer mehr Unternehmen die Marktbedeutung des Designs als Wettbewerbs- und Innovationsfaktor erkennen. Zu diesem Erfolg tragen die vielfältigen öffentlichen Maßnahmen der Design-Initiative der deutschen Wirtschaft sowie des Bundes, der Länder und Design-Institutionen bei. ———— Der diesjährige Wettbewerb hat deutlich gemacht: Es kann sich sehen lassen, was die deutsche und europäische Wirtschaft in punkto Design zu bieten hat! Möge der *Designpreis* den Gewinnern, aber auch denen, die diesmal noch nicht zu den Ausgezeichneten gehören, Ansporn sein, Design konsequent und vielleicht noch offensiver als bisher als Wettbewerbsfaktor einzusetzen. ———— Ich freue mich, dass auch in diesem Jahr die prämierten Produkte und Kommunikationsleistungen auf einer Wanderausstellung einer breiten internationalen Öffentlichkeit vorgestellt werden.

communication services which, over the past five years, have won a regional, national, or international award, come from the European Economic Area, and are available on the German market, are entitled to take part. They are nominated through the Ministers of Economics, Economics Senators of the federal states, and on behalf of the *Federal Ministry of Economics and Technology*, by the Chair of the *German Design Council.* ———— Over 900 nominations were submitted for the 2007 award. This high level of participation is gratifying proof that an increasing number of companies is recognizing the market importance of design as a factor in competition and innovation. The numerous public measures from the Design Initiative of German Industry as well as the German government, the federal states and design institutions all play a part in this success. ———— This year's competition has revealed one thing: German and European can be proud of what is produces in the field of design. May the *Design Award* be an incentive both to the winners as well as to those who this time were not among those commended, to make use of design as a competition factor consistently and perhaps more forcefully than in the past. ———— It gives me enormous pleasure that this year, once again, the winning products and communications services are being presented to a wide international audience in a touring exhibition.

EMOTIONEN — DIFFERENZIERUNG DURCH DEN PREIS DER PREISE Über Design gibt es nach wie vor sehr unterschiedliche Vorstellungen. Weit verbreitet ist jene, dass Design ein Produkt zwar verteuert, aber seinem Nutzer einen emotionalen Mehrwert bietet, der den höheren Preis durchaus rechtfertigt. Avancierter ist hingegen jenes Verständnis, dass das Design den Herstellungspreis eines Produktes kaum – zumindest nicht negativ – beeinflusst, wohl aber ein entscheidendes Differenzierungsmerkmal von Produkten und effektives Positionierungsinstrument von Unternehmen im globalen Wettbewerb darstellt. Der Schlüssel zu einer solchen Erkenntnis liegt in einer professionellen, strategischen Beschäftigung mit Design, die die Disziplin als integralen Bestandteil einer umfassenden Unternehmenspolitik betrachtet und auf langfristigen, ganzheitlichen Zielen beruht. ———— Differenzierung durch Design zu ermöglichen, bedeutet hier nicht weniger als ein markenspezifisches, ganzheitliches Erscheinungsbild zu definieren. In der Automobilindustrie beispielsweise ist so in Folge der technischen Angleichung von Produktleistungen auf technologisch

EMOTIONS — DIFFERENTIATING WITH THE AWARD OF AWARDS Ideas about design vary greatly, today just as much as in the past. One very widespread idea is that, although design increases the price of a product, it offers its user an emotional added value, which thoroughly justifies the higher price. However, more advanced is the view that design hardly has an influence, or at least not a negative one, on the production price of a product, but probably represents a decisive distinguishing feature of products and effective positioning instrument of companies in the competitive global market. The key to a realization such as this lies in engaging with design in a professional, strategic manner, considering discipline as an integral part of a comprehensive corporate policy and looking towards long-term, holistic goals. ———— Here, enabling differentiation through design means nothing less than defining a brand-specific, holistic image.

Designpreis der Bundesrepublik Deutschland 2007

Design Award of the Federal Republic of Germany 2007

Andrej Kupetz Geschäftsführer des Rat für Formgebung
Managing Director of the German Design Council

hohem Niveau und schwindenden Möglichkeiten, über die Preisgestaltung Wettbewerbsvorteile zu erzielen, das spezifische Markendesign der Produkte zum wichtigsten Differenzierungsmerkmal überhaupt geworden. ——— Aber auch für andere Branchen – und das zeigt der Wettbewerb um den höchsten deutschen *Designpreis* erneut – gilt: Je stärker das Preis-Leistungsverhältnis in einer Branche ausgeglichen ist, desto entscheidender ist die Rolle jener Faktoren zu bewerten, die geeignet erscheinen, für die Kunden erkennbare Differenzierungsmerkmale im jeweiligen Wettbewerbsumfeld zu schaffen. Wenn Produkte oder Dienstleistungen funktional identisch sind, müssen für die Kunden andere Merkmale kreiert werden, um sie für sich einzunehmen. Die Qualität der emotionalen Ansprache – über Design, Kommunikation, Service – bildet für den Kunden die Grundlage zu entscheiden: Will ich oder will ich nicht. ——— Gerade für die emotionale Differenzierung und die Schaffung eines positiven Images spielen *Designpreise* eine entscheidende Rolle. Sie kommunizieren die Designkompetenz von Unternehmen gegenüber Kunden, Partnern und nicht zuletzt gegenüber den Mitarbeitern eines Unternehmens selbst. Sie sind nicht nur für Unternehmen als Marketinginstrument von nicht zu unterschätzender Bedeutung, sondern auch für Designer eine der wenigen Möglichkeiten überhaupt, Marketing für sich zu betreiben.

In the car industry, for example, the specific market design of a product has become the most important distinguishing feature as a consequence of similar product performance on a high technological level and decreasing possibilities of creating a competitive advantage through price. ——— Yet for other sectors, too, – and this has again been shown by the competition for the top *German design award* – we can say: The more the price-performance ratio in a sector is balanced, the more decisive is the role of valuing those factors that appear to be suitable and creating recognizable distinguishing characteristics for customers in the relevant competitive environment. If products or services are functionally identical, other features need to be created for the customer in order to win them over. For the customer, the quality of the emotional address – through design, communication and service – forms the basis for their decision: Do I want it or not. ——— It is precisely in emotional differentiation and the creation of a positive image that design awards play a decisive part. They communicate companies' design expertise to customers, partners and not least their own employees. Their significance as a marketing instrument must not be underestimated, not only for companies, but also for designers, as they present one of the few opportunities for them to market themselves. After all, customers of any age will see design prizes as positive image factors. In addition, consumers then know that products that have been awarded design prizes are of a particular quality. ——— For us therefore it is all

Schließlich werden *Designpreise* von Kunden über alle Altergruppen hinweg als positive Imagefaktoren wahrgenommen. Auch die Botschaft, dass designprämierte Produkte über eine besondere Qualität verfügen, ist bei den Konsumenten angekommen. ———— Umso wichtiger ist es für uns, den *Designpreis der Bundesrepublik Deutschland* für seine Nutzer zu profilieren und seine Bedeutung im internationalen Rahmen zu erhöhen. Entscheidend ist die dabei Qualität der Ausschreibung und die breite Kommunikation der Ergebnisse. Dass wir hier auf dem richtigen Weg sind, zeigen uns nicht zuletzt die Ergebnisse einer unabhängigen GFK-Studie, die dem *Designpreis der Bundesrepublik* in punkto Image, Seriosität, Unabhängigkeit und Integrität sowie hinsichtlich der Qualität der Teilnehmer Bestnoten ausgestellt hat. ———— Genauso bedeutend ist es, die Ausschreibung mit den Bedürfnissen ihrer Zielgruppen abzugleichen und – wenn nötig – zu aktualisieren. Mit den im Jahr 2000 vorgenommenen Änderungen bei Deutschlands offizieller Designauszeichnung – die der *Rat für Formgebung* seit immerhin 1969 für das Bundesministerium für Wirtschaft durchführt – vom *Bundespreis Produktdesign* zum *Designpreis der Bundesrepublik Deutschland* lag zum Beispiel der Gedanke zu Grunde, dass nicht mehr allein das Produktdesign betrachtet werden kann, sondern alle kommunikativen Maßnahmen eines Unternehmens einer Bewertung zu unterziehen sind, um ein ganzheitliches Verständnis für die Qualität der Nominierungen zu entwickeln. ———— Umso mehr freut es uns, dass im Jahr 2007 mit über 200 Ein-

the more important to make a profile of the *Design Award of the Federal Republic of Germany* for its users and to increase its significance in the international arena. The quality of the competition organisation and widespread communication of the results are decisive factors. The fact that we are on the right track here is illustrated not least by the results of an independent GFK study, which has given the *Design Award of the Federal Republic* top marks in terms of image, respectability, independence and integrity as well as the quality of participants. ———— Equally as important is to bring the organisation of the competition in line with the needs of its target groups and, if necessary, to update them. For example, the changes introduced in 2000 to Germany's official design award, which the *German Design Council* has been in charge of for the Federal Ministry of Economics since 1969 and where the *Federal Award for Product Design* became the *Design Award of the Federal Republic of Germany*, were essentially the result of the realization that considering product design alone was no longer enough. Instead, all a firm's communicative measures had to be evaluated in order to develop a complete understanding of the quality of the nominations. ———— We are even more delighted that, with over 200 entries (from a total of 940) in the area of communicative design, the number of participants in 2007 compared to

sendungen (von insgesamt 940) im Bereich Kommunikationsdesign der Anteil gegenüber dem Vorjahr noch einmal erheblich gestiegen ist. Auch die Tatsache, dass 2007 die internationale Jury bei 25 zu vergebenden Preisen erstmals sieben Arbeiten aus dem Bereich Kommunikationsdienstleistungen mit dem *Designpreis der Bundesrepublik Deutschland* ausgezeichnet hat, zeigt uns, dass der Preis zunehmend den ganzheitlichen Anspruch des Designs repräsentiert. ——— Eines wird sich aber auch in Zukunft nicht ändern: Die Positionierung des *Designpreises der Bundesrepublik Deutschland* als Preis der Preise. Nach wie vor werden die Teilnehmer zum Preis nominiert. Nur, wer bereits mindestens eine Auszeichnung errungen hat, ist berechtigt, am Wettbewerb um den *Designpreis der Bundesrepublik* teilzunehmen. Jede Nominierung – die durch die Wirtschaftsminister der Länder und im Auftrag des Bundesministers für Wirtschaft durch den *Rat für Formgebung* erfolgt – bedeutet dabei bereits eine Auszeichnung für sich. Denn die Nominierten haben ihre überragende Designqualität bereits in vielen hochkarätigen Wettbewerben unter Beweis gestellt. Diese Preise sind für uns bedeutende Qualifizierungsinstrumente, ohne die das Verfahren um den *Designpreis der Bundesrepublik Deutschland* nicht funktionieren kann. Im Interesse der klaren Differenzierung unserer Auszeichnung und im Interesse ihrer Nutzer wird auch in Zukunft die Zahl der Preisträger mit 20 Preisen in Silber und 5 Preisen in Gold beim *Designpreis der Bundesrepublik Deutschland* überschaubar bleiben.

the prior year has once again risen considerably. In addition, the fact that in 2007 the international jury, with 25 awards to bestow, first selected seven works from the field of communication services to receive the *Design Award of the Federal Republic of Germany* shows us that the award is increasingly coming to represent design's holistic claim. ——— Yet one thing will never change, not even in the future, namely, the status of the *Design Award of the Federal Republic of Germany* as the award of all awards. Participants are still nominated for the award, as before. Only, those who have already won at least one award are entitled to enter the competition for the *Design Award of the Federal Republic*. Each nomination, which is made by the economics ministers of the Länder and on behalf of the *Federal Minister of Economics* by the *German Design Council*, is therefore already an award in itself. This is because those nominated have already proven the outstanding quality of their design in numerous top-class competitions. These awards are important qualifying instruments for us, without which the procedure of the *Design Award of the Federal Republic of Germany* cannot function. In the interests of clearly differentiating our award and in the interests of its users, in the future too, the number of winners of the *Design Award of the Federal Republic of Germany* will remain clear, with 20 silver prizes and five gold prizes.

v. l. n. r. James Irvine, Bodo Hacke (Bundesministerium für Wirtschaft und Technologie), Reinhard Binder, Nicolette Naumann, Torsten Neeland, Andrej Kupetz (Geschäftsführer des Rat für Formgebung), Robert Thiemann, Barbara Friedrich, Herbert H. Schultes

Bericht der Jury
Report of the Jury

Herbert H. Schultes Vorsitzender der Jury
Chairman of the Jury

DER DESIGNPREIS DER BUNDESREPUBLIK DEUTSCHLAND ist die höchste offizielle Design-Auszeichnung des Landes und wird vom *Rat für Formgebung/German Design Council* im Auftrag des *Bundesministeriums für Wirtschaft und Technologie* verliehen. ——— Seit dem Jahr 2006 wird der *Designpreis der Bundesrepublik Deutschland* jährlich verliehen. Neben Produkten und Leistungen aus dem Bereich Kommunikationsdesign wird auch eine Persönlichkeit oder eine Personengruppe für ihre Leistungen im Bereich Design ausgezeichnet.

THE DESIGN AWARD OF THE FEDERAL REPUBLIC OF GERMANY (hereafter: *German Design Award*) is the nation's highest official design prize. It is conferred by the *German Design Council* on behalf of the *Federal Ministry of Economics and Technologies.* ——— Since 2006, the German Design Award is bestowed annually. Apart from honouring product and communication design, it is also awarded to a person or group of persons for achievements in the area of design.

Die im Jahr 2003 durch den *Bundesminister für Wirtschaft und Arbeit* für vier Jahre eingesetzte Jury repräsentiert die Bereiche, Industrie, Handel, Design, Medien und Lehre. Die Mitglieder der Jury sind: PROF. TANJA DIEZMANN, *Hochschule Sachsen-Anhalt, Dessau* und *pReview*, Berlin // BARBARA FRIEDRICH, *Architektur & Wohnen*, Hamburg // NICOLETTE NAUMANN, *Messe Frankfurt* // REINHARD BINDER, *Design und Kommunikation*, Berlin // KONSTANTIN GRCIC, Designer, München // JAMES IRVINE, Designer, Mailand / London // TORSTEN NEELAND, Designer, London // HERBERT H. SCHULTES, Designer, Fürstenfeldbruck // ROBERT THIEMANN, *FRAME*, Amsterdam. ———— Den Vorsitz führt, auf Vorschlag des Bundeswirtschaftsministers, der Designer HERBERT H. SCHULTES. Die Jurymitglieder Tanja Diezmann und Konstantin Grcic konnten wegen anderer Verpflichtungen nicht an der Jurysitzung, die am 28. und 29. September 2006 in Frankfurt am Main stattfand, teilnehmen. ———— Im Jahr 2007 wird der *Designpreis* zum dritten Mal verliehen. Es wurden insgesamt 940 Produkte nominiert. Gemäß den Richtlinien des *Designpreises* erfolgten die Nominierungen durch die Wirtschaftsministerien und Wirtschaftssenatoren des Bundes und der Länder beziehungsweise durch die von ihnen beauftragten Institutionen. Voraussetzung ist, dass die vorgeschlagenen Produkte bereits auf dem Markt angeboten werden und bereits eine Auszeichnung in einem regionalen, in einem anderen nationalen oder

The Jury appointed in 2003 for four years by the *Federal Ministry of Economics and Labour* consists of ten persons, who represent industry, trade, design, the media and academia. The jury members are: PROF. TANJA DIEZMANN, *Anhalt Institute of Further Education, Dessau, Saxony-Anhalt*, and *pReview*, Berlin // BARBARA FRIEDRICH, *Architektur & Wohnen*, Hamburg // NICOLETTE NAUMANN, *Messe Frankfurt GmbH* // REINHARD BINDER, *Design und Kommunikation*, Berlin // KONSTANTIN GRCIC, Designer, Munich // JAMES IRVINE, Designer, Milan / London // TORSTEN NEELAND, Designer, London // HERBERT H. SCHULTES, Designer, Fürstenfeldbruck // ROBERT THIEMANN, *FRAME*, Amsterdam. ———— The Chairman – appointed on the advice of the *Federal Economics Minister* – is designer HERBERT H. SCHULTES. Owing to previous engagements the jury members Tanja Diezmann and Konstantin Grcic were unable to attend the meeting on September 28 and September 29, 2006 in Frankfurt / Main. ———— In 2007, the *German Design Award* was bestowed for the third time. A total of 940 products were nominated. In accordance with the *Design Award* guidelines, the nominations were made by the economics ministries and senate departments of economics at national or federal state level, or by an institution they commissioned. In order to qualify

internationalen Designwettbewerb erhalten haben. ———— Ein Projektteam des *Rat für Formgebung* hatte die Informationen der Hersteller über die eingereichten Produkte aufgearbeitet und den Mitgliedern der Jury formgerecht zur Verfügung gestellt. Auch während der Jurysitzung stand das Team für technische Fragen zur Verfügung. ———— Wie auch in den vorangegangenen Jahren kam es vor, dass Jurymitglieder an der Entwicklung eines eingereichten Produkts als Designer oder Berater beteiligt waren. Diese Situation wurde in bewährter Weise behandelt: Das betroffene Jurymitglied nahm nicht an der Produktdiskussion teil und war auch bei der jeweiligen Abstimmung nicht anwesend. Hierdurch wurde die Neutralität der Jurierung gewahrt. ———— Gestaltungsqualität, Gebrauchswert, technische und ökologische Qualität sowie die Gesamtkonzeption sind die wichtigsten Kriterien für die Prämierung zum *Designpreis der Bundesrepublik Deutschland.* Außerdem spielen Produkt-, Design- und Marketingstrategie des jeweiligen Unternehmens eine große Rolle für die Beurteilung der Produkte. Im Jahr 2007 zeichnet die Jury fünf Produkte mit einem Preis in Gold aus, 20 weitere Produkte erhalten einen Preis in Silber. Darüber hinaus wird dem Designer Dieter Rams der *Designpreis der Bundesrepublik Deutschland* für sein bisheriges Lebenswerk verliehen.

for the competition the proposed products must already be available on the market, and must already have won a prize in a regional, national or international design competition. ———— A project team of the *German Design Council* processed the information from the manufacturers on the submitted products, and made it available to jury members in proper form. This team was also present during the jury meeting to answer technical questions. ———— As was the case in previous years, in several instances jury members were involved in the development of a submitted product either as a designer or consultant. Such situations were dealt with in the usual manner: The jury member in question was excluded from the participating in the product discussion and was not present during voting for that product. This procedure ensured the neutrality of the judging process. ———— The key criteria considered in judging products competing for the *German Design Award* are design quality, practical value, technical and ecological quality as well as the overall conception. However, the manufacturing firm's product strategy, design strategy and marketing strategy also play a major role. In 2007 the jury selected five products for a Gold award, while 20 other products received a Silver. Designer Dieter Rams received the *German Design Award* for his life's work.

Designpreis Persönlichkeit: Dieter Rams

Verfechter des Vernünftigen — Championing the rational

Ein Besuch bei Dieter Rams

A visit to Dieter Rams

der Jahre ganze Produktgattungen verändert haben: die für den Privatgebrauch ergonomisch richtigere Schrägstellung des Griffs bei den Haartrocknern; die Ersetzung der gemuldeten Tasten durch Pastillen bei den Rechnern, weil das neue elektronische Innenleben eine leichtere und schnellere Bedienbarkeit ermöglichte; die Verbindung von harten und weichen Kunststoffen, erstmals durchgesetzt bei den Rasierern (damit sie nicht mehr vom Wachbecken rutschen); die Induktionszündung beim „Cylindric"-Feuerzeug – „ohne meine Sturheit wäre das nicht so gemacht worden". ——— Dieter Rams der „Ingenieur-Designer"? Das Etikett ist zweifellos zu kurz gegriffen. Schließlich ist es nicht die technische, sondern die ästhetische Meisterleistung, die die *Braun*-Geräte der Ära Rams wie auch sein ab 1957 entwickeltes Vitsoe-Möbelsystem (heute sdr+) zu Leuchttürmen der Designgeschichte macht. Und damit, jenseits aller rationalistischen Kriterien, die exzeptionelle Sensibilität des Designers Dieter Rams. ——— Proportionen, Konturen und Übergänge, die feinen Farbdifferenzen und Loch-, Noppen-, Linienraster und nicht zuletzt das so funktionsbedachte wie „grafisch" durchgestaltete Ordnungssystem der Phono-Bedienelemente – alles vermittelt ein Erlebnis von so perfekter Harmonie und Schönheit, dass man geradezu ins Schwärmen kommen kann. Heute noch. Nicht von ungefähr sind diese Geräte nahtlos vom Gebrauchs- zum begehrten Sammlerobjekt avanciert. Dass die Preise sich dennoch im zwei- bis

11
Portables Radio T 52; mit Einschub auch als Autoradio, 1961
Portable radio T 52; with case can be used as a car radio, 1961

12
Phono-Kombination TP1, 1959 („First Walkman")
Phonographic combination TP1, 1959 ("first Walkman")

11

12

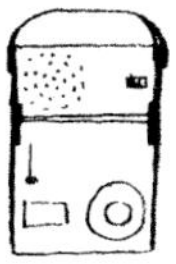

needed to assert himself, but also in Rams' ability to repeatedly challenge both himself and others to look for and to extend the boundaries of the feasible. ——— Let's talk about innovation. In his "ten theses" which all centre around the primacy of the quality and durability of utility items which is the first commandment. In conversation he underlines "I have always attempted to keep technical matters in the foreground. If a product is to have a meaning it must include something new – new technology or new utility value". ——— Examples by means of which the *Braun* designers have changed entire product genres over the years: putting a slant on the handle of hair dryers for private use, the more ergonomically correct option; replacing recessed keys with raised ones on calculators because the new electronics inside them made it easier and quicker to use them; combining hard and soft plastics, first achieved with razors (so that they no longer fall off the wash-basin); induction lighting in the "Cylindric" cigarette lighter – "if I hadn't been so stubborn they would never have made it the way they did". ——— Dieter Rams the "engineer-designer"? Doubtless, the tag is too short. After all, it is his aesthetic masterpieces, not the technical ones, that have made beacons of design history out of the *Braun* appliances dating from the Rams era, such as his Vitsoe furniture system (today sdr+), produced as of 1957. And accordingly, beyond all rationalist criteria, it is the exceptional sensitivity of the designer who is Dieter Rams. Proportions, contours and transitions, the subtle colour differences, the fine

niederen dreistelligen Bereich halten, hat einen einfachen Grund: Es gibt noch erstaunlich viele. Was wiederum nicht nur an den hohen Stückzahlen liegt, sondern auch daran, dass *Braun*-Geräte seltener als andere ausgediente Konsumartikel in der Mülltonne landeten. Wenn Rams von den Partnern und Widerparts spricht, die (außer den Designern) im komplexen Prozess der industriellen Produktgestaltung eine Schlüsselrolle spielen, sind die Rollen klar verteilt. Der Widerpart hieß und heißt postum noch Marketing. Das Vernünftige gegen das leicht Verkäufliche zu verteidigen, hat er zeitlebens als eine Herausforderung für sich begriffen und über seinem Aufstieg vom Designer zum Design-Manager, der nicht dem Marketing, sondern direkt dem Vorstand unterstellt ist, als exemplarische *Braun*-Firmenstruktur propagiert (was Kompromisse und Niederlagen freilich nicht ausschließt). ——— Die Partner suchte und fand er indessen in der Technik. Von Anfang an. Manche Flasche Chantré habe es ihn gekostet, erzählt er gern, die Techniker auf seine Seite zu bringen – sei es, um die Produktqualität zu optimieren (anfangs habe das Äußere der Geräte „ja mehr versprochen, als das Innere halten konnte"), sei es, um die Realisierbarkeit von Ideen zu beweisen. Letztlich beantwortet die Nähe zu den Entwicklungs- und Produktionstechnikern sogar die Frage, warum er *Braun* nie verlassen hat, um als freier Designer Fähigkeiten, Kenntnisse und Namen auf eigene Rechnung zu vermarkten. „Nur wenn Du eng mit den Tech-

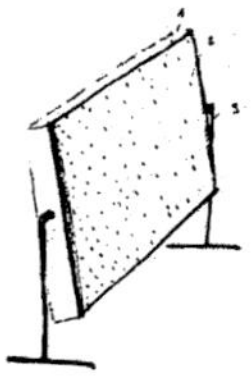

13

13
Elektrostatischer Lautsprecher, 1960
Electrostatic loudspeaker, 1960

patterns of holes, knobs and lines and, not least, the arrangement of the operating elements on his phonographic equipment, perfectly designed both in terms of functions and of optics, everything makes for an experience of the kind of perfect harmony and beauty that sends people off into rhapsodies. Even today. And it is no coincidence that these appliances have moved seamlessly from their place as utility objects to being sought-after collector's items. And there is a simple reason for the fact that the prices are still in the two to low three figure region. There are still a surprisingly large number of them about. Something which is not, in turn, only a result of the large numbers of them that were produced, but also because *Braun* appliances tended to end up in the garbage can less often than other out-of-service consumer articles. When Rams talks about his partners and the opposition who (apart from the designers) play a key role in the complex process of industrial product design, the different parts are clearly assigned. The opposition was and still is even posthumously Marketing. Throughout his life he has always seen it as a personal challenge to defend what is reasonable against what can be sold easily and has broadcast this as something exemplary in the *Braun* company structure, seen in his rise from designer to design manager, reporting directly to the Executive Board rather than the Marketing Department (something that admittedly does not exclude compromises and defeats). ——— Meanwhile, he looked for and found his partners in the Technology Department. From the very beginning. As

nikern zusammen arbeitest, kannst Du den Erfolg erreichen, den Du Dir wünschst. Sonst bleibst Du irgendwo unterwegs hängen." Topaktuelles Beispiel: Jasper Morrison und seine *Rowenta*-Geräte. Da habe Morrison „zwar hervorragendes Design gemacht, aber die Techniker waren noch nicht so weit". ———— Als Dieter Rams zu *Braun* kam, war der Weg, den das Design gehen sollte, „bereits fixiert" (F. Eichler). Die Marksteine – Einfachheit, Funktionalität, Systemgestaltung – hatten die Partner von der HfG, allen voran Hans Gugelot, eingepflockt. Doch Ulm war weit weg, Gugelot kam in der Regel einmal monatlich vorbei, die Produktentwicklung lag in Frankfurt. Die Chance für Dieter Rams, sich in der täglichen Arbeit zu qualifizieren, zu profilieren, selbst Meilensteine zu setzen. 1959 gelang ihm das mit gleich zwei Entwürfen: dem Taschenempfänger TP 1 mit aufsteckbarem Plattenspieler und mit der Phonoanlage „studio", der ersten aus stapelbaren Einzelkomponenten bestehenden Phonoanlage. Mit Gerd A. Müller hatte er mittlerweile einen jungen Kollegen ins Team geholt, der Zeichen im Bereich Küchengeräte (u. a. mit dem Küchenmaschinen-Klassiker KM 3, 1957) und Rasierer setzte – für die Firma schon damals mindestens ebenso wichtig wie die publizistisch prominenteren „Radios" und à la longue sogar ein kommerzieller Kernbereich, der das Ende der Phonoproduktion (1990) bis heute überlebte. 1962 wurde Rams Leiter der kleinen, aber stetig wachsenden und längst Marken-profilierenden

14, 15
Alternativentwürfe zu T 1000 Weltempfänger, 1962
Alternative design for T 1000 world receiver, 1962

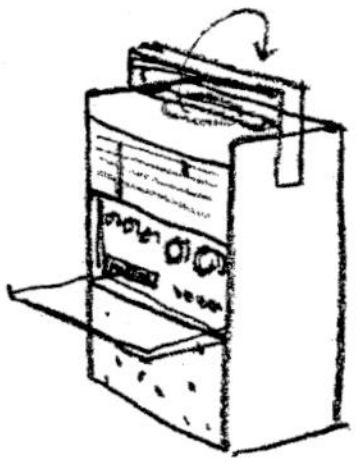

14

15

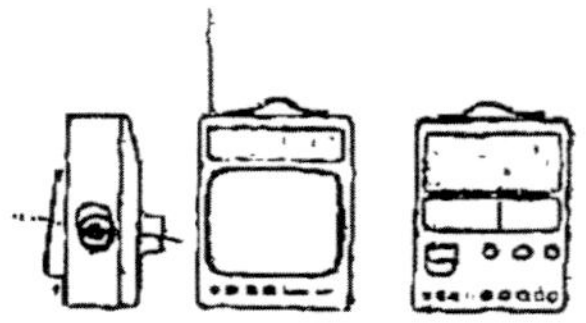

he likes to report, it cost him quite a few bottles of Chantré to get the technical staff on his side – whether he was trying to optimize the product quality (in the beginning, the outside of the appliances "promised more than their insides were capable of"), or to prove the feasibility of ideas. In the final analysis, his closeness to the development and production engineers even explains why he never left *Braun* to market his skills, knowledge and name independently as a freelance designer. "It is only when you collaborate closely with the engineers that you can achieve the success that you're looking for. Otherwise, you just get stuck somewhere on the way." One up-to-the-minute example: Jasper Morrison and his Rowenta appliances. In Rams' opinion "although [Morrison] produced an outstanding design, the engineers were not as far advanced". ———— When Dieter Rams arrived at *Braun*, the approach to design was "already fixed" (F. Eichler). The cornerstones – simplicity, functionality, system design – had been laid down by the partners from the Ulm School of Design, in particular Hans Gugelot. But Ulm was far away. As a rule, Gugelot stopped by once a month and product development was in Frankfurt. This was Dieter Rams' chance to prove himself in his daily work, to distinguish himself and to set his own milestones. And he achieved this in 1959 with two designs: the TP 1 pocket receiver with attachable record player and with the "studio" sound system, the first sound system consisting of stacking individual components. By now, he had taken on board a young colleague, Gerd A. Müller,

Design-Abteilung. ——— Für etliche Jahre bezog er auch als Entwerfer einzelner Produkte einen dominierenden Platz. Ab den 70ern taucht sein Name in den Design-Autoren-Angaben dann immer seltener auf. Mit der Expansion von Firma und Gestaltungsaufgaben, und gar nach dem Rückzug von Erwin Braun, der das Unternehmen 1967 an den amerikanischen Gillette verkauft hatte, kündigte sich eine neue Rolle für den Team-Leiter an: Design-Manager. Was das für seine Rolle im Team respektive die konkrete Entwurfsarbeit bedeutete, war lange und ist wohl noch immer ein Kapitel für sich, an dem sich mancher Archivar und Philologe abarbeiten kann. Unangefochten ist dabei seine Leistung als Primus inter Pares insofern, als er, wie Dietrich Lubs erzählt, die Mitarbeiter „an der langen Leine" führte: „Wir konnten frei arbeiten." Freilich ohne persönlich in Erscheinung zu treten. ——— Dieter Rams, der Anreger, Mentor – und machtbewusste Boss ... ? Als Rams 1980 im Berliner IDZ mit einer Einzelausstellung geehrt werden sollte, durfte das Team mit Rams-Unterstützung erstmals Mann für Mann in Erscheinung treten. Allerdings erst, nachdem „die", wie der Freund Lubs sich erinnert, „opponiert" und „dem Dieter Druck gemacht" haben. Die Ausstellung hieß dann nicht „Dieter Rams", sondern „Dieter Rams et". Manches Zuschreibungsschildchen musste dennoch nachträglich überklebt werden. Die Kaffemühle KSM 1/ KSM 11 von 1967 zum Beispiel: Damals als Opus von Dieter Rams ausgewiesen, figuriert sie im

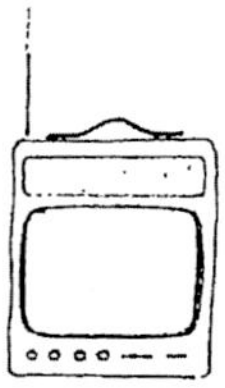

16
Alternativentwurf zu T 1000 Weltempfänger, 1962
Alternative design for T 1000 world receiver, 1962

who left his mark in the field of kitchen appliances (including that classic food processor, the KM 3, 1957) and razors – even then, as far as the company was concerned, at least as important as the "radios" although the latter received more publicity, and, in the long term, even a more commercial core market that survived the end of audio production (1990) and is still going today. In 1962 Rams became the head of the small but constantly growing design department that had long been causing the brand to stand out from the crowd. ——— He also occupied a dominant place as the designer of individual products for a number of years. But then, as of the 1970s, his name started to appear less and less frequently in the designer credits. As the company began to expand and with it, its design assignments, and then after the retirement of Erwin Braun who sold the company to US manufacturer Gillette in 1967, the team leader took on a new role, that of design manager. What this meant in terms of his role in the team with regard to concrete design work was, for a long time and doubtless still is a story in itself, one for the archivists and philologists. But what is uncontested is his achievement as primus inter pares inasmuch as, as Dietrich Lubs tells it, he allowed his staff "a long leash": "we were allowed to work freely." And this without making his presence felt. Dieter Rams, the source of inspiration, the mentor – and boss, fully aware of his power …? In 1980 when Rams was to be honoured with a solo exhibition in the IDZ, Berlin the team was allowed for the first time, with Rams' support, to appear them-

2005 erschienenen Bernd Polster-Band zum 50-jährigen *Braun*-Design-Jubiläum als Arbeit von Reinhold Weiss. Auch Dieter Rams ließ sie außen vor, als er im selben Jahr in Kyoto seine One-man-Ausstellung „Less but better. Die Welt von Dieter Rams" bestückte. In einer Zen-Tempelanlage übrigens – was nun nicht gerade bescheiden anmutet, aber dennoch verblüffend diskret und stimmig gelang. Und en passant einmal mehr unterstreicht, was jeder merkt, der sich gelegentlich außer Landes begibt: Von Japan bis in die USA ist sein Name fast mehr bekannt und der Mann mehr geachtet als zuhause. Tendenz steigend. ——— Dieter Rams, der Zeitgenosse. Als er Anfang 2006 auf der Kölner Möbelmesse, als „Altmeister" gegen drei junge Designer ins Feld geschickt, ein so genanntes „ideales Haus" konzipieren sollte, machte er daraus eine Hommage an das Beständige und bestückte sein japanisch inspiriertes „Haus" mit Möbeln von Charles Eames bis Konstantin Grcic. Einmal mehr ein Bekenntnis zu seinem Credo „Gutes Design ist möglichst wenig Design". Dass die Botschaft nicht so recht zündete, geht in der Tat zu einem Gut Teil auf das Konto der miesen Rahmenbedingungen. Aber nur? ——— Vielleicht, argwöhne ich, fehlte ihm selbst da auf einmal das Team, das sonst so oft von ihm zu Höchstleistungen gebracht wurde? Ganz sicher offenbarte sich da aber auch, wie weit die seit den 90er Jahren revitalisierten Bezugsgrößen Sachlichkeit, Bescheidenheit, Vernunft sich mittlerweile gewandelt haben. Für Rams

17, 18
Alternativentwürfe zu T 1000 Weltempfänger, 1962
Alternative design for T 1000 world receiver, 1962

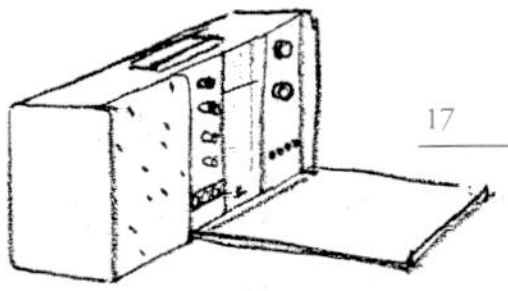

17

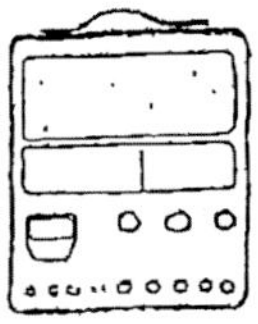

18

selves, man for man. However, as his friend Lubs remembers, this was only after "they rebelled and put pressure on Dieter". The exhibition was thus not entitled "Dieter Rams" but rather "Dieter Rams et." It did, however, involve some of the attribution labels having to be covered over with new ones. The coffee grinder KSM 1/KSM 11 from 1967, for example: displayed back then as the work of Dieter Rams, appeared in the 2005 volume by Bernd Polster on the 50th anniversary of *Braun* design, as the work of Reinhold Weiss. Dieter Rams also left them out in the cold when in the same year he put together his one-man show in Kyoto, "Less but better. The World of Dieter Rams." This, incidentally, was held in a Zen temple complex – which does not exactly sound modest, but nevertheless was surprisingly discrete and coherent. And en passant, once again, underlines what anybody who ever leaves the country on occasion notices: From Japan to the United States, his name is virtually better known and the man more respected than he is at home. The trend is on the increase. ——— Dieter Rams, the contemporary. At the beginning of 2006 at the Cologne Furniture Fair, as an old master fielding off three young designers, when he was supposed to conceive a so-called "ideal house" he turned the occasion into a homage to sustainability and equipped his Japanese inspired 'house' with the furniture by designers ranging from Charles Eames to Konstantin Grcic. Yet another affirmation of his credo, "Good design is as little design as possible". The fact that the message did not really catch on

bedeutete das Aufkommen der „neuen Einfachheit" – auch wenn „mir das zu sparsam gewesen ist" – eine späte Bestätigung. „Wir sind nicht weit auseinander", kommentiert er Ansatz und Designauffassung von Jasper Morrison oder Konstantin Grcic. ———— Grcic sieht da sehr wohl persönliche und Generations-Unterschiede: Rams' „Anspruch auf Verbindlichkeit" und sein Versuch, „Werte zu etablieren", sind ihm nicht nur als Designer, für den die Autorenschaft immer ein substanzieller Teil der Arbeit ist, zutiefst fremd. Auch die Verhältnisse ringsum spielen eine Rolle: Ich bin heute ja konfrontiert mit und „verwöhnt von dem Zustand, dass viele Systeme und Wertvorstellungen nebeneinander existieren und funktionieren", sagt er. Nichtsdestoweniger weist er wie viele seiner Generation Rams eine spät entdeckte Vorbildrolle zu: Wie „unnachgiebig und fast penetrant er auf Qualität gepocht hat, das ist schon einzigartig und unglaublich stark". ———— Die seltsamste Hommage übrigens, die ich in den zahllosen Würdigungen von Dieter Rams gefunden habe, stammt von Ettore Sottsass. Als „sehr konzentriert" (was jedermann unterstreichen kann), „sehr geduldig" (was sicher nicht jeder Mitarbeiter erlebt hat) und sehr weise" (was Rams mit grandseignoralem „das wäre ich gerne" kommentiert) rühmt der italienische Design-Individualist da den deutschen Design-Programmatiker. Um gleich drauf abgründig ironisch in frage zu stellen, ob die geschätzten Qualitäten der Rams-Opera wirklich aus der These

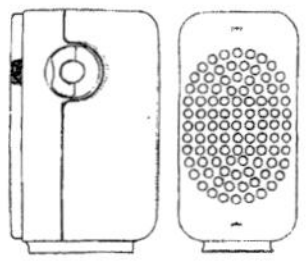

19

20

19
Weiterentwicklung Weltempfänger mit Kunststoffgehäuse, 1970
Improved world receiver with plastic casing, 1970

20
Portable Radiokassetten-Kombination; parallel verschiebbar, 1978
Portable radio cassette combination; parallel moveable, 1978

was actually due in good part to the miserable general conditions. But was it that alone? Perhaps, I suspect that he suddenly missed the presence of the team which he so often pushed to a level of high performance? Without a doubt here too, it was revealed to what degree the revitalized reference values – practicality, modesty, rationality – had changed since the 1990s. For Rams the rise of "new simplicity" implied – even if "this was too sparing for my liking" – a late confirmation. "We are not far apart", he commented on the approach and design concept of Jasper Morrison and Konstantin Grcic. ———— Grcic, however, certainly sees personal and generational differences: Rams' "claim of commitment" and his attempt "to establish values" are deeply alien to him, not just as a designer for whom authorship is always a substantial part of work. The surrounding context also plays a role. Today I am confronted with and "spoiled by the fact that several systems and ideas of values exist side by side and function", he says. Nevertheless, like many of his generation, he attributes Dieter Rams with a role model position which was discovered late. Just how "unflinchingly and almost obtrusively he emphasised quality is indeed unique and incredibly strong". ———— The most extraordinary homage, incidentally, which I have found among the countless appraisals of Dieter Rams, comes from Ettore Sottsass. Here, the Italian design individualist praises the German programmatic designer as "very concentrated" (something everyone can identify with), "very patient" (certainly something not experienced by

„Gutes Design ist wenig Design“ abzuleiten sind, ja ob „dieser Gemeinplatz“ überhaupt „von Dieter oder irgendwelchen Journalisten stammt“ – „und es ist mir auch egal“. ———— Natürlich ist das weder fraglich noch gar egal (ein Fremdwort im Denksystem von Dieter Rams). Als Designer, Manager, Hochschullehrer (von 1981 bis 1995 an der Hochschule der Künste in Hamburg) und noch immer aktiver Vortragsreisender für die Überzeugung einzutreten, dass Design „eine ethische und moralische Dimension“ hat und dazu beitragen muss, dass wir wegkommen „von der Produktkultur des Überflusses, der Verschwendung, der Billigkeit im Wortsinne, aber auch im übertragenen Sinne“, ist ein integraler Bestandteil des Lebenswerks von Dieter Rams. Für ihn persönlich geradezu ein Grundgesetz, dem eigentlich auch keiner widersprechen kann (auch wenn sich nicht alles Design nach seinem Schwarz-Weiß-Schema beurteilen lässt). ———— Aber ist es auch eine Handlungsanweisung? So sehr seine eigenen Produkte sich an seinen Prinzipien messen lassen, so sehr muss man einräumen, dass überragendes Design à la Rams mehr ist als das Nachbuchstabieren eines Regelwerks. Ob es ihm gefällt oder nicht: Wenn das Leitbild Dieter Rams bis heute unangefochtene Zeichen setzt, dann eher mit den tadellosen Dingen als mit den entschiedenen Worten. Wobei, zugegeben, beides zu tief in seiner Persönlichkeit verwurzelt ist, als dass es sich wirklich auseinander dividieren ließe.

21
Entwurf für Mixgerät, 1967
Design for mixer, 1967

21

all his co-workers) and “very wise” (which Rams commented in great style “I would love to be”). Only to challenge in inscrutable irony whether the esteemed qualities of the Rams opera really derive from the statement “Good design is little design”, and if “this platitude” really originated from Dieter or perhaps some journalist” – which “is all the same to me”. ———— Of course this is neither questionable not is it “all the same” (a foreign word in the mind of Dieter Rams). As designer, manager, university lecturer (from 1981 to 1995 at the Hochschule der Künste in Hamburg) and still an active guest lecturer who stands for the conviction that design has an “ethical and moral dimension” and that we need to participate in moving away from “the product culture of superabundance, waste, cheapness in the literal as well as metaphorical sense” it is an integral part of the life work of Dieter Rams. For him personally, some kind of constitution which really no one can contradict (even if not everything in the world of design can be judged in black and white terms). But could this also be a directive for action? As much as his own products can be measured against his own principles, one must admit that outstanding design à la Rams amounts to more than the re-spelling of a set of rules. Whether he likes it or not: If the Dieter Rams model still, up to the present day, indisputably sets standards then it is rather with faultless things than with resolute words. Although admittedly, both are too deeply rooted in his personality to be able to really separate.

Die Preisträger Gold
The Awards Gold

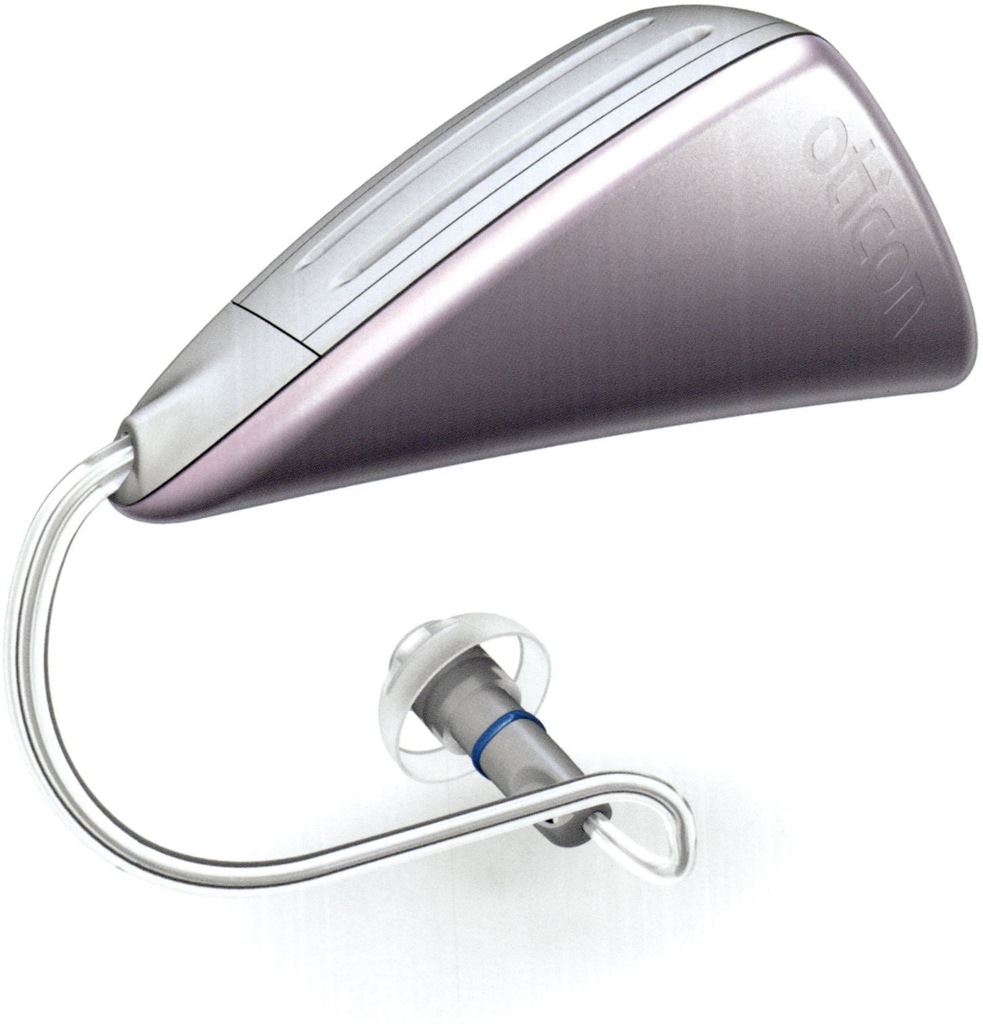

Oticon Delta *Hörsystem*
Hearing device

BEI EINEM HÖRGERÄT GEHT ES IN ERSTER LINIE DARUM, ein Objekt zu schaffen, dass formal weitgehend in den Hintergrund tritt. Gutes Design ist meistens, wie in diesem Fall, unsichtbar. Denn nur ein Gerät, das allen technischen Anforderungen entspricht und keine physische Präsenz hat, vermittelt dem Benutzer das Gefühl von Sicherheit und Unabhängigkeit. Diese Kriterien wurden bei diesem Hörsystem erfüllt. Im Zuge von starken zukünftigen sozialen Veränderungen haben sich *Oticon* und *Lockenwitz* insbesondere der Generation 50+ gewidmet und einen innovativen Beitrag geleistet. Der Entwurf überzeugt durch seine fein dimensionierte, zurückhaltende Form, die sich dem Ohr optimal anpasst und der Erscheinung eines Schmuckstücks nahe kommt.

MOST IMPORTANT WITH A HEARING DEVICE is to create an object whose formal aspect largely disappears into the background. Good design is usually invisible, as in this case. Users only have a feeling of security and independence if the device which complies with all technical demands has no physical presence. This hearing device has fulfilled the criteria. In the wake of strong social changes, *Oticon* and *Lockenwitz* have devoted themselves in particular to the over-50 generation thereby making an innovative contribution. The design is convincing in its finely dimensioned, low-key form which fits the ear optimally and looks a little like a piece of jewellery. — TORSTEN NEELAND

Jurybegründung

Statement of the Jury

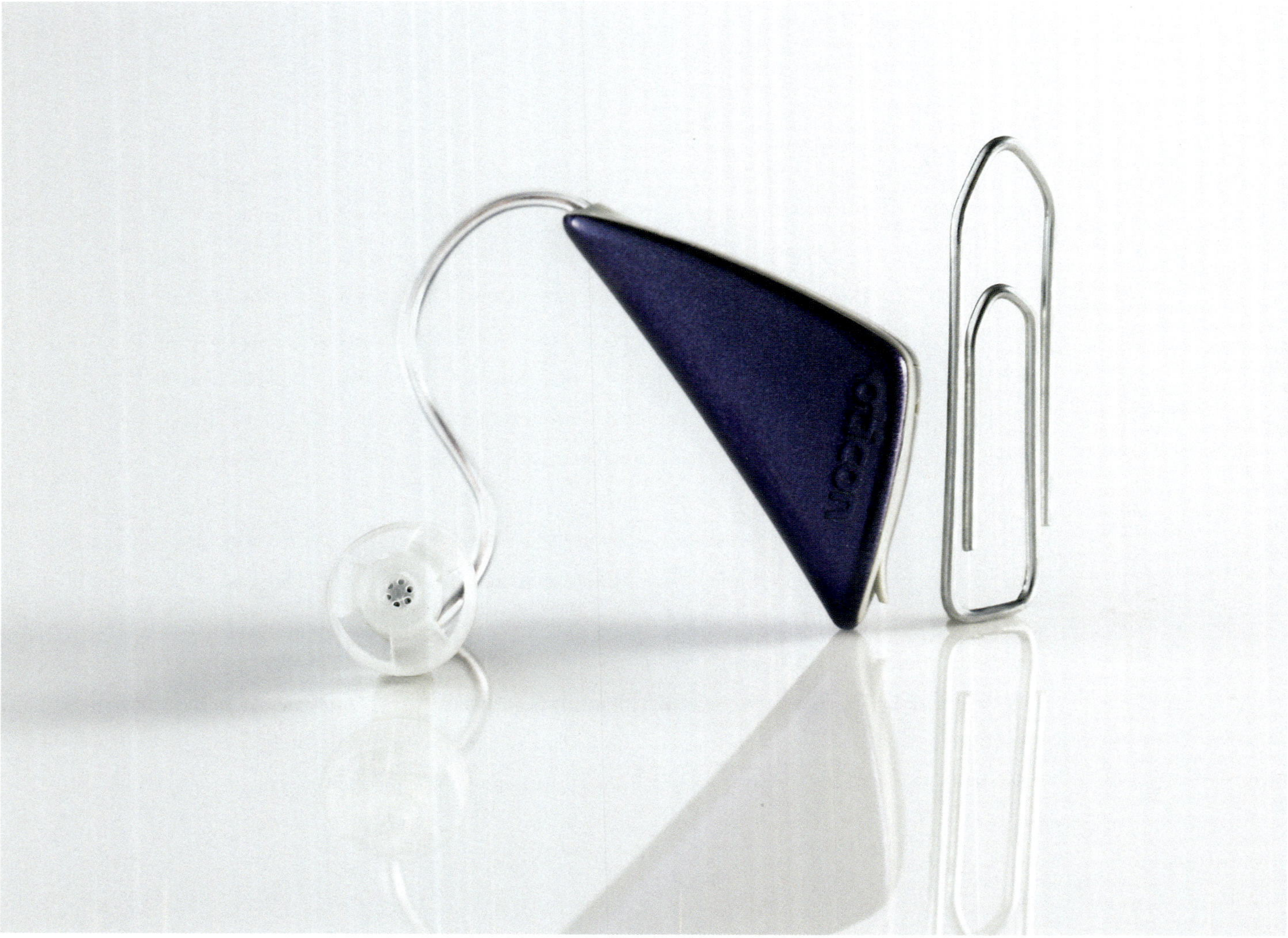

Unternehmen Company

Oticon A/S Kongebakken 9 — 2654 Smørum — Denmark — www.oticon.com

Design

lockenwitz i/d Christian Lockenwitz — www.lockenwitz.dk //
Werksdesign / Inhouse Michael Frank Petersen, Claus Würfel, Søren Skjærbæk, Lars Tuborg Jensen, Søren Colding

DELTA KOMBINIERT INNOVATIVES DESIGN mit der neuesten Hörgerätetechnologie: künstlicher Intelligenz. Das System analysiert permanent alle akustischen Umgebungen, in denen es sich befindet – und stellt sich kontinuierlich automatisch darauf ein. Es arbeitet ununterbrochen an der jeweils besten Balance zwischen Sprachverständlichkeit, gutem Klang und reduziertem Lärm. So ermöglicht es selbst in schwierigen Situationen gutes Sprachverstehen. *Deltas* Design wurde durch die besondere Platzierung des Lautsprechers ermöglicht. Der Hörer sitzt – separat vom Gehäuse – am Ende einer dünnen Kabelverbindung in einem weichen Fixierschirm direkt im Ohr. Der Fixierschirm passt sich der Innenohrform genau an, lässt sich offen anpassen und sichert dadurch eine exzellente natürliche Klangqualität sowie ein angenehmes Tragegefühl. Das kleine *Delta*-Hördreieck, in dem der Verstärker sitzt, verschwindet nahezu unsichtbar hinter dem Ohr. Die dreieckige Form macht die optimale Platzierung zweier, qualitativ hochwertiger Richtmikrofone möglich, die aus jeder Richtung automatisch die dominierende Stimme erkennen. Sobald keine Sprache vorhanden ist, sorgt ein ausgeklügeltes Lärm-Unterdrückungssystem für komfortablen Klang. Das *Delta*-Gehäusedesign macht eine weitere Revolution in der Hörgerätebranche möglich: das flexible Wechseln des farbigen Gehäuses. Der modulare Aufbau des Gerätes erlaubt dem Hörgeräteakustiker – den Wünschen des Trägers entsprechend – das *Delta*-Design zu ändern.

DELTA COMBINES ITS UNMATCHED DESIGN with the latest in hearing aid technology – Artificial Intelligence. *Delta's* AI-enabled system constantly works to deliver the best possible combination of sound quality, speech understanding and comfort in all situations. *Delta's* digital processor makes millions of calculations per second to attenuate noise, accentuate speech and thereby enhance people's ability to communicate – even in challenging situations. *Delta's* unique design has been made possible by placing its speaker unit away from the shell into the "Dome" that is placed in the ear canal – giving room to create a groundbreaking new shape while at the same time deliver superior performance. The location of the speaker offers a number of benefits: combined with an open fitting, it is the key to preserving the user's existing hearing capabilities to deliver excellent sound quality in the most natural way, and offers unprecedented high frequency performance. The soft Dome itself fits comfortably within the curve of the individual ear and incorporates an advanced wax protection system. *Delta* is designed to fit discreetly behind the ear, and is virtually invisible from the back. Its concave shape ensures a comfortable fit on the ear. The triangular form enables the optimal placement of two, high quality directional microphones that can cancel out up to three noise sources from different directions and focus on the important speech signals. The triangular *Delta* design makes another novelty possible: the flexible exchange of its shell colours. The modular set up of the device allows hearing care professionals to change the look of the instruments in no time – depending on the users' preference.

Archetun *Kochgeschirr*
Non-stick aluminium cookware

AN DER KOCHTOPFSERIE ARCHETUN gefiel der Jury die archtetypische, reduzierte Form, die durch die Doppelwandigkeit „schwer“ wirkt, dann jedoch überraschend leicht ist. Neben der einprägsamen Form sind in der Gestaltung zahlreiche Funktionsanforderungen auch durch optimalen Materialeinsatz gelöst.

WHAT THE JURY LIKED ABOUT THE COOKING POT SERIES, ARCHETUN, was the archetypical, reduced form which although it appears "heavy" due to the double wall, is actually surprisingly light. In addition to the expressive form numerous other functional demands are solved by the optimal use of material. — NICOLETTE NAUMANN

Jurybegründung
Statement of the Jury

Unternehmen Company

TVS Spa Via G. Galilei, 2 — 61033 Fermignano (PU) — Italy — www.tvs-spa.it

Design

STUDIO MATTEO THUN Matteo Thun — www.matteothun.com

EINE IKONE SCHLICHTEN UND ESSENTIELLEN DESIGNS: *Archetun* TVS birgt in der Reinheit seiner strengen Formen hochwertigste Gebrauchszweckmäßigkeit. Dies dank seines dickwandigen Körpers aus einer druckgepressten Aluminiumlegierung in glattem Profil, für eine praktische Reinigung und eine optische Sauberkeit, dank der thermostabilen Handgriffe aus wärmegehärtetem Material für einen bequemen Griff und dank der wertvollen kratzfesten wie abreibungsfesten Antihaftbeschichtung Teflon® Platinum DuPont. Teflon® Platinum ist die bisher gegen Kratzer und Abreibung widerstandsfähigste Beschichtung, die je von DuPont hergestellt wurde, da sie sogar den Gebrauch von üblichen metallischen Küchengeräten und Scheuerschwämmen übersteht. Aufgebracht in drei dicken Schichten, enthält sie mikroskopisch kleine Teilchen aus harten Mineralien. Metallische Gegenstände gleiten über diese harten Teilchen, ohne so die Oberfläche, in der sie enthalten sind, beschädigen zu können. Dadurch bleibt die Antihaft-Eigenschaft dieser Beschichtung bestehen.

AN ICON OF SIMPLE, MINIMAL DESIGN, TVS *Archetun* encapsulates superior ease of use within its clean-cut geometry. This is thanks to its body in thick die-cast aluminium alloy with a continuous profile for visual and real cleanliness, to its comfortable heat-resistant handles in soft-grip thermosetting material, to the exceptionally high-quality DuPont Teflon® Platinum. *Archetun* TVS was born of special fusions of aluminium and the top-quality non-stick durability of Teflon® Platinum DuPont. In real terms, this means you get optimal, even cooking in the shortest time thanks to the aluminium's excellent thermo diffusion properties. What's more, the washing up is incredibly easy – and no scratches or dents thanks to highly durable, long-lasting mineral particles in the non-stick surfaces.

Pluma *Butangasflasche*
Butane gas cylinder

BUTANGASFLASCHEN WAREN EINMAL FÜRCHTERLICHE OBJEKTE. Die Griffe schnitten einem in die Finger. Sie machten hässliche scheppernde Geräusche, wenn sie von Lastwagen entluden wurden. Die Farbe der Flaschen blätterte gewöhnlich ab. Sie waren sehr schwer. Heutzutage kann sogar die bescheidene Gasflasche ein schönes Objekt sein. Das *Pluma*® Produkt ist das Ergebnis eines radikalen Ansatzes, der neue Technologien und feinfühliges Design verbindet, um ein Hochleistungsprodukt mit einer erstklassigen Ergonomie zu entwickeln. Ein klassisches Beispiel von ausgezeichnetem Design, das sich positiv auf das Alltagsleben auswirkt.

THERE WAS A TIME WHEN BUTANE GAS CYLINDERS WERE TERRIBLE OBJECTS. The handles would cut your fingers. They would make terrible clanging sounds when unloaded from trucks. The paint on the cylinder would be chipped. They were very heavy. Today even the humble butane gas cylinder can be a beautiful object. The *Pluma*® product is the result of a radical approach combining new technology and sensitive design to produce a high performance product with superior ergonomics. A classic example of brilliant design having an impact on peoples daily lives. — JAMES IRVINE

Jurybegründung
Statement of the Jury

Unternehmen Company

Galp Energia Rua Tomás da Fonseca, Torre C — 1600-209 Lisboa — Portugal — www.galpenergia.com

Design

Brandia Creating Rui Sampaio de Faria (Design Director) — José Carlos Mendes (Graphic Design) — Teresa Costa Reis — Rita Norton Sampaio — Tiago Mendes — www.brandia.net

PLUMA® – DAS PORTUGIESISCHE UND SPANISCHE WORT FÜR FEDER – ist eine Butangasflasche für den Hausgebrauch. Sie wiegt leer 7,5 kg und besteht aus zwei Teilen: dem eigentlichen Gasbehälter und einer äußeren Hülle. Der Behälter enthält einen inneren Tank aus Stahlplatten mit einer Dicke von weniger als 1 mm, gestanzt und vollständig verkleidet mit einem Netz aus Twintex®, einer Zusammensetzung aus Polypropylen und zusammenhängenden Glasfasern. Abgeschlossen wird die Oberfläche des Tanks durch einen Schutzfilm aus einem weißen Gel, wodurch *Pluma®* eine höhere Widerstandsfestigkeit und größere Sicherheit erhält, als herkömmliche Gasflaschen. Die äußere Schutzhülle besteht aus vier verschiedenen Teilen – Korpus, innerem Deckel und zwei identischen Griffen, die mit großer Dichte in Polyethylen eingespritzt sind. *Pluma®* wird gemäß der internationalen Vorschrift N12245 hergestellt und ist TÜV zertifiziert. Sie vereinigt verschiedene Vorteile in einem Produkt: die Sicherheit und die Widerstandsfestigkeit herkömmlicher Eisenflaschen mit der Leichtigkeit der Flaschen neuer Generationen aus gemischten Materialien. Das exklusive Design bietet eine höhere Funktionstüchtigkeit und Ergonomie als die anderer Konkurrenzprodukte.

PLUMA® (PLUMA IS THE PORTUGUESE AND SPANISH WORD FOR A FEATHER) is a butane gas cylinder for domestic use. It weighs 7.5 kg when empty and is composed of two main parts: the gas container and the exterior skin. The container comprises a pressed steel plate liner less than 1 mm thick, entirely coated with layers of Twintex® (polypropylene compound) fibre and continuous glass fibres, surface finished with a white Gel coat, giving it a strength and safety performance greater than those of traditional cylinders. The outer skin consists of a full-jacket made up of four separate parts – body, base and two identical handles formed by injection moulding in High Density – HD – Polyethylene. The *Pluma®*, which is constructed to international standard N12245 and TUV certificated, has the advantage of combining in a single product the safety and strength of traditional steel cylinders with the lightness of the new generation of cylinders made from composite materials. Its unique design offers higher functionality and ergonomics than those of any rival product.

POC Skull Comp Helmet *Skihelm*

Ski helmet

DER NAME DIESES PRODUKTES SPRICHT FÜR SICH. Das Wort „Schädel" (= engl. skull) erinnert daran, dass unser Gehirn in etwas ruht, dass einer großen Eierschale ähnelt. Der POC Helm ist ein ernst zu nehmendes Objekt für eine ernst zu nehmende Aufgabe. Den Schutz unseres Gehirns. POC hat ein schützendes Meisterstück konzipiert und produziert. Technologie als funktionale Kunst. Der Helm ist so leicht, dass man kaum merkt, dass man ihn auf dem Kopf hat. Skihelme werden nie wieder sein, was sie einmal waren.

THE NAME OF THIS PRODUCT SAYS IT ALL. The word "Skull" reminds one that our brains are contained in something which resembles a big egg shell. The POC helmet is a serious piece of equipment for a serious task. Protecting our brains. POC has conceived and produced a protective masterpiece. Technology as functional art. The helmet is so light that one hardly notices that one is wearing it. Ski helmets will never be the same again. — JAMES IRVINE

Jurybegründung

Statement of the Jury

Unternehmen Company

POC Sweden AB Skogsövägen 22 — 133 33 Saltsjöbaden — Sweden — www.pocski.com

Design

Werksdesign/Inhouse POC Design Team

BISLANG GAB ES ZWEI GRUNDTYPEN VON HELMEN; den herkömmlichen Hartschalenhelm mit einer dicken Außenschale und einem Kern aus EPS und den leichteren gegossenen Helm mit einer dünneren Schale, in die EPS eingespritzt ist. Diese Helme hatten beide ihre Schwachpunkte. Der Hartschalenhelm ist schwer und prallt bei einem Unfall vom Boden ab, was zu Verletzungen an Hals und Gehirn führen kann. Der gegossene Helm schützt nicht zuverlässig bei eindringenden Gegenständen. ——— POC hat deswegen eine neue Generation von Helmen entwickelt, die die Vorteile der Hartschalenhelme und der gegossenen Helme hinsichtlich des Schutzes des Kopfes verbinden. Das Ergebnis ist ein Helm mit fünf entscheidenden Vorteilen:

1. geringeres Abprallen vom Boden im Falle eines Sturzes
2. höhere Widerstandskraft bei eindringenden Gegenständen
3. besseres Energieaufnahmevermögen
4. besserer Sitz
5. geringeres Gewicht

In der Verbindung der einzigartigen spitz zulaufenden Schale, dem patentierten Schutz vor Durchdringung in Form einer Aramidmembran und der energieaufnehmenden Supracor-Auspolsterung ist dieser Helm für jene gemacht, die auf Abfahrtspisten, extremen Steilhängen oder beim X-Ski-Fahren keine Kompromisse eingehen wollen.

UNTIL NOW THERE HAVE BEEN TWO MAIN TYPES OF HELMETS; the traditional hard-shell helmets with a thick outer shell and an inner core of EPS, and the lighter, in-mould helmet with a thinner shell into which EPS is injected. These helmets both had their weaknesses. The hard-shell helmet is heavy and bounces against the ground when you crash, which may harm your neck and brain. The in-mould helmet is vulnerable to penetration. ——— POC has developed a new generation of helmets that provide the advantages of both the hard-shell and in-mould types of head protection. This results in a helmet with five major advantages:

1. Less bouncing of the ground in a fall
2. higher penetration resistance,
3. better energy absorption properties,
4. better fit,
5. lower weight.

Combining the unique tapered shell, the patented aramid-membrane penetration barrier and an Supracor energy-absorbing padding, this helmet is aimed at those who demand the best – downhill, steep skiers or Ski Cross.

DB Type *Exklusives Schriftsystem für die Deutsche Bahn*
Exclusive system of corporate typefaces for Deutsche Bahn

DAS NEUE SCHRIFTENSYSTEM DER BAHN, bestehend aus sieben funktional differenzierten Einzelschriften, ist ein Meisterwerk für mehr Informationskultur in Deutschland. Die Bahn prägt seit jeher landauf, landab mit einer sehr hohen Präsenz das öffentliche Bild in unseren Städten mit. Durch das neu geschaffene Schriftensystem entsteht einerseits ein erneuertes, kultiviertes Markenbild, andererseits erleichtern die einzelnen Schriftschnitte in den unterschiedlichsten Anwendungen vom Fahrplan über Wegeleitsysteme bis zur Werbedrucksache die Informationsaufnahme für die Kunden. Herausragende Funktionalität und kultivierte Anmutung stehen der deutschen Alltagskultur gut zu Gesicht.

THE NEW SYSTEM OF TYPEFACES FOR DEUTSCHE BAHN, which consists of seven functionally differentiated individual fonts, is a masterstroke, helping to further information culture in Germany. For a long time now the Bahn's very visible presence has characterised the public image in our cities up and down the country. The newly designed system of typefaces creates a renewed and refined brand image. At the same time the individual fonts, in the most varied applications from timetables to signage to printed advertisements, make it easier for customers to absorb information. Excellent functionality along with a cultivated impression improve the look of everyday German culture. – REINHARD BINDER

Jurybegründung
Statement of the Jury

Die neue Schrift der DB muss alles *können*, was die *alte* Hausschrift auch konnte, aber viel mehr. 28 pt DB Sans

Eine exklusive Hausschrift ist nicht nur der Marke eigen, sie kann auch *speziell* auf die technischen und kommunikativen Aufgaben des Unternehmens zugeschnitten werden. Diese Aufgaben sind bei der Bahn so vielfältig wie sonst selten. Vom Formular und Fahrplan über Zeitungen und Zeitschriften bis zu Werbung und Leitsystem muss die Schrift nicht nur sympathisch für die Marke stehen. Sie hat auch ganz *spezifische* Aufgaben zu lösen, soll überall verfügbar sein und so gut ausgebaut, dass es keine Entschuldigungen geben kann, sie nicht einzusetzen. 24 pt

Eine serifenlose Grotesk kann diese verschiedenen Inhalte nie angemessen darstellen. Nimmt man *verschiedene* andere Schriften hinzu, die jeweils besser geeignet wären, verblasst der gewünschte geschlossene Eindruck, den jedes Unternehmen und jede Marke vermitteln will. 18 pt

DB Type ist ein Schriftsystem, das alle nötigen Schriftarten aus einem Formrepertoire entwickelt, den einzelnen Typen dabei aber genügend *formale Eigenständigkeit* läßt. So entsteht Selbstähnlichkeit über Medien und Zielgruppen hinweg, aber nie Uniformität. 12 pt

Der Ton macht die Musik

Gedrucktes wirkt vertrauenerweckend: Niemand will einen handgeschriebenen Fahrplan lesen. Wenn es jedoch auf die *emotionale* Ansprache ankommt, auf die Aufladung einer Werbebotschaft, auf die weichen Werte der Marke, dann reicht keine gewöhnliche „Druckschrift".

Nicht umsonst nutzen immer mehr Marken die Assoziationen, die Schriften bei uns auslösen. Sie sind tief verwurzelt in unseren Gewohnheiten und über das Unterbewusstsein *gerade deshalb* wirksam.

Bücher werden immer noch aus *Antiqua* gesetzt, das verspricht Kultur und hochwertigen Inhalt.

Schnelle Information hingegen braucht aktuelle, schmale, eilige Typen.

Die Zeitung verlangt nach robusten Zeichen, die wie eine *Antiqua* wirken, aber soviel mechanischen Missbrauch aushalten wie eine Grotesk.

DB Head Black, Sans Regular, Serif, Sans Condensed Bold, News

Unternehmen Company

Deutsche Bahn AG Potsdamer Platz — 10785 Berlin — Germany — www.bahn.de

URW ++ Poppenbütteler Bogen 36 — 22399 Hamburg — Germany — www.urwpp.de

Design

SpiekermannPartners Erik Spiekermann, Christian Schwartz — www.spiekermannpartners.com

DIE NEUE SCHRIFT DER DB – heute Anbieter von weit mehr als Schienenverkehr – muss alles können, was die alte Hausschrift auch konnte, aber darüber hinaus noch viel mehr. Eine exklusive Hausschrift ist nicht nur der Marke eigen, sie kann auch speziell auf die technischen und kommunikativen Aufgaben des Unternehmens zugeschnitten werden. Diese Aufgaben sind bei der Bahn so vielfältig, wie sonst selten. Vom Formular und Fahrplan über Zeitungen und Zeitschriften bis zu Werbung und Leitsystem muss die Schrift nicht nur sympathisch für die Marke stehen. Sie hat auch ganz spezifische Aufgaben zu lösen, soll überall verfügbar sein und so gut ausgebaut, dass es keine Entschuldigungen geben kann, sie nicht einzusetzen. ———— Eine serifenlose Grotesk kann diese verschiedenen Inhalte nie angemessen darstellen. Nähme man andere Schriften hinzu, die jeweils besser geeignet wären, verblasste der gewünschte geschlossene Eindruck, den jedes Unternehmen und jede Marke vermitteln will. DB Type ist deshalb ein Schriftsystem, das alle nötigen Schriftarten aus einem Formrepertoire entwickelt, den einzelnen Typen dabei aber genügend formale Eigenständigkeit lässt. So entsteht Selbstähnlichkeit über Medien und Zielgruppen hinweg, aber nie Uniformität.

THE NEW TYPEFACE CHOSEN BY DEUTSCHE BAHN, the German railways and logistics group, ought to be able to do everything the old one did and more. An exclusive corporate typeface is not merely unique to its brand, it can also be tailored to fit the company's technical and communications requirements. With an enterprise like Deutsche Bahn, these requirements are as diverse as they get. From forms and timetables, newspapers and magazines through to advertising and signage, the typeface has to do more than just represent the brand in a sympathetic manner. It must also fulfil specific functions, be available everywhere and be made well enough to never grant excuses for not employing it. ———— One sans serif font cannot adequately represent this diversity of content. However, using additional fonts from other sources to cover individual functions would detract from the overall impression that every brand wishes to create. With this in mind, DB Type is a system of typefaces that derives each font from a single formal repertoire whilst allowing each one the freedom of form they require. In this way, the fonts are all similar – independent of medium or target group – but never uniform.

Die Preisträger Silber
The Awards Silver

Individual 32 *LCD-TV*

LCD-TV

INDIVIDUAL 32 STEHT FÜR INDIVIDUALISIERUNG UND WANDELBARKEIT im hochwertigen Bereich der LCD Fernseher. Oberflächen, Materialien, Grundfarben des Gehäuses und der Seiten können je nach Bedürfnis des Verbrauchers mit der Inneneinrichtung abgestimmt werden. Es geht um die Frage, in welchem Verhältnis der Benutzer heutzutage zu einem Produkt steht. *Individual 32* kann perfekt auf dem Sideboard, Regal, Boden, an der Wand oder frei im Raum integriert werden. Der Jury gefiel insbesondere die Überlegung, das Gerät mit Hilfe von abgestimmtem Zubehör auf unterschiedliche Raumbedingungen anzupassen.

INDIVIDUAL 32 STANDS FOR INDIVIDUALIZATION AND FLEXIBILITY in the field of high-quality LCD televisions. Surfaces, materials, basic housing colours and sides can be adjusted to match interior designs, according to the user's needs. The key question is to what extent a user is loyal to a product. *Individual 32* can be perfectly integrated on the sideboard, shelf, floor or wall or can be free-standing. The jury were particularly taken with the idea of being able to adapt the device to different spatial conditions with the aid of matching accessories. — TORSTEN NEELAND

Jurybegründung
Statement of the Jury

Unternehmen Company

Loewe AG Industriestraße 11 — 96317 Kronach — Germany — www.loewe.de

Design

Phoenix Design www.phoenixdesign.de

DER KUNDE KANN AUF SEINE WÜNSCHE HIN sein Gerät entsprechend technischer Ausstattung, Oberflächen, Materialien und Aufstelllösungen individuell zusammenstellen. Das Produkt bietet die Wahlmöglichkeiten zwischen acht verschiedenen Gehäuse-Farben (auf Wunsch auch Sonderfarben) und Aluminium Eloxalen mit einer Vorsatzglasscheibe, plus frei wählbarer Kombinationen mit seitlich eingesetzten Dekorintarsien verschiedenster Farben, Metalle und Hölzer. Mit dem abgestimmten Zubehör ist jedes Produkt perfekt auf dem Sideboard, Regal, Boden, an der Wand oder auch mit der Decken-Boden-Stange frei im Raum aufstellbar. Zur optimalen Bildausrichtung sind die Geräte neigbar und manuell oder optional auch motorisch verstellbar. Die technische Ausstattung ist ebenso wählbar und garantiert Zukunftssicherheit und Flexibilität dank integriertem DVB-T / DVB-C-Empfang. Die Funktion DR+ mit Dual-DVB ermöglicht es, ein digitales Programm anzusehen und gleichzeitig ein anderes digitales Programm aufzuzeichnen.

THIS CONCEPT ALLOWS THE CUSTOMER TO CUSTOMISE his own set when it comes to technical features, surfaces, materials and set-up options. The product is available with a choice of eight different housing colours (special colours also available on request) and anodised aluminium finishes with a glass pane and choice of combinations featuring decorative marquetry inlaid in the sides in a wide variety of colours, metals and woods. With the matching accessories, each product fits perfectly on the sideboard, shelf, floor, wall or even placed anywhere in the room using a floor-to-ceiling pole. For perfect picture alignment, the sets can be tilted as well as optionally adjusted either manually or electrically. The technical features can also be optionally selected and are guaranteed future-proof and flexible thanks to integrated DVB-T / DVB-C reception. The DR+ function with dual DVB allows you to watch one digital programme while you record another at the same time.

Loewe IFA 2005

Messestand

Fair stand

LOEWE STEHT FÜR SCHICKE AUDIOVISUELLE HIGH-END-GERÄTE IM MARKT. Die Messestanddesigner *Braun Wagner* arbeiteten mit *Zintzmeyer & Lux* zusammen, um diese Qualitäten auf einen Messestand für die IFA zu übertragen. Veranstaltungen dieser Art setzen das Publikum einer derartig bombastischen Reizüberflutung aus, dass der flashgeplagte Messebesucher geradezu nach architektonischer Klarheit dürsten und diese als wahre Oase empfinden muss. In einer weißen Welt werden nicht nur die Vorteile der *Loewe*-Produkte angestrahlt, sie erstrahlen im wörtlichen als auch bildlichen Sinne. Vor einer großen, mit Pastelltönen erleuchteten Wand, die die gesamte Länge des Messestandes einnahm, wurde die *Loewe* Produktlinie gezeigt, während auf einem Laufsteg das Neueste vom Neuen präsentiert wurde. Zudem bot der Messestand ein inszeniertes häusliches Umfeld, in dem der Besucher die Wirkung der *Loewe* Geräte erleben konnte, als wäre er in den eigenen vier Wänden.

LOEWE STANDS FOR STYLISH AUDIOVISUAL EQUIPMENT AT THE HIGH END OF THE MARKET. *Braun Wagner's* stand designers collaborated with *Interbrand Zintzmeyer & Lux* to translate these qualities into a trade-fair stand for the IFA, the world's largest consumer electronics show. Events of this kind bombard visitors with so many stimuli that the use of architectural purity as a concept must have drawn flash-fatigued fairgoers like an oasis. In a white world, *Loewe* products were not only shown to advantage: they shone, literally and figuratively. A large pastel-illuminated wall extending the entire length of the stand displayed *Loewe's* product line, while a catwalk presented the latest of the latest. The stand also included a domestic setting in which visitors could experience the effects of *Loewe* equipment as if they were at home. — ROBERT THIEMANN

Jurybegründung
Statement of the Jury

Unternehmen Company

Loewe AG Industriestraße 11 — 96317 Kronach — Germany — www.loewe.de

Design

Braun Wagner www.braunwagner.de

Interbrand, Zintzmeyer & Lux www.interbrand.de

DER MESSEAUFTRITT VON LOEWE bei der IFA 2005 demonstrierte die Positionierung der Marke als führenden Premiumanbieter von visionären individuellen Flat-TV-Lösungen mit höchstem ästhetischem Anspruch. Als architektonisches Grundkonzept des über 2000 m^2-Standes wurde die „weiße Bühne“ gewählt. Eine intelligente Kombination von Architektur, Design und Kommunikation vermittelte Transparenz, Größe und Vielfalt in beeindruckend minimalistischer Ästhetik. Die Reduktion der Materialität und Tonalität schaffte Raum für hochwertige Produkterlebnisse und rückte das Produkt in den Mittelpunkt – speziell die Weltneuheit *Loewe Individual*, der erste Flat TV, der nach individuellen Bedürfnissen gestaltet werden kann. Eine 65 m lange und 7 m hohe Kommunikationswand „Wonderwall“ trennte Publikums- vom Händlerbereich, öffnete sich jedoch temporär über Lamellen und schaffte somit erweiterte Wahrnehmungsebenen für beide Seiten. Analog zum Aufbau der „Wonderwall“ wurde den Besuchern ein Themen-„Catwalk“ angeboten, der zum interaktiven Kennenlernen der *Loewe*-Produktpalette einlud.

THE LOEWE FAIR STAND at IFA 2005 demonstrated the brand's positioning as leading premium manufacturer of visionary, individual flat-TV solutions with maximum aesthetics. The underlying architectural principle of the more than 2,000 m^2 stand was the “white stage”. An intelligent combination of architecture, design and communication steps transported size, transparency and diversity with impressively minimalist aesthetics. The reduced materiality and tonality let the products be the centre of attention – especially the highlight *Loewe Individual*, the first flat-TV that can be customized in terms of material, housing and inset colours as well as set-up solutions.
A 65 m long and 7 m high communications wall, the “Wonderwall”, separated the public and dealer zones. By temporarily opening six rotating lamella an extended level of perception for both sides was achieved. In accordance with the “Wonderwall” a “theme catwalk” was created for interactive exploration of the Loewe product range.

Stash, retail design

Inneneinrichtung für ein Taschengeschäft
Interiordesign for a handbagshop

WIE HOLT MAN DAS BESTE AUS EINER KLEINEN LADENFLÄCHE HERAUS, in der Taschen präsentiert werden sollen. Dieser Frage sah sich der Niederländer Maurice Mentjens gegenüber, als man ihn fragte, ob er das Ladendesign für ein Taschengeschäft in Maastricht entwickeln wolle. Indem er die rückwärtige Wand mit Spiegeln verkleidete, vergrößerte er den Innenraum optisch um ein Vielfaches. Die Wände und Decken ließ er feuerrot streichen, wobei die Übergänge abgerundet wurden, so dass der Eindruck entstand, dass sie miteinander verschmelzen. Die entscheidende Idee war jedoch so einfach wie großartig: indem er viele Taschen mittels Magneten an Metallwänden befestigte, ermöglicht er es dem Besucher, eine Tasche abzunehmen, anzuschauen und sie entweder zu kaufen oder wieder an der Wand zu befestigen. Keine zerbrechlichen Aufhänger, keine raumgreifenden Regale und die ständige Aufräumarbeit für das Personal entfällt auch, denn egal wo die Tasche wieder an der Wand angebracht wird, der Laden ist immer tipptopp.

HOW TO MAKE THE MOST OF A FAIRLY SMALL RETAIL SPACE that's to be stashed with bags? It's a question that faced Dutchman Maurice Mentjens on being asked to design a retail interior for the sale of contemporary bags in Maastricht. By cladding the rear wall with mirrors, he visually enlarged the interior many times over. And by having ceilings and walls merge into one another and painting them fire-engine red, he created a striking architectural element. But his most defining idea was both basic and brilliant: by attaching scores of bags to metal walls with magnets, he's allowed the visitor to remove a bag, inspect it, and buy it or stick it back on the wall. No breakable hangers, no space-consuming shelving and no personnel constantly tidying up, because no matter where a bag goes back on the wall, the shop is always shipshape. — ROBERT THIEMANN

Jurybegründung
Statement of the Jury

Unternehmen Company

Stash Wyckerbrugstraat 45 — 6121 EB Maastricht — Netherlands — www.s-t-a-s-h.com

Design

Maurice Mentjens Design Maurice Mentjens — www.mauricementjens.com

STASH IST EIN TASCHENGESCHÄFT mit einem neuen Displaysystem. Normalerweise stehen Taschen in Verkaufsräumen auf Regalen. Taschen, die getragen werden, stehen jedoch nicht, sondern hängen. Darum wollte Maurice Mentjens die Modelle in *Stash* hängend präsentieren. Recht bald entstand die Idee, die Taschen mit Magneten an Stahlplatten zu „kleben". Auf diese Weise können die Kunden jeden Artikel von der Wand nehmen und wieder „ankleben". Die gesamte Wandoberfläche des eher kleinen Raumes (33 m^2) wird damit optimal zur Präsentation benutzt. ——— Über den Sockeln für die Blickfänger hängen mitten im Ladenraum zwei stählerne, zur Decke hin gewölbte Displayelemente. Dadurch vergrößert sich nicht nur nochmals die Präsentationsoberfläche, es ist zugleich ein eindrucksvoller architektonischer Kunstgriff. So entstehen zwei Gänge mit einem großen Spiegel an der Rückwand, durch den das Geschäft wesentlich größer (tiefer) scheint. ——— Das Motiv der Displayelemente und Sockel wird im Schaufenster wiederholt. Einerseits erfolgt dadurch die räumliche Trennung von Schaufenster und Verkaufsraum, andererseits schützt diese Einteilung die Verkaufstheke – deren Rückseite ansonsten vom Schaufenster aus einsehbar wäre – vor allzu neugierigen Blicken. Durch die rote Wand- und Deckenbeschichtung übt das Geschäft eine enorme Anziehungskraft auf Passanten aus. Der Fußboden, die Verkaufstheke und die Sockel gehen durch die graue PU-Beschichtung nahtlos ineinander über.

STASH SELLS MODERN BAGS. Most shops display bags perched on a shelf. But bags do not "perch" when they are worn; they hang. For this reason, Maurice Mentjens wanted a hanging display for the bags in *Stash*. The idea was quickly born to work with magnets, enabling the bags to be "stuck" to steel plates. Customers could then remove each bag from the wall to look at it and then stick it back in place. In this way, optimum use could be made of the entire wall surface of the fairly limited space (33 m^2) for the presentation of the bags. ——— In the middle of the shop are two steel coves suspended from the ceiling with pedestals underneath for the eye-catchers. This increases the presentation surface area still further, as well as adding a strong architectural element. This creates two passageways with a large mirror wall at the end, making the shop appear much larger (longer). ——— The cove and pedestal theme can also be seen in the shop window with a smaller space in between. This adds more intrigue, as it distinguishes the shop window from the shop itself and also restricts the view in to the counter (which has its back facing the window) from outside the shop. The red coating on the walls and ceiling makes the shop very appealing to passers-by. The floor, counter and pedestals fit in seamlessly due to their grey PU coating.

Circle *Möbel*
Furniture

DER ENTWURF DES HOLLÄNDISCHEN ARCHITEKTEN Ben van Berkel von UN-Studio für die Möbelmarke *Walter Knoll* ist eine mutige Umsetzung seiner Interpretation des Kreises. Zerschnitten in Viertelkreiselemente können sie beliebig zu einer Sitzlandschaft zusammengesetzt werden: ein optischer „Knaller" – auch in den präsentierten Stoffqualitäten und Farbkombinationen. Wüsste man nicht, dass die „schnittige Sitzlandschaft" von dem schwäbischen Traditions- und Qualitätspolstermöbler *Walter Knoll* gefertigt wird, würde man sie wohl einer italienischen Marke zuschreiben. Ein maßgeblicher Grund, *Circle*, das sich als zeitgemäßes „Lifestyle"-Möbel für Bars, Lounges, Designhotels etc. vorstellt, mit einem *Designpreis der Bundesrepublik Deutschland* auszuzeichnen.

THE DESIGN BY DUTCH ARCHITECT Ben van Berkel from UN Studio for the furniture brand, *Walter Knoll*, is a courageous implementation of his circle interpretation. Cut into quarter circle elements it can be assembled as desired to form a seating landscape: an optical hit – also in terms of the fabric quality and colour combinations. If you did not realise that the "sleek seating landscape" was manufactured by the Swabian maker of traditional quality furniture, *Walter Knoll*, you would assume it belonged to an Italian brand. An important reason why *Circle* which presents itself as a contemporary lifestyle furniture for bars, lounges, design hotels etc. be awarded a *German Design Award*. — BARBARA FRIEDRICH

Jurybegründung
Statement of the Jury

Unternehmen Company

Walter Knoll AG & Co. KG Bahnhofstraße 25 — 71083 Herrenberg — Germany — www.walterknoll.de

Design

UNStudio Ben van Berkel — www.unstudio.com

VIER TEILE BILDEN CIRCLE, je zwei einen Halbkreis, gemeinsam eine kreisrunde Plastik. Die Neigung des Rückens und die Sitztiefe verändern sich fließend vom aufrechten Sitzen bis zum entspannten Liegen. Konstant bleibt die Höhe. Der geschlossene Kreis oder die nur aus drei Teilen bestehende Variante – zum Begehen geöffnet – bilden einen eigenen Raum: Sitzen im Innenkreis. Das Design umfasst auch Varianten des nach außen gerichteten Sitzens: Sitzen im Außenkreis. Aus der Verbindung von konvexen und konkaven Formen entstehen schwungvolle Skulpturen. *Circle* entstand im Zusammenhang mit dem Entwurf für das *Mercedes-Benz Museum* in Stuttgart, dessen Konzept die Designidee anregte. „Charakteristisch für das UN Studio ist das Vereinen einer immensen Vielfalt von Möglichkeiten und Assoziationen in einer kraftvollen Geste, ohne eindimensional oder symbolisch zu wirken."

CIRCLE IS MADE UP OF FOUR PARTS – two of its parts making up a semicircle, all of them creating the round sculpture. The slope of the back and the seating area make up an unbroken flow from sitting upright to lying in relaxation. The height is the one thing that remains constant. The closed or the open circle, consisting of only three parts, allowing access, creates its own space: seating in the inner circle. Change the configuration and the design allows one to sit on the outside: seating in the outer circle. The connection of convex and concave shapes results in sweeping sculptures. *Circle* was inspired by the design of the *Mercedes-Benz Museum* in Stuttgart, Germany. "Clenching a great diversity for options and references within a powerful, but not unambiguous or symbolic gesture, is characteristic for UN Studio."

MIURA *Barhocker*
Barstool

BEI DIESEM BARHOCKER GIBT ES NICHTS, WAS NICHT INNOVATIV IST. Ein gewöhnlicher Barhocker ist dagegen ein Sammelsurium aus Komponenten und Materialien. *Miura* reicht eines. Und das ist nicht irgendein Material: verstärktes Spritzgusspolypropylen ist leicht, stabil, und, was genauso wichtig ist, nicht teuer. Mit der eigenartigen, eckigen und auf den ersten Blick eher ungemütlich anmutenden Form, ist *Miura* ein Star im provozierenden Œuvre von Konstantin Grcic, ein Mann, dessen Vorstellungswelt nicht wenige seiner Designkollegen inspiriert hat. Im Gegensatz zu vielen anderen geht es Grcic jedoch nicht nur um Form: der Stuhl ist wunderbar bequem und zudem auch stapelbar. Eine Tour de force.

NAME ONE THING ABOUT THIS BAR STOOL THAT ISN'T INNOVATIVE. Whereas an ordinary bar stool is a clutter of components and materials, *Miura* manages with just one. And this is not just any material: reinforced, injection-moulded polypropylene is lightweight, strong and, equally important, inexpensive. With its curious, angular and, at first glance, rather uncomfortable looking form, *Miura* plays a starring role in the provocative œuvre of Konstantin Grcic, a man whose universe has been the inspiration of many a fellow designer. Unlike others, however, Grcic is not all about form: the stool sits like a dream and is stackable in the bargain. A tour de force. — ROBERT THIEMANN

Jurybegründung
Statement of the Jury

Unternehmen Company

PLANK Plank Collezioni Srl — Via Nazionale 35 — 39040 Ora — Italy — www.plank.it

Design

Konstantin Grcic Industrial Design Konstantin Grcic — www.konstantin-grcic.com

DER BARHOCKER MIURA ist als Monoblock aus verstärktem Polypropylen gefertigt und für den Einsatz im Wohn- sowie Objektbereich für innen und außen geeignet. Die innovative Spritzguss-Technologie sowie der Anspruch einen stapelbaren Barhocker für die Bereiche „Privat" und „Objekt" zu entwickeln, haben zur charakteristischen Form des Hockers *Miura* geführt. Optisch zeichnet sich der Hocker durch seine dynamische Form aus, bei der Konstruktion und Ergonomie einen ausdrucksstarken Einklang ergeben. Das Thema Ökologie wurde bei der gesamten Planung stets berücksichtigt – *Miura* ist recycelbar. Dimensionen: Höhe: 81 cm; Sitzhöhe: 78 cm; Tiefe: 40 cm; Breite: 47 cm; Gewicht: 4,2 kg.

THE MIURA STOOL is made of a single block of reinforced polypropylene and suitable for domestic and mass use, indoors and outdoors. The characteristic shape of the *Miura* stems from the innovative injection moulding technology as well as the wish to develop a barstool for commercial and domestic environments. Visually speaking, this stool has a truly dynamic form, with the structure and the ergonomics developing a powerful and expressive harmony. Ecology was an important consideration throughout the design – *Miura* is recyclable. Dimensions: seat height: 78 cm; depth: 40 cm; width: 47 cm; weight: 4.2 kg.

KAA *Duscharmaturen*
Shower taps

DIE KAA DUSCHARMATUR VON AGAPE überzeugte die Jury durch den innovativen Einsatz des Antiallergiematerials Silikon. Die Funktionsflächen der Armaturen wölben sich durch den Wasserdruck nach außen. Der Entwurf besticht durch seine Simplizität und sympathische Farbigkeit. Er erfüllt die Hauptaufgabe guten Designs, nämlich die Dinge neu zu sehen.

THE KAA SHOWER FITTING BY AGAPE convinced the jury with its innovative use of the anti-allergic material, silicon. The functional surfaces of the fittings curve outwards under water pressure. The design appeals for its simplicity and congenial colouration. It fulfils the principle task of good design, namely seeing things in a new light. — HERBERT H. SCHULTES

Jurybegründung

Statement of the Jury

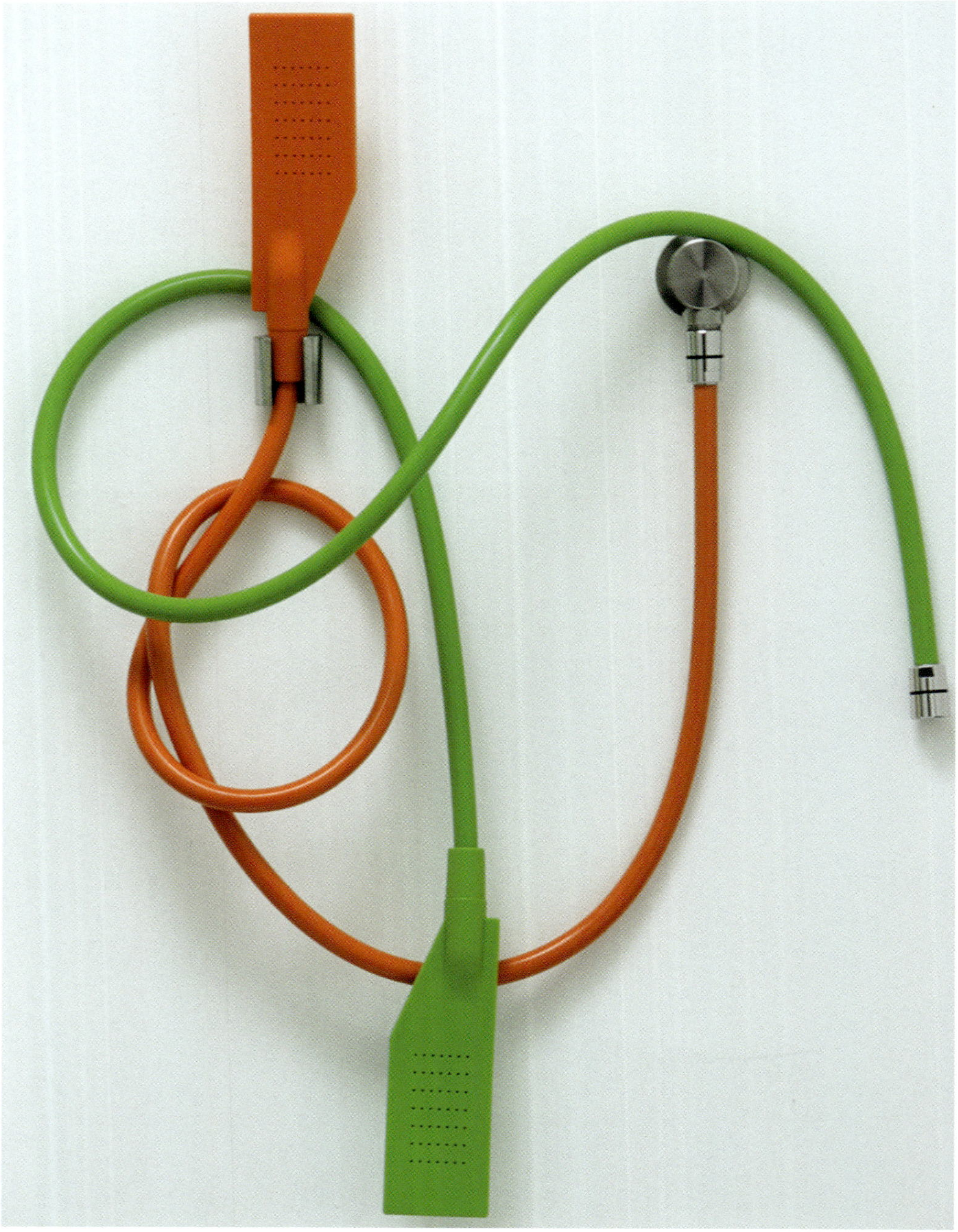

Unternehmen Company

agape Via Po Barna 69 — 46031 Bagnolo San Vito — Italy — www.agapedesign.it

Design

Giulio Gianturco www.giuliogianturco.com

DAS KONZEPT SAH DEN ENTWURF eines Duschkopfes und einer Handbrause vor, die unempfindlich gegen Kalkablagerungen sind, eine weiche Haptik mit guter Griffergonomie verbinden und möglichst leicht sind. ——— Seit vielen Jahren werden, um das Entfernen von Kalkablagerungen zu vereinfachen, Duschköpfe mit einer Silikongummischicht mit weichen Wasserdüsen überzogen, da auf diese Weise die Kalkreste einfach mit der Hand abgewischt werden können. Silikongummi ist für die Verwendung in Badezimmereinrichtungen ideal, da es unempfindlich gegen Kalkablagerungen, weich anzufassen ist und eine ausgezeichnete Griffigkeit gewährleistet. Somit war dies der Ausgangspunkt für das Projekt. ——— Das Projekt wurde in Zusammenarbeit mit der technischen Abteilung von AGAPE entwickelt und führte zur Umsetzung von KAA, das dank der Verwendung von antiallergenem Silikongummi, die Anmutung einer leichten Massage vermittelt, um das Wohlbefinden zu erhöhen. Die Farbpalette (Grau, Orange, Grün) und die Verpackung runden das Projekt ab.

THE CONCEPT INVOLVED DESIGNING a shower head and a hand-held shower that would be resistant to limescale deposits, soft to the touch, with a good grip and, if possible, lightweight. ——— For years now, in order to facilitate the removal of limescale build-up, the bodies of shower heads have been fitted with a silicone rubber membrane with soft water nozzles, which can be cleaned by simply wiping with the hand. Silicone rubber is ideal for use in bathroom applications, as it is resistant to limescale, soft, and offers excellent grip, and so this was the starting point for the project. ——— The project was developed in collaboration with AGAPE's technical department, and led to the realization of KAA, which, thanks to the employment of anti-allergy silicone rubber, evokes the softness of massage to restore personal well-being. The range of colours (grey, orange, green) and packaging complete the project.

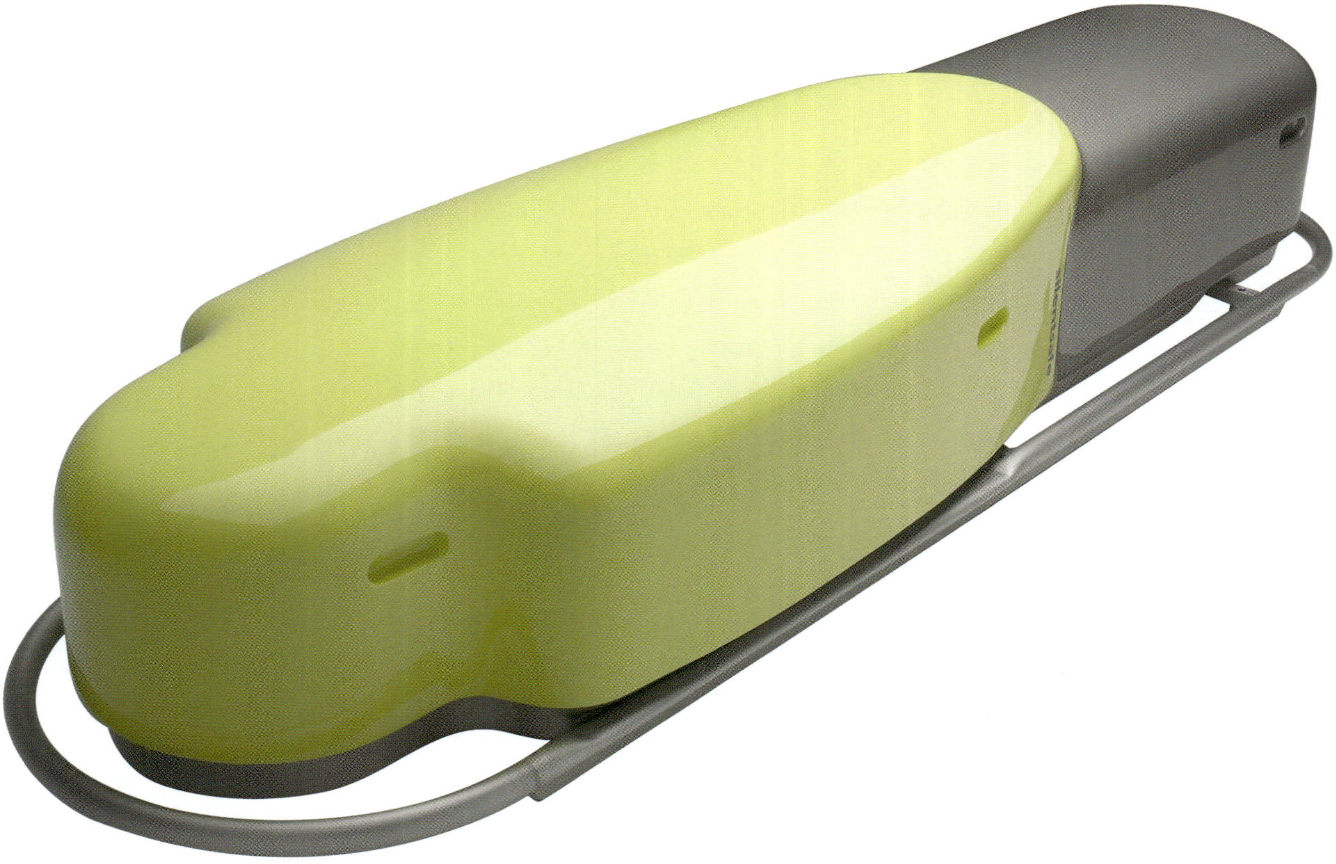

Silentsafe

Leichentransport- und Bergesystem

System to transport and salvage dead bodies

AUF DEN ERSTEN BLICK ÜBERZEUGTE dieses Leichentransport- und Bergesystem die Jury vor allem durch die gelungene Umsetzung von funktionalen Anforderungen in eine moderne Ästhetik. So leitet sich die Form einerseits aus den Anforderungen für Leichen unterschiedlichen Alters, Gewichts und unterschiedlicher Größe nutzbar zu sein und dabei von einer Person gehandhabt werden zu können, ab; zugleich übersetzt sie aber auch piktogrammartig Ergonomie in Form. Auch die kräftige Farbgestaltung ist zunächst der Nutzung in verschneitem Gelände geschuldet, stellt jedoch zugleich eine mutige Entscheidung für ein Leichentransportsystem dar.

AT FIRST GLANCE this corpse transport and rescue system convinced the jury in particular for its successful conversion of functional demands into modern aesthetics. On the one hand the form derives from the practical demands of transporting corpses of different age, weight and size single-handedly. On the other it translates ergonomics into form, rather like a pictogram. The strong colour design is also due to usage in various terrains, but at the same time represents a courageous decision for a hearse system. — NICOLETTE NAUMANN

Jurybegründung
Statement of the Jury

Unternehmen Company

Starmed GmbH Im Lehrer Feld 30 — 89081 Ulm — Germany — www.starmed.eu

Design

Reform Design Matthias Wagner — www.refom-design.de

Werksdesign/Inhouse René Stern

DAS NEUARTIGE LEICHENTRANSPORT- UND BERGESYSTEM besteht aus einem in Leichtbauweise hergestellten, hochwertigen Transportsarg und einer entnehmbaren, speziellen Schaufeltrage. Trotz kompakter Abmessungen bietet der *Silentsafe* viel Platz, auch für große oder schwergewichtige Leichen. Das wird zum einen durch die besondere Ausformung der Schaufeltrage erreicht, die die Schultern anhebt und die Leiche schmäler werden lässt. Zum anderen ermöglicht die anspruchsvolle Körperform des *Silentsafe* eine optimale Raumnutzung bei geringen Außenabmessungen. Wenn die Beinsegmente der Schaufeltrage angewinkelt werden, können auch etwas längere Körper gelagert werden, da sich dadurch die Gesamtlänge um einige Zentimeter reduziert. Der *Silentsafe* ist so konzipiert, dass eine Überführung ebenso wie das Umlagern von Tisch zu Tisch von einer Person durchgeführt werden kann. Die Arbeitsergonomie für den Bestatter wurde deutlich verbessert. So können Leichen dank der Schaufeltrage mühelos zum Beispiel vom Boden aufgenommen werden. Die Materialien, die besonders reinigungsfreundlich sind, und glatte, desinfektionsmittelbeständige Oberflächen garantieren optimale Hygiene.

THE NOVEL BODY TRANSPORTATION AND RESCUE SYSTEM consists of a high-quality coffin of light construction and a removable special scoop stretcher. Despite compact dimensions, the *Silentsafe* offers plenty of room even for larger or heavier bodies. On the one hand, this is achieved thanks to the special shape of the scoop stretcher, which lifts the shoulders and makes the body narrower. On the other hand, the sophisticated body shape of the *Silentsafe* enables optimal space utilization where there is little room. When the leg segments of the scoop stretcher are set at an angle, it can also be used for longer bodies, because this reduces the total length by a few centimetres. The *Silentsafe* is designed so that one person is able to pass it on or transfer it from one table to another. Work ergonomics have been considerably improved for the undertaker. For example, bodies can easily be lifted up from the floor thanks to the scoop stretcher. The materials, which are particularly easy to clean, and smooth, disinfectant-resistant surfaces guarantee optimal hygiene.

Rosenbauer Panther 8x8 *Flughafen-Feuerwehrfahrzeug*

Airport fire engine

DAS NEUE 8X8 MODELL PANTHER mit acht angetriebenen Rädern, einem Einsatzgewicht von 40 Tonnen, 14.500 l Löschmitteln und 140 km / h Geschwindigkeit überzeugt vor allem durch die straffe Gliederung in Ladevolumen, Fahrerhaus und Antriebsrädern. Raffiniert gelungen ist die Gestaltung des Fahrerhauses durch den Einsatz großzügig dimensionierter Panorama- und Seitenscheiben und der nach hinten schräg geneigten Trennungslinie zwischen Fahrerhaus und Löschmitteltank. Das futuristische Aussehen im Frontbereich entsteht vor allem durch die keilförmige Ausbildung der Panoramascheibe und des extrem angewinkelten Unterbodenbleches. Diese ausgeprägte und formal anspruchsvolle Linienführung erzeugt eine hohe Wiedererkennbarkeit.

THE NEW 8X8 PANTHER MODEL with eight driven wheels, a read-to-go weight of 40 tons 14,500 litres extinguishing agent and 140 km / h speed is especially convincing for its tight division into loading volumes, driver's cab and driving wheels. The design of the driver's cab succeeds ingeniously with large sized panorama and side windows and the dividing line between driver's cab and extinguishing agent tank which slopes towards the back. The futuristic appearance in the front section is created primarily by the wedge-shaped construction of the panorama window and the extremely bent undercarriage plate. The distinct and formally demanding lines give it high recognition value.

— HERBERT H. SCHULTES

Jurybegründung

Statement of the Jury

Unternehmen Company

Rosenbauer International AG Paschinger Straße 90 — 4060 Leonding — Austria — www.rosenbauer.com

Design

Spirit Design / Innovation and Branding Daniel Huber, Tomas Maly — www.spiritdesign.at

ZIEL DER NEUENTWICKLUNG des *Rosenbauer Panther* war es, Design und Technik auf die künftigen Anforderungen der Flughafenfeuerwehren auszurichten. Schwerpunkt waren Ergonomie, dynamisches Design, Sicherheit, moderne Werkstoffe, Leistungsfähigkeit der Löschtechnik sowie die Anpassung an neue Umweltvorschriften. Mit der kraftvollen Formensprache betont Spirit Design die Attribute Verlässlichkeit, Sicherheit, Kraft und Dynamik. Die Kabine des *Panther* verfügt über eine Panorama-Windschutzscheibe aus Verbundsicherheitsglas und transparente Türen. Durch die weit nach oben und unten gezogenen Front- / Seitenscheiben überblickt der Fahrer seine Umgebung optimal.

THE AIM OF THE *Rosenbauer Panther* 8x8 redevelopment project was to gear the vehicles' design and engineering to airport fire brigades' future requirements. The focus was placed on ergonomics, dynamic design, safety, modern materials, high performance fire-fighting systems, and adaptation to new environmental regulations. Spirit Design has used a vigorous stylistic idiom to emphasise such attributes as reliability, safety, power and dynamism. The cab has a panoramic windscreen made of safety glass, and transparent doors. The front and side panels extend a long way up and down, giving the driver an unobstructed wide-angle view of the surroundings.

Fendt 936 Vario *Traktor*

Tractor

DIE NEU GESTALTETE, TRANSPARENTE MOTORHAUBE, die zahlreichen gut durchdachten Beleuchtungselemente und die großzügige transparente X5 Kabine, die ein Gefühl von Offenheit vermittelt, sind die ersten ungewöhnlichen Gestaltungsmerkmale, die beim näheren Betrachten des Traktors *Fendt 936 Vario* angenehm auffallen. Ergonomische Gesichtspunkte wurden bei der Gestaltung der Kabine ebenso einbezogen wie Überlegungen zum Thema Fahrsicherheit und Fahrkomfort. Das Beispiel *Fendt 936 Vario* zeigt, dass dieser Großtraktor im Hochleistungsbereich nicht nur den neuesten technischen Anforderungen für eine professionelle Landwirtschaft entspricht, sondern auch formal hohen ästhetischen Ansprüchen im Spitzensegment gerecht wird.

THE NEWLY DESIGNED, TRANSPARENT MOTOR COVER with its numerous well-conceived lighting elements and the generous transparent X5 cabin which gives a feeling of openness are the first unusual design characteristics which on closer inspection of the *Fendt 936 Vario* tractor are really a very pleasant surprise. Ergonomic factors as well as driving safety and comfort considerations were integrated in the design of the cabin. The *Fendt 936 Vario* shows that this large tractor in the high performance field not only complies with the technical demands of professional farming but also a high level of formal aesthetic demands in a top-notch market. — TORSTEN NEELAND

Jurybegründung
Statement of the Jury

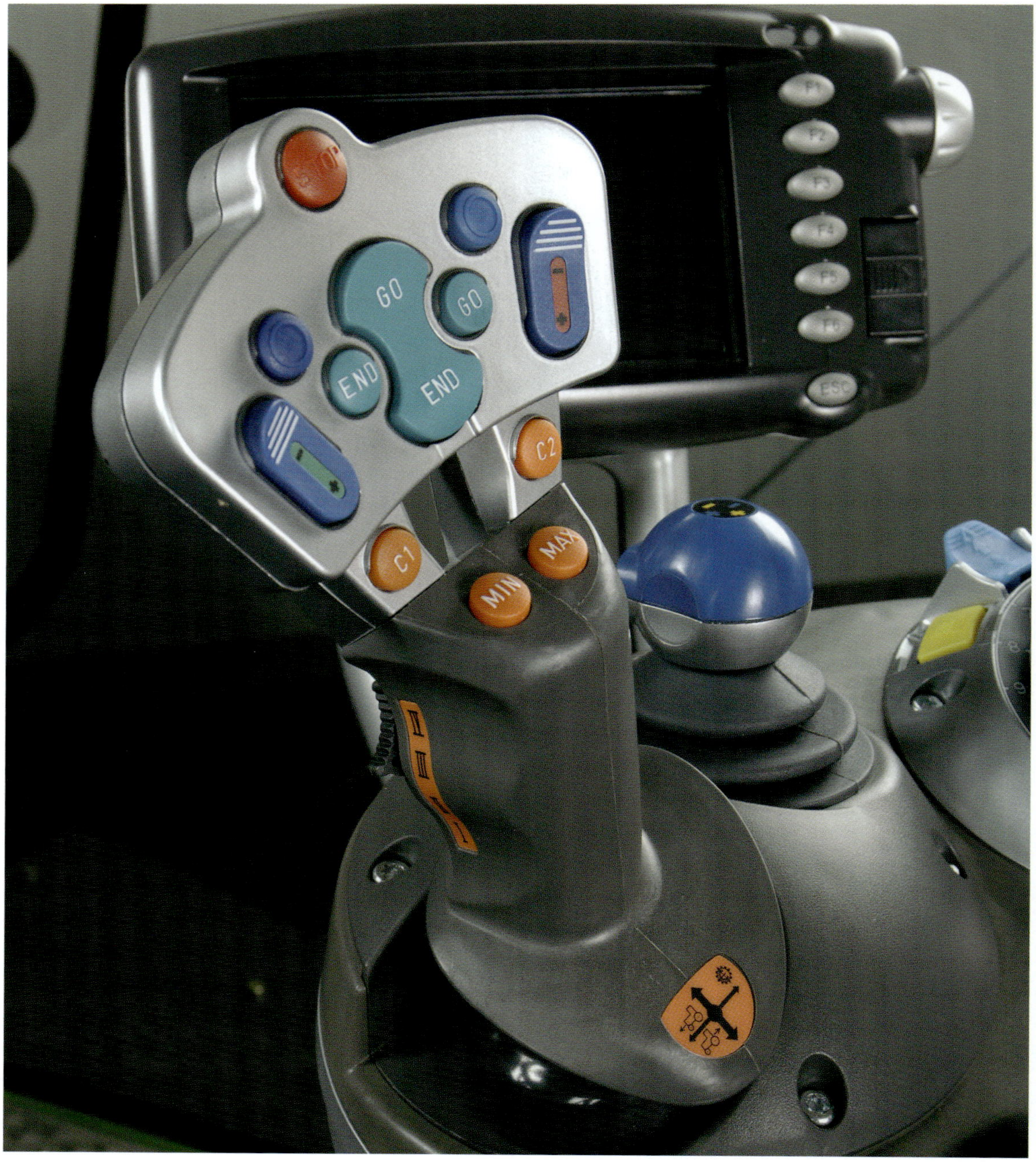

Unternehmen Company

AGCO GmbH Fendt — Johann-Georg-Fendt-Straße 4 — 87616 Marktoberdorf — Germany — www.fendt.com

Design

Werksdesign/Inhouse Johann Epple, Tom Maier

DER FENDT 936 VARIO bietet als stärkster und schnellster Standardtraktor leistungsstarke, zeitsparende und kostenorientierte technische Lösungen für Feld und Straße im Gesamtpaket. Charakteristisch für diesen neuen *Fendt* Großtraktor sind neben den technischen Daten, das ausdrucksstarke Erscheinungsbild, bei dem Form und Funktion konsequent und optimal aufeinander abgestimmt sind. Die sich zur Kabine hin stark verjüngende und nach vorne abfallende Motorhaube signalisiert durch den dominanten Kühlergrill und die seitlichen Lufteinlässe eindrucksvoll die Kraft und Leistungsfähigkeit des Fahrzeugs. Das neue Beleuchtungskonzept des *936 Vario* gewährleistet eine 360° Ausleuchtung der Fahrzeugumgebung. ——— Die komplett neu entwickelte, 70 dB leise, serienmäßig klimatisierte Kabine X5 bietet dem Fahrer hervorragende Rundumsicht und höchsten Komfort. Das Variocenter im Inneren umfasst als Arbeitsplatz des neuen *936 Vario* alle Bedienelemente des Schleppers zentral am Lenkturm und auf der Multifunktionsarmlehne des Fahrersitzes. Diese Armlehne bietet außerordentliche Verstellwege, wodurch das bequeme Auslösen sämtlicher Arbeitsfunktionen ohne Relativbewegung zwischen Fahrer und Bedienelement ermöglicht wird. Der optional aktivgefederte Sitz lässt sich inklusive Armlehne insgesamt 20° nach links und rechts schwenken. Zusätzlich kann bei der optional angebotenen Rückfahrvariante der gesamte Fahrerstand (Lenkturm, Sitz und Armlehne) um 180° im Uhrzeigersinn gedreht werden. Jeder Fahrer kann sich damit für jeden Einsatz seinen Arbeitsplatz ideal einrichten.

THE MOST POWERFUL AND SPEEDIEST of standard tractors, the *Fendt 936 Vario* offers high-performance, time-saving and cost-oriented technical solutions for the field and the road as a complete package. Alongside its technical capabilities, the new large tractor by *Fendt* boasts a strongly expressive appearance with shapes and function in consistent and optimum harmony. The engine hood tapers severely towards the cab and falls away sharply at the front. Its prominent radiator grill and its air inlets on the sides impressively accentuate the vehicle's power and its capability. The *936 Vario's* new lighting concept allows for the area around the vehicle to be illuminated over a radius of 360°. ——— Completely redesigned, only 70 dB loud and equipped with air-conditioning as standard, the X5 cab affords the driver an outstanding all-round view and absolute comfort. As the workstation of the new *936 Vario*, the Variocenter inside the tractor houses all the operating elements centrally on the steering columns and on the multifunction armrest on the driver's seat. This armrest offers amazing adjustment options, meaning that the driver can trigger all work functions without relative movements between himself and the controls. The seat is optionally available with active springs and can, together with the armrest, be rotated by as much as 20° to the left or to the right. Additionally, a version with an optional reverse function allows the entire driver unit (steering column, seat and armrest) to be rotated 180° clockwise. As a consequence, every driver can set his workplace to the ideal position for himself and for any job.

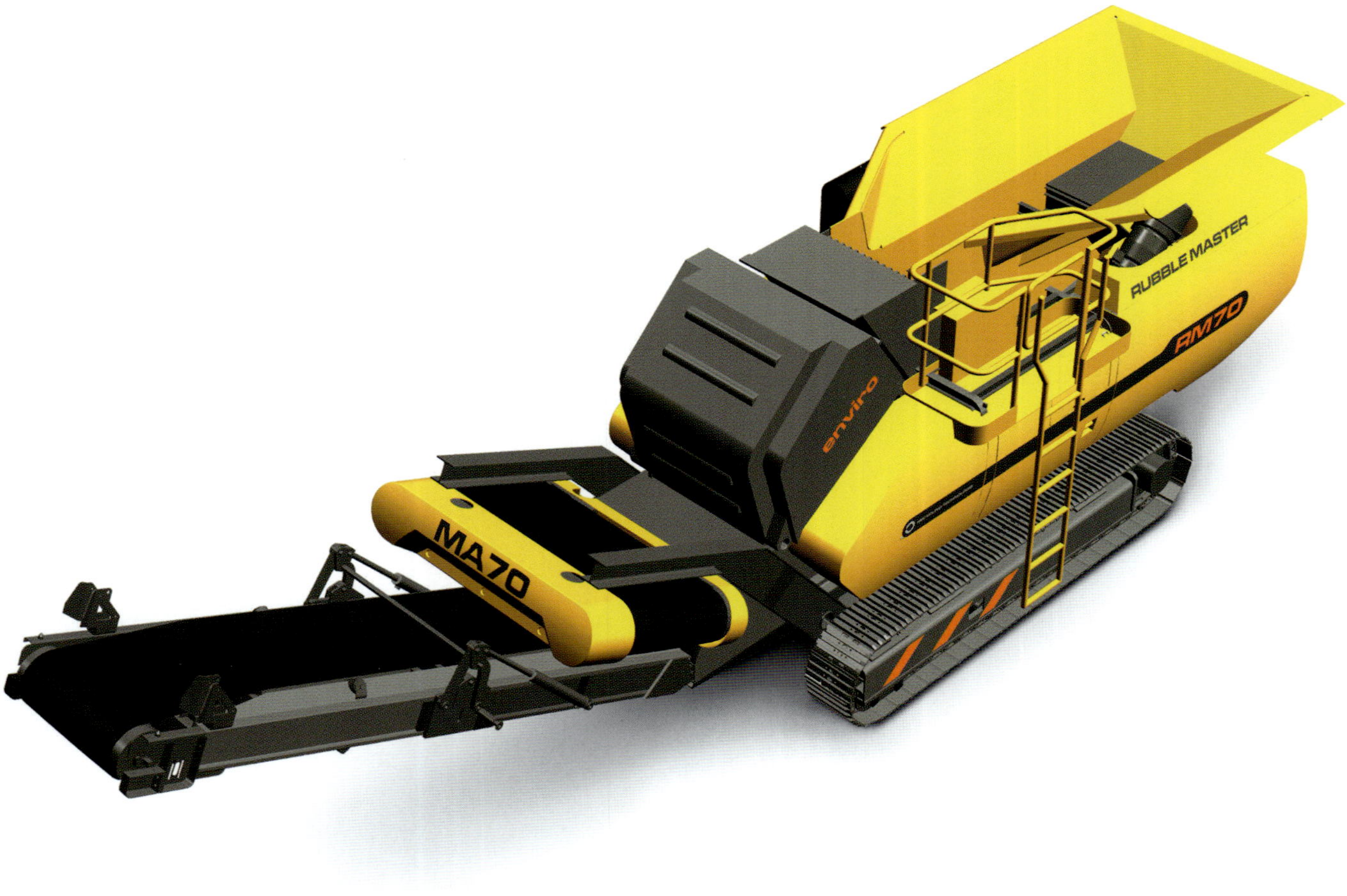

RM70 Compact Recycler

Bauschutt-Recyclinganlage
Rubble recycling plant

DER RM 70 COMPACT RECYCLER ist eine selbstfahrende Bauschutt-Recyclinganlage mit hoher Brecherleistung. Dank der niedrigen Emissionswerte bei Lärm und Staub ist der kompakte RM 70 der ideale Innenstadtbrecher, optimal für kleine und mittlere Baustellen in sensiblen Gebieten. Sie ist wohltuend klar und minimalistisch gestaltet, eindeutig in statische und dynamische Baugruppen unterteilt. Durch die Überwölbung der seitlichen Gehäuseflächen wird beim Aufprallen von beispielsweise zurückfallendem Gestein eine Verbeulung vermieden. Unabhängig von der klaren, formal anspruchsvollen Linienführung ist hier eine Ikone für diesen Typ Baumaschinen entstanden.

THE RM 70 COMPACT RECYCLER is a self-operating recycling system for building waste, with strong crushing power. Thanks to the low emission values for sound and dust the compact RM 70 is the ideal inner city crusher, optimal for small and medium building sites in sensitive areas. It features a pleasingly clear minimalist design and is divided clearly into static and dynamic components. The excessive curving of lateral casing elements helps to avoid dents caused by falling stone. Independent of the clear, formally demanding lines it has become an icon for this type of construction machine.

— HERBERT H. SCHULTES

Jurybegründung

Statement of the Jury

Unternehmen Company

HMH Engineering-Consulting-Trading GmbH Im Südpark 196 — A-4030 Linz — Austria — www.rubblemaster.com

Design

idukk industrial design union kittler kurz partner Reinhard Kittler, Heinrich Kurz, Gerald Fasching — www.idukk.at

RM 70 IST EINE SELBSTFAHRENDE BAUSCHUTT-RECYCLINGANLAGE mit hoher Brecherleistung. Durch die speziell niedrigen Emissionswerte betreffend Staub, Lärm und Abgasen sowie den kompakten Abmessungen eignet sich RM 70 optimal für den Einsatz im innerstädtischen Raum. Mit nur 20,5 to Transportgewicht ist die Maschine ohne Sondergenehmigung transportierbar und aufgrund des – für derartige Maschinen niedrigen – Bodendrucks auf jeder Baustelle leicht zu manövrieren. RM 70 zeichnet sich durch besondere Servicefreundlichkeit aus. Das Ground-Based-Anordnungskonzept ermöglicht Bedienungs- und Wartungstätigkeiten vom Boden aus. Die Maschinenverkleidung wird von großflächig überwölbten Stahlblechverkleidungen bestimmt, die sich zum Förderband hin verjüngen. ——— Die Maschine wirkt dadurch kraftvoll und dynamisch. Neben der formalen Funktion übernehmen die Verkleidungsteile die Schallführung / Schalldämmung bzw. durch die Überwölbung entsprechende Stabilisierungsfunktion gegenüber auftreffendem Baumaterial, Steingut etc.

RM 70 IS A SELF-PROPELLING DEMOLITION RUBBLE FACILITY with a high breaking capacity. With unusually low dust, noise and exhaust fume emissions plus its compact dimensions, RM 70 is ideal for use in inner-city areas. With a transport weight of only 20.5 tons the machine does not require a special transport permit and, thanks to its low – for this type of machine – load pressure it is easy to manoeuvre on any construction site. RM 70 is extremely service-friendly. Its ground-based array system means that it can be operated and serviced from the ground. The covering of the machinery is dominated by vaulted steel sheeting that tapers off towards the conveyor belt. ——— This feature lends the machine a powerful, dynamic appearance. Alongside its formal function the covering acts as a noise conductor / absorber and the vaulting performs a stabilizing function if it is hit by stones, rubble etc.

PLAYMOBIL Transport & Verkehr Hafen
PLAYMOBIL Cargo Harbor

Spielzeug
Toys

KLAR SCHIFF FÜR MARITIMES ROLLENSPIEL! An der *Playmobil-Hafenwelt* gefiel der Jury besonders, dass Kinder spielerisch an Themen der modernen Arbeitswelt herangeführt werden. Längst ist *Playmobil* international ein Begriff für unverwüstliche und interaktive Spielwelten. Die Hafenwelt besticht durch eine sachliche Gestaltung, die die Funktionalität der einzelnen Spieloptionen veranschaulicht. Die Detailfülle macht Kinder vertraut mit den Rollen der verschiedenen Akteure im Frachthafen; erste physikalische Erkenntnisse sind unaufdringlich ins Spiel integriert.

CLEAN SWEEP FOR MARITIME ROLE PLAY! In the case of *Playmobil's Harbor World* the jury were especially impressed with how it playfully leads children to themes in the modern working world. *Playmobil* has long since been an international concept for an indestructible and interactive play world. The harbour series captivates with its down-to-earth design, which visualizes the function of individual play options. The wealth of detail serves to familiarize children with the roles of the various actors in a freight harbour; first physical knowledge is integrated unobtrusively in the game.
— NICOLETTE NAUMANN

Jurybegründung

Statement of the Jury

Unternehmen Company

geobra Brandstätter GmbH & Co. KG PLAYMOBIL — Brandstätterstraße 2–10 — 90513 Zirndorf — Germany — www.playmobil.de

Design

Werksdesign/Inhouse

IN DER PLAYMOBIL-HAFENWELT sind die Kinder und der multifunktionale Hafenkran (Höhe: 57 cm) ein eingespieltes Team. Der Kranaufsatz des vierbeinigen Giganten lässt sich um 360 Grad drehen. Sein Arm hebt mit Power jede Last. Mit den Kurbeln an der Seite ist das Be- und Entladen ganz einfach: Schaufel öffnen – Kohle fassen – Schaufel schließen und schon ist die Ladung sicher im großen Frachtschiff (Länge: 60 cm) verstaut. Im Inneren des schwimmfähigen Superfrachters (aufrüstbar mit Unterwassermotor) findet sich Platz für Frachten aller Art. Die Arbeiter mit Hubwagen, Kleinstapler und Sackkarre sowie die „Seebären" an Bord packen kräftig mit an. Dank der schwimmenden Leucht-Bojen mit Blinkfunktion hält der Ozeanriese sicher den Kurs und ist bereit für eine geballte Ladung Spielspaß. Ein schnittiges Zollschiff ergänzt die vielseitige Spielwelt in PLAYMOBIL-Design und -Qualität.

THE CHILDREN AND THE MULTI-FUNCTIONAL HARBOUR CRANE (height: 57 cm) are an efficient team. The crane superstructure of the four-legged giant can be rotated around a full circle. Its powerful arm can lift any load. With the cranks on the sides loading and unloading are completely easy: Open the tip – grab coal – close the tip and store the freight safely away in the Cargo Ship (length: 60 cm). There is room for freight of any kind within the floating super freighter (which can be equipped with an underwater motor). The Dock Worker with pallet truck, mini forklift or hard cart and the sailors on board work hard to get the job done. Thanks to the floating and blinking light buoys, the huge ship stays safely on course and is ready for a full load of play fun. The floating harbour police complete the play world.

BMW Cruising Bike

Fahrrad

Bicycle

MAN WÜRDE DENKEN, dass die Möglichkeiten des Designs beim Fahrrad mehr oder weniger ausgeschöpft sind. Dennoch werden Forscher nicht müde, das ultimative, vom Menschen angetriebene Zweirad weiterzuentwickeln. Das BMW *Cruising Bike* ist insofern noch ein weiterer Schritt in die Zukunft. Durch den Einsatz von Technologien, die man bislang in der Welt des Zweirads noch nicht zu sehen bekommen hat, hat BMW ein hervorragendes Fahrrad entwickelt, bei dem die Technologie eine neue Ästhetik definiert, die nicht nur Styling ist. Die Form ergibt sich aus der Dynamik der verwendeten Materialien und hat als solche eine ihr ganz eigene Schönheit. Das Design beschränkt sich auf das Wesentliche, ganz so wie ein Fahrrad sein soll.

ONE WOULD THINK that there is little left to design in a bicycle. However the research for the ultimate human powered two-wheeler continues. The BMW *Cruising Bike* is yet another step into the future. Utilising technologies so far unseen in the bicycle world they have produced a superb machine where the technology defines new aesthetics which are not just styling. The form comes from the dynamics of the use of the materials and as such has an inherent beauty. The design is essential just like a bicycle should be. — JAMES IRVINE

Jurybegründung
Statement of the Jury

Unternehmen Company

BMW Group Knorrstraße 147 — 80788 München — Germany — www.bmw.com

Design

BMW Group Design Designworks USA — Studio München

BEIM BMW CRUISE-BIKE sorgt die Rahmengeometrie, die auf dem neuesten Stand der Technik ist, für optimalen Sitzkomfort. Das Fahrrad besticht durch seinen dynamischen Auftritt und sein Design. Durch die Verwendung von ungewöhnlichen Schmiedeteilen werden fließende Übergänge an den Verbindungsstellen ohne sichtbare Schweißnähte erreicht: Hydroforming sorgt hier für ein ungewöhnliches Rahmendesign, das einzigartig in der Fahrradkonstruktion ist. Es ermöglicht eine präzise Linienführung sowie konvexe und konkave Oberflächen. ——— Das Fahrrad ist mit einem 24-Gang Alivio Schaltwerk von Shimano ausgestattet. Damit wird Cruisen auch in hügeligem Gelände zum Vergnügen. Der unschlagbare Vorteil des BMW *Cruise-Bikes* liegt in seiner Vielseitigkeit. Die Suntour NEX-4000 Teleskopgabel gleicht Unebenheiten einfach aus. Über die notwendige Bremskraft muss man sich mit dem Cruiser auch keine Sorgen machen. V-Brakes von Shimano stoppen schnell; die Bremszüge verlaufen integriert im Rahmen.

WITH ITS SATE-OF-THE-ART TECHNOLOGY, the geometry of the BMW *Cruise-Bike's* frame offers maximum sitting comfort. The bicycle stands out for its dynamic appearance and design. Its unusual forging makes for smooth transitions at the joints without visible welding seams, here, hydro-forming creates an unusual frame design that is unique in bicycle construction. This makes both precise line management and convex and concave surfaces possible. ——— The bicycle comes equipped with a 24-gear Alivio control unit by Shimano which makes cruising a pleasure, even in hilly terrain. The unbeatable advantage of the BMW *Cruise-Bike*, however, is its versatility. The Suntour NEX-4000 telescopic fork compensates for uneven surfaces. And there is no need to worry about whether the Cruiser will demonstrate the necessary braking capabilities. Shimano's V-Brakes stop quickly; the brake cables are integrated into the frame.

+F50 TUNIT *Fußballschuh*

Football shoe

ES IST 2006 UND SPORT IST MEHR ALS ENTSPANNUNG. Sport ist ein großes Geschäft. Worte, die nicht nur auf die großen Athleten zutreffen, sondern auch auf die Hersteller von Sportartikeln, die sich im Wettbewerb um das technologisch fortgeschrittenste und coolste Produkt gegenseitig aus dem Rennen zu schubsen versuchen. Lederfußballschuhe zum Beispiel werden bald der Vergangenheit angehören. Der +F50, der sich dem Fuß wie ein Handschuh anpasst, ist ein Hightech-Produkt, das zugleich äußerst leicht ist und aus widerstandsfähigen Materialien wie Mikrofasern besteht und sich durch ein dynamisches Design auszeichnet. Das Obermaterial ist in einem breiten Farbspektrum erhältlich und kann persönlich ad infinitum ausgewählt werden. Der Unterbau wiegt weniger als 75 Gramm und die Stollen gibt es in drei verschiedenen Ausführungen. Die Materialien, der Anblick und die Möglichkeit in passgenauen Schuhen Fußball zu spielen, machen den +F50 zu einem Produkt des 21. Jahrhunderts, das Leistung, Komfort und Individualität verbindet.

IT'S 2006, AND SPORT IS MORE THAN RELAXATION. Sport is big business. Words that apply not only to professional athletes but also to manufacturers of sports equipment, who continually jostle and shove in the race for the most technologically advanced – and the coolest – product. Leather football shoes, for example, will soon be a thing of the past. The +F50, which fits the foot like a glove, is a high-tech product that combines lightweight, durable materials such as microfibres with a dynamic design. Uppers that come in a wide range of colours can be customized ad infinitum. The chassis weighs less than 75 grams, and three types of studs are available. Materials, looks and the opportunity to play ball in customized footwear make the +F50 a 21st century product that integrates performance, comfort and individuality. — ROBERT THIEMANN

Jurybegründung
Statement of the Jury

Unternehmen Company

adidas AG Adi-Dassler-Straße 1 — 91074 Herzogenaurach — Germany — www.adidas.com

Design

Werksdesign/Inhouse adidas Global Design — Matthias Bieling, Senior Designer Football

DER +F50 TUNIT IST EINE NEUENTWICKLUNG im X-EDGE-Segment der adidas-Fußballschuhe. Das Ziel war es, einen Fußballschuh zu entwickeln, der den Spieler begeistert und motiviert und seinen Teil dazu beiträgt, diesen bei Höchstleistungen zu fördern und zu unterstützen. Das Design des +F50 TUNIT ist modular aufgebaut. Es ermöglicht dem Spieler, den Schuh seinem persönlichen Style anzupassen und für unterschiedliche klimatische Bedingungen und Bodenverhältnisse zu konfigurieren. Fließend-dynamische Grafik- und Konstruktionslinien bilden eine Symbiose mit kristallenen-facettierten Strukturen. Das bestimmende Grafikelement im Vorfuß schwingt zusammen mit den drei Streifen dynamisch um den gesamten Schuh und bildet ein zusammenhängendes, umschließendes 360°-Gefüge, das aus jedem Betrachtungswinkel erkennbar ist. Alle grafischen Elemente bilden sich aus dynamischen „Echo-Linien", die an Wellenbilder im Wasser erinnern und Bewegungsfluss und Leichtigkeit vermitteln. Der visuelle Focus liegt auf dem Vorfuß. Der Schuh erinnert – unterstützt von einem eigens für diesen Schuh entwickelten Leisten – mit seinen Proportionen an einen Sprintschuh und signalisiert einen starken Vorwärtsdrang sowie Schnelligkeit. Das facettierte Material bildet eine hochfunktionale „zweite Haut", die den gesamten Schuh überzieht und durchdringt. Das Material ist anschmiegsam und gleichzeitig pflegeleicht, abriebfest und langlebig. In Kombination mit einer neuartigen Konstruktion ermöglicht es eine extrem gute Passform, besonders wichtig für ein optimales Ballgefühl. Die Schnürsenkelabdeckung bietet zusätzlich eine saubere und homogene Kontaktfläche und vereinfacht somit die Kontrolle über den Ball.

THE +F50 TUNIT IS A NEW DEVELOPMENT in the X-EDGE segment of adidas football boots. The goal was to create a football boot that excites and motivates the player, helping as well as supporting the player for maximum performance. The +F50 TUNIT has a modular design. It allows players to adjust the boot to their personal style and to tune it according to varying climate or ground conditions. Flowing dynamic graphic and construction lines form a symbiosis with crystal-facetted structures. The dominating graphic element at the forefoot, together with the three stripes, dynamically swings around the entire shoe to form a coherent, encompassing a 360° arrangement which can be recognised from all angles. All graphic elements are comprised of dynamic "echo lines" which resemble waves in the water and communicate the flow of movement as well as lightness. The visual focus is on the forefoot. Supported by a last specifically developed for this shoe, the boot bears a resemblance to a sprinter's shoe in its proportions, thereby signalling strong forward propulsion and speed. The facetted material performs like a highly functional "second skin" which encompasses and penetrates the entire boot. The material has a snug fit and at the same time is easy to clean, abrasion resistant and durable. The combination with a new kind of construction allows for a glove-like fit, which is especially important for optimal feeling for the ball. The lace cover additionally offers a clean and homogenous point of contact, facilitating ball control.

FALKE Second Layer *Pullover*

Pullover

DASS DAS ÜBER 110 JAHRE ALTE SAUERLÄNDISCHE UNTERNEHMEN den Weg vom mittelständischen Strickwarenhersteller zur internationalen Modemarke geschafft hat, verdankt es sicher weitsichtiger Entscheidungen der in vierter Generation von Familienmitgliedern besetzten Geschäftsführung. Mit der Produktion der intelligenten, nahtlosen Sportbekleidung, die Sportlern in Extrembedingungen als Körperklimaregulierung dient, zeigt das Textillabel FALKE seine Zukunftsorientierung: Nur mit hochwertiger Hightech-Produktion ist eine qualitative Unterscheidung zu asiatischer Billig-Konkurrenz zu erzielen. Der FALKE Sportpullover überzeugt in Material und Design.

THE FACT THAT AN OLD SAUERLAND COMPANY which has been in business for more than 110 years has succeeded in transforming from a medium-sized knitwear manufacturer into an international fashion brand, is surely due to the broad-minded decisions of the management consisting of fourth generation family members. With the production of the intelligent, seamless, functional sports pullover which serves as a body climate regulator for sportsmen in extreme conditions, the textile label FALKE demonstrates its future orientation: Only with high quality, hi-tech production is it possible to achieve a qualitative difference to cheap Asian competitors. The FALKE sports pullover is convincing in both material and design. — BARBARA FRIEDRICH

Jurybegründung
Statement of the Jury

Unternehmen Company

FALKE KG Oststraße 5 — 57392 Schmallenberg — Germany — www.FALKE.com

Design

N3twork Dennis Hinze — www.n3twork.de

MIT SEINER EIGENEN SPORTRANGE, FALKE Ergonomic Sport System, hat sich das Traditionsunternehmen FALKE in den vergangenen Jahren als Innovationsführer im Bereich funktioneller Sportbekleidung etabliert. Das gilt auch für den prämierten FALKE Sportpullover für Active Sports, der in konsequenter Fortsetzung und Ergänzung zu dem bestehenden Sportprogramm entwickelt wurde. Der in Seamless-Stricktechnik verarbeitete Hightech-Pullover folgt dem grundlegenden Prinzip der Ergonomie und verbindet auf perfekte Weise höchste Funktionalität und hochmodisches Design. Als absoluter Vorreiter im Markt bietet FALKE hiermit erstmalig Funktionalität auf Basis der Rundstricktechnik, die eine optimale, körpernahe Passform für ideale Beweglichkeit garantiert. Die innovativen Techniken, wie die anatomisch konstruierte Rippenstruktur, gestricktes Plush auf der Innenseite sowie die Intarsien in Kombination mit funktionalen Garnen bilden die Grundlage für die Entwicklung von ultimativer Form und Funktion des FALKE Sportpullovers. Dabei werden die natürlichen Körperkonturen hervorgehoben und die Definition der Muskulatur durch die ungewöhnliche Nahtführung akzentuiert – Ästhetik und Funktion in vollendetem Design.

IN RECENT YEARS, the FALKE company with long-standing tradition has established itself as an innovative leader in the functional sports wear sector with its own FALKE Ergonomic Sport System range. This also encompasses the award-winning FALKE high-tech sports pullover for active sports, which has been developed as a logical continuation and supplement to the existing sports range. The seamless knitted high-tech pullover is based on ergonomic principles, and perfectly combines functionality and a highly fashionable design. As a market leader, FALKE provides functionality based on circular knit technology for the first time, ensuring perfect, body-contoured fit for optimal movement. The innovative technologies such as the anatomical rib structure, knitted plush on the inside and inlays form the basis for the development of the ultimate shape and function of the FALKE sports pullover. The natural body contours are emphasised and the muscle definition is accentuated – combining aesthetics and function in one perfect design.

me Markus T *Brille*

Spectacles

WENIGE GRAMM SCHWER, sitzen die Brillengestelle *me* aus TMi-Kunststoff und Titan auf der Nase. Der westfälische Augenoptiker und Betriebswirt Markus Temming hat für seine Modelle das Prinzip der „Reduktion auf das Notwendigste" auf die Spitze getrieben: ein zum Patent angemeldetes Stecksystem verbindet die superleichten, form-schlichten Kunststoffrahmen und -bügel mit Titan-Verbindungselementen. Ein Hauch von Brille, der überzeugte.

THE GLASSES FRAME *me* made from TMi plastic and titan sits on the nose, weighing just a few grams. In this model, the Westphallian optician and graduate in business administration, Markus Temming, has carried the principle of "reduction to the essentials" to the extreme. A connector system which has registered a patent, combines the super light, simple-form, plastic frame and earpiece with titanium joining elements. The glasses are as light as a breeze and convincing too. — BARBARA FRIEDRICH

Jurybegründung

Statement of the Jury

Unternehmen Company

deSIGN Markus T. GmbH Isselhorster Straße 374 — 33334 Gütersloh — Germany — www.markus-t.com

Design

Werksdesign/Inhouse Markus Temming

UNAUFDRINGLICHE KONTUREN in der Seitenansicht, futuristische Ansätze von vorne betrachtet. *me Markus T* ist die einzigartige Komposition aus Titan und TMi-Kunststoff, angemeldet zu zwei Patenten. Ein Design, das markante Dynamik mit Harmonie verbindet. Mit der Kollektion *me Markus T* sollte etwas Neues, nie da Gewesenes geschaffen werden, umgesetzt durch innovative Technik kombiniert mit anspruchsvollem Design. Entwickelt wurde ein Kunststoff kombiniert mit einem Mix verschiedenster Materialien, darunter Kohlefaser und Glasfaser. Diese spezielle Zusammensetzung, verarbeitet in einem einzigartigen System, setzt neue Maßstäbe. Das flache Kunststoff-Mittelteil ist mit seinem einzigartigen Materialquerschnitt eine Besonderheit dieser Brille. Von vorn betrachtet markant und stark akzentuiert, verwandelt sich das Mittelteil in der Seitenansicht in eine dünne Ebene und wartet mit der gewohnt filigranen Optik von *Markus T* auf. Ergonomisch geformt und mit weniger als drei Gramm Gewicht ist *me Markus T* eine der leichtesten Kunststoffvollrandbrillen der Welt.

UNOBTRUSIVE CONTOURS when seen from the side, futuristic bases from the front. *me Markus T* is a unique composition of titanium and TMi – plastic, and two patents are pending. A single concept that connects a striking design with harmony. The idea behind the collection *me Markus T* is to create something new, unprecedented, and realised using innovative technology combined with high quality design. A plastic was developed combined with a mixture of different kinds of materials, among them carbon fiber and fiberglass. This special composite, processed in a singular system, sets new standards. With its unique cross-section of materials, the flat plastic middle forms the characteristic feature of these glasses. Seen from the front, it is striking and strongly accented, from the side the middle section is merely a thin line illustrating the familiar filigree appearance of *Markus T.* Ergonomically shaped and weighing less than three grams, *me Markus T* is one of the lightest plastic glass frames in the world.

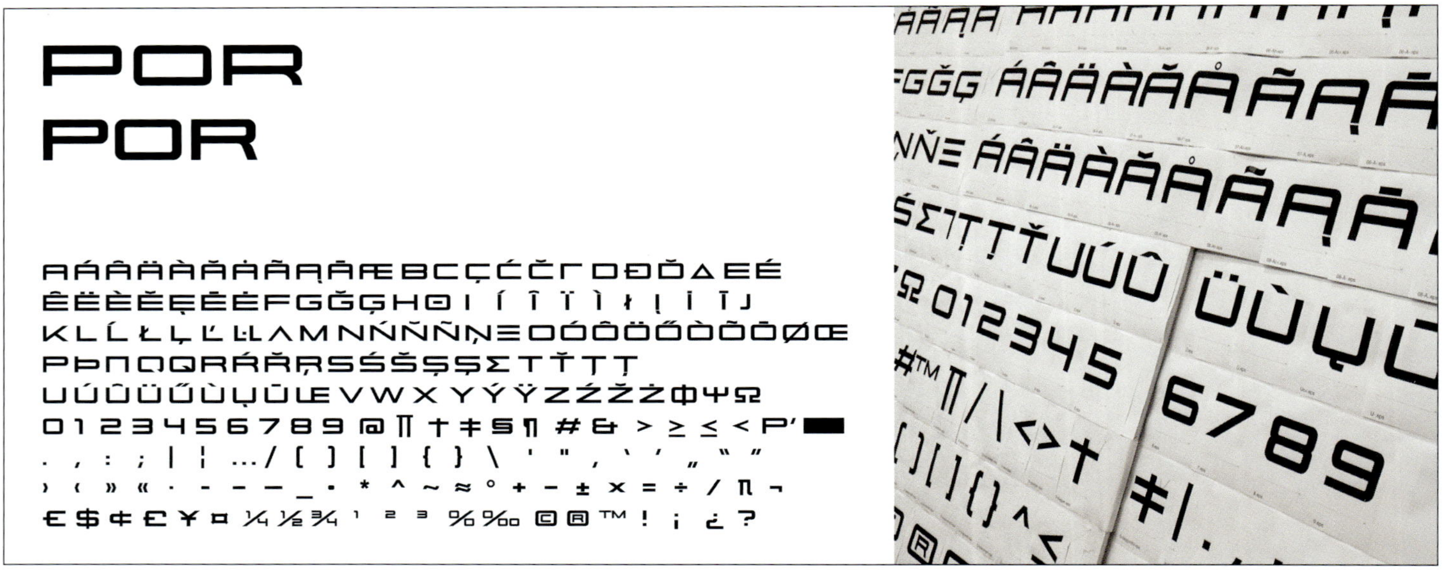

Erscheinungsbild für Porsche Design
Corporate Design for Porsche Design

Corporate Design

Corporate Design

DAS PORSCHE DESIGN ERSCHEINUNGSBILD ist ein herausragendes Beispiel für klare Linien, reduzierte Ästhetik und konsequente Umsetzung. Von der Verpackung über die Medien bis hin zum Point of Sale ist die exklusive, eigenständige Note sofort erkennbar. Und die raffinierten Details und Materialeinsätze stellen die Marke für technische Ingenieursprodukte in ein hochwertiges, emotional erlebbares Umfeld. Die eigenständige Schrift, das monochrome Farbklima und die technisch anmutende Bildsprache verbinden Technikkompetenz und Designqualität zu einem starken, profilierten Auftritt.

THE PORSCHE DESIGN IMAGE is an excellent example of clear lines, minimalist aesthetics and thorough implementation. The exclusive, independent flair is immediately recognizable – from packaging and media to the point of sale. Also, refined details and use of materials place this brand of technically engineered products in a qualitative setting with an emotional element. The independent font, the monochromatic colour atmosphere and the technical image language combine design quality and technical competence to create a strong, profiled appearance. – REINHARD BINDER

Jurybegründung

Statement of the Jury

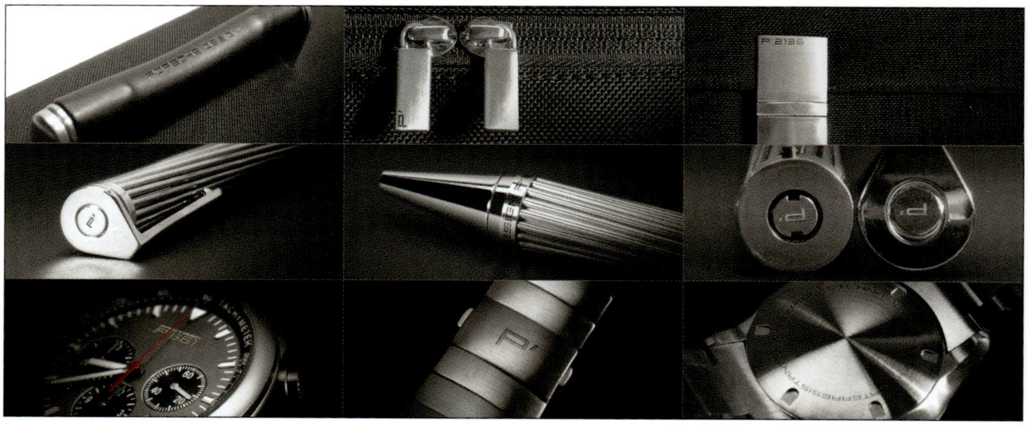

Unternehmen Company

Porsche Lizenz-und Handelsgesellschaft mbH & Co. KG Porschestraße 1 — 74321 Bietigheim-Bissingen — Germany — www.porsche-design.com

Design

KMS Team GmbH Knut Maierhofer, Patrick Märki, Dirk Koy, Christina Baur, Sandra Ehm — www.kms-team.de

PORSCHE DESIGN GESTALTET MODERNE und reduzierte Designprodukte, deren besonderer Mehrwert in ihrer überlegenen Funktionalität liegt. Für den internationalen Relaunch der Marke war ein Erscheinungsbild zu konzipieren, das die Strahlkraft von *Porsche* (Automobil) nutzt, zugleich aber die Eigenständigkeit der Marke *Porsche Design* unterstreicht. Ein zentrales Element der Arbeit ist die eigens gezeichnete *Porsche Design-Schrift*, die sich an den *Porsche*-Schriftzug anlehnt, allerdings schmaler und lesbarer ist. Der daraus gesetzte Namenszug „PORSCHE DESIGN" stellt zugleich das Logo dar. Als Hausfarben wurden Schwarz, Weiß, Titan und Silber definiert; die ersten beiden stehen für Purismus und Reduktion, die technoiden Metalltöne unterstreichen sowohl die Wertigkeit als auch den „Engineering-Ansatz" der Marke. ——— Auch die neue, auf dem Schema „P plus vier Ziffern" basierende Kennzeichnungssystematik für die Produkte erinnert an technische Nomenklaturen. Das Package-Konzept nimmt darüber hinaus direkten Bezug auf *Porsche*. Als Grundform für alle Formate wurde ein Rechteck mit dem Seitenverhältnis (2:5) ausgewählt – den Proportionen des 911ers.

PORSCHE DESIGN CREATES MODERN DESIGN PRODUCTS that are particularly valuable due to their superior functionality. A corporate image had to be conceived for the international relaunch of the brand, that would exploit the aura of *Porsche* (as an automobile), while also underscoring the autonomy of the *Porsche Design* brand. ——— As a central element a new font was designed, following the *Porsche* lettering very closely, but boasting significantly improved legibility. The chosen colours, black, white, titanium and silver, also bring to mind the car brand, thus emphasizing the techno-strong positioning of *Porsche Design*. For the packaging a formal raster grid was advanced. In spatial terms it is based on the proportions of the horizontal projection of *Porsche's* model 911 (2:5) and conceptually on a consistent universal naming system. Every product appears in a package with a geometric form derived from the "original rectangle". The ID number is always a four-digit number preceded by a "P".

Volkswagen Corporate Design Essentials

Corporate Book

Corporate book

DAS VOLKSWAGEN CORPORATE DESIGN BOOK wirkt wie ein erfrischendes Kaleidoskop des *Volkswagen* Markenbildes. Informativ, vor allem aber inspirierend und gestalterisch vorbildlich werden hier die unterschiedlichsten Aspekte des *Volkswagen*-Auftritts und der Kommunikation nachvollziehbar vermittelt. Hier zeigt sich: Design macht Sinn, macht sogar Spaß, und eben unverwechselbar. Ein vorbildliches Markenbuch über Design wird selbst zum Design. Ein schönes Stück Markenidentität zum Anfassen und Begreifen. Zur Nachahmung wärmstens empfohlen.

THE VOLKSWAGEN CORPORATE DESIGN BOOK is like a refreshing kaleidoscope for the *Volkswagen* brand image. Informative, particularly inspiring and with an exemplary design the various aspects of *Volkswagen's* presentation and communication are presented comprehensibly. You can see here that design makes sense, it can even be fun and distinctive. An exemplary brand book on design is itself good design. A fine piece of brand identity to get your hands on. More like it ... would be my recommendation. — REINHARD BINDER

Jurybegründung

Statement of the Jury

Unternehmen Company

Volkswagen AG Postfach 19060 — 38436 Wolfsburg — Germany — www.volkswagen.de

Design

MetaDesign AG www.metadesign.de

VOLKSWAGEN IST WELTWEIT MIT AUTOMOBILEN vom Fox bis zum Phaeton präsent. Egal, welches Land, egal, welches Modell: Die Marke *Volkswagen* spricht immer eine klare Sprache und vermittelt ihren Kunden überall das gleiche Gefühl. Dafür sorgt ein nachhaltiges internationales Management der Marke. MetaDesign entwickelte und implementierte ein flexibles Corporate Design-System und begleitet *Volkswagen* seit elf Jahren bei der kontinuierlichen Führung, Pflege und Weiterentwicklung der Marke. ———— Die *Volkswagen* Corporate Design-Essentials sind ein besonderes Kommunikations- und Motivationsinstrument innerhalb der Markenkommunikation des Automobilkonzerns und richten sich an Führungskräfte und externe Agenturen. In ihnen werden auf sinnlich nachvollziehbare Weise die Markentonalität und die Gestaltungsgrundlagen der Marke *Volkswagen* erläutert. Die Leser entwickeln so ein Gefühl für die Marke und das Corporate Design, werden sich der Identität des Unternehmens bewusst und spüren, was die Marke besonders macht.

VOLKSWAGEN HAS AN INTERNATIONAL PRESENCE with cars from the Fox to the Phaeton. No matter what the country, no matter what the car, the *Volkswagen* brand speaks a clear language and generates the same emotion among customers. This is ensured by effective international brand management. MetaDesign developed and implemented a flexible Corporate Design system and, for the last eleven years, has assisted *Volkswagen* in the ongoing management and optimization of its brand. ———— *Volkswagen* Corporate Design Essentials is a special communications and motivational tool that targets both *Volkswagen* executives and external agencies. The book explains the tonality and basic design principles of the *Volkswagen* brand in a way that appeals to all the senses. Users gain a feel for both the brand and Corporate Design. They are made aware of the identity of the company and those things that make the brand unique.

Ascii

AIDS-Präventionskampagne

AIDS Prevention campaign

WIE SPRICHT MAN HEUTE DAS JUNGE URBANE PUBLIKUM AN, wenn es um das Thema Aids-Prävention geht? Für Anfänger – auf Vorträge verzichten, nicht langweilig sein und sich kurz fassen. Die Idee ist, die Botschaft kurz, aber drastisch zu formulieren und die Sprache der Kids in der Zielgruppe zu sprechen. Man sieht ein Paar beim beschleunigten Vorspiel, das unvermeidlich auf einen One-Night-Stand hinauszulaufen scheint. Sexuell eindeutige Bilder, die aus schwarz-weißen Buchstaben und Symbolen zusammengesetzt sind, fesseln die Aufmerksamkeit. Der Höhepunkt erfolgt, als die Kamera auf einen Geschlechtsakt zoomt, der vollständig aus Zeichen der Zeichencodierung ASCII besteht, die die finale Botschaft des Films zeigen: Schütze Dich vor HIV. Die Idee, grafische Zeichen zu verwenden, um diese „pixelierte" Warnung zu transportieren, ist großartig: die Bilder sind ausreichend abstrakt, um nicht auf Pornoniveau abzusinken, während sie einen flüssigen Übergang von der animierten Szene zur Textbotschaft garantieren.

HOW DO YOU ADDRESS TODAY'S URBAN YOUTH on the subject of AIDS prevention? For starters – skip the lecture, don't be dull, and keep it short. The idea is to tell it like it is, make the message brief but bold, and speak the language of the kids in your target group. We see a couple that fast-forwards through foreplay and ends up starring in a one-night stand. Sexually explicit images composed of black and white letters and symbols grab and hold our attention. The climax comes as the camera zooms in on a sex act that consists completely of ASCII characters, which spell out the film's final message: arm yourself against HIV. The idea of using graphic characters to convey this "pixelated" warning is brilliant: the images are sufficiently abstract to keep them from sinking to the level of porn, while guaranteeing a fluid transition from animated scene to textual message. — ROBERT THIEMANN

Jurybegründung
Statement of the Jury

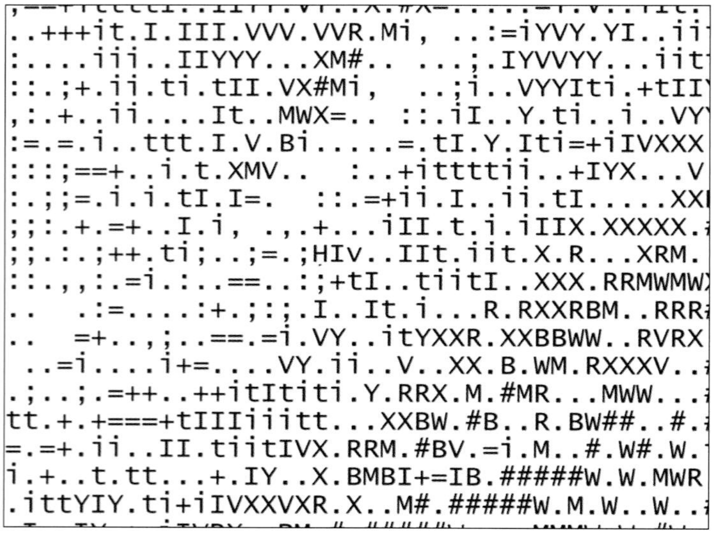

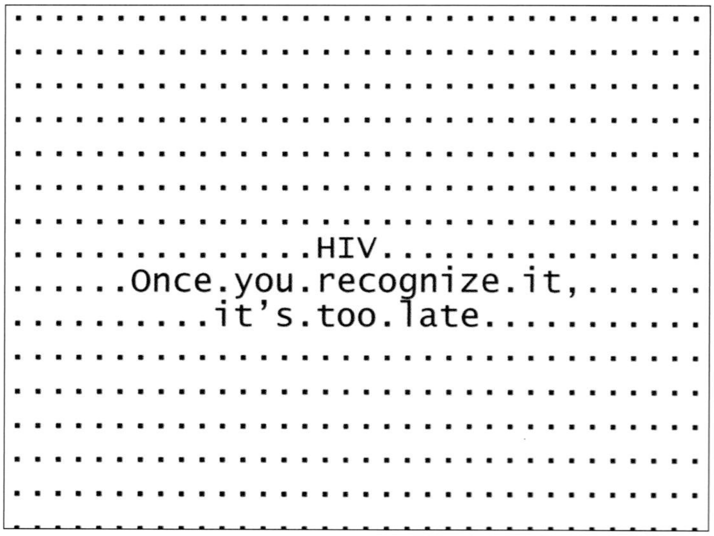

Unternehmen Company

BBDO Düsseldorf GmbH Königsallee 92 — 40212 Düsseldorf — Germany — www.bbdo.de

Design

Werksdesign/Inhouse Marco Obermann, Jean-Pierre Gregor, Concetta Milione, Michael Ohanian, Raphael Milczarek, Hans-Holger Pollack

DIE PRÄVENTIONSKAMPAGNE „Ascii“ besteht aus zwei Spots. Diese Spots warnen aufmerksamkeitsstark vor dem leichtsinnigen Umgang mit AIDS. Wir sehen jeweils ein homosexuelles und ein heterosexuelles Paar bei einem One-Night-Stand. Die Spots laufen im ungesehenen ASCII-Look, bei dem jedes der Bilder ausschließlich aus Buchstaben und Zeichen besteht. Am Ende der Spots macht ein Zoom in die Buchstaben deutlich, dass einer der Partner den HI-Virus in sich trägt. Die Botschaft lautet: Once you recognize it, it's too late – Wenn Du es bemerkst, ist es zu spät.

THE “Ascii” PREVENTION CAMPAIGN consists of two commercials which warn, in an attention grabbing way, of the dangers of being careless with regards to AIDS. We see a homosexual and a heterosexual couple during a one night stand. The commercials are shown in the unseen ASCII look in which each of the images is solely made up of letters and figures. At the end of the commercial, a zoom in on the letters makes it clear that one of the partners has HIV. The message is: “Once you recognize it, it's too late.”

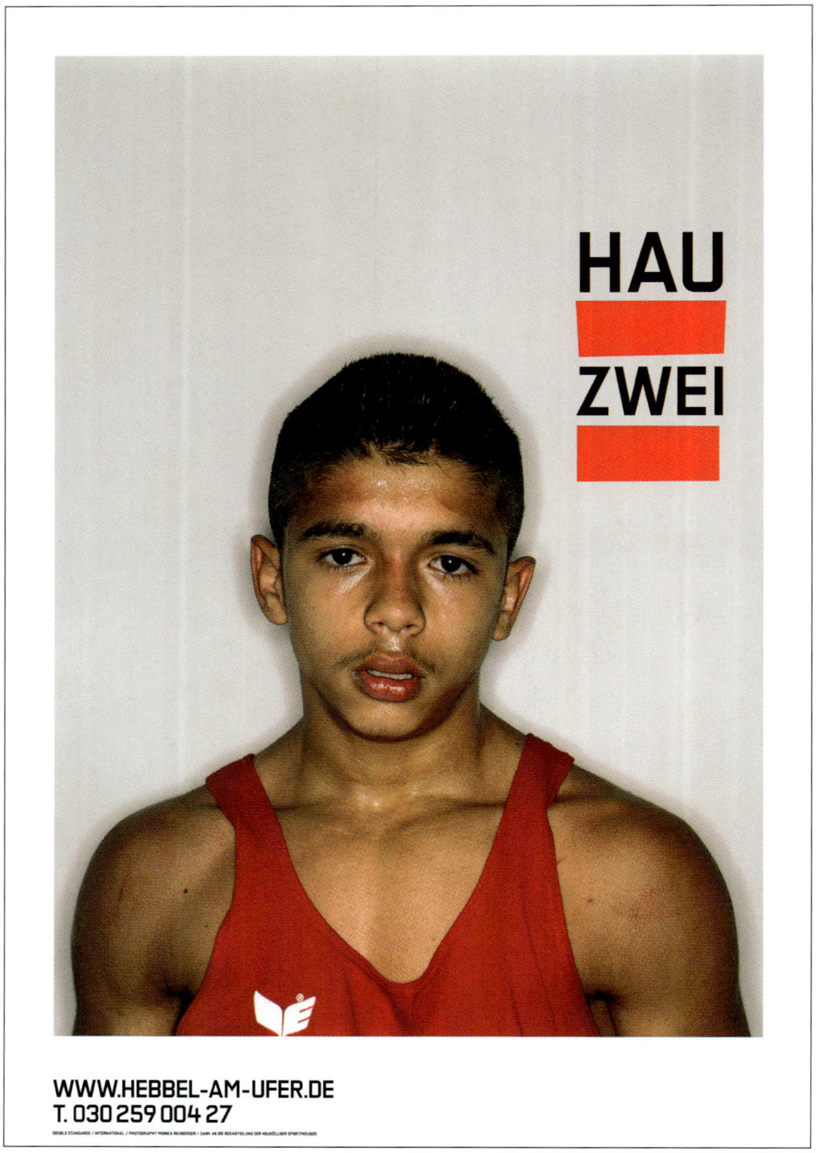

HEBBEL AM UFER – HAU123 – Kampagne
HEBBEL AM UFER – HAU123 – CAMPAIGN

Plakatkampagne
Poster campaign

AN DER KAMPAGNE GEFIEL DER JURY der für eine kulturelle Institution ungewöhnliche Ansatz, der mit seinem überraschendem Wortwitz wesentlich dazu beigetragen hat, das HAU in einer Stadt, in der Kulturinstitutionen in ungewöhnlichem Maße einem Distinktionsdiskurs ausgesetzt sind, neu zu positionieren.

WHAT THE JURY LIKED ABOUT THE CAMPAIGN was the unusual approach adopted by a cultural institution whose surprising play on words played a considerable part in the fact that HAU was able to reposition itself in a city in which cultural institutions are subject to an unusual extent to discourse on distinction. — NICOLETTE NAUMANN

Jurybegründung

Statement of the Jury

Unternehmen Company

Hebbel am Ufer Hallesches Ufer 32 — 10963 Berlin — Germany — www.hebbel-am-ufer.de

Design

Double Standards Chris Rehberger, Todd Schulz, Monika Rehberger — www.doublestandards.net

BERLIN-KREUZBERG: Seit September 2003 steht das Theaterkombinat mit seinen Spielstätten HAU 1, HAU 2 und HAU 3 unter der künstlerischen Leitung von Matthias Lilienthal. Durch die Zusammenlegung der ehemals drei Theater können internationale Gastspiele, Vorstellungen der Berliner und (inter)nationalen Tanzszene und innovative Theaterprojekte gleichzeitig entwickelt und gezeigt werden. Lilienthal umschreibt mit seinem Credo, „eine hysterische Sehnsucht nach Realität" zu haben, wesentliche Eckpfeiler des Programms. So hat er das Thema Migration offensiv in den Spielplan eingebaut und bietet Themenwochenenden zum „Nahen Osten", zu „Beirut" oder zu politischen Transformationsprozessen im Osten an. Das HAU arbeitet kontinuierlich mit dem Regieteam RIMINI PROTOKOLL, das Alltagsexperten statt Schauspieler auf der Bühne zu Wort kommen lässt und dadurch Collagen frappierender Gegenwartsfragmente schafft. ——— HAU steht für Hebbel am Ufer und ist Drohgebärde und Motivationsgeste zugleich. „Sie sind jung und verschwitzt, ihre Gesichter grün und blau geprügelt. Sie wirken erschöpft, aber glücklich: die Boxerinnen- und Boxermodels, die sich für die Werbekampagne des neuen HAU fotografieren ließen" – schrieb Eva Behrend über die Startkampagne. Als Gesichter des Bezirks repräsentieren die Boxer die Öffnung des Theaters hin zu den urbanen Themen Migration und Minorität, welche nicht nur im HAU quasi vor der Tür stehen.

BERLIN-KREUZBERG: with its three venues, HAU 1, HAU 2 and HAU 3, this theatre combine has been headed by Matthias Lilienthal as Artistic Director since September 2003. Amalgamating what used to be three theatres has allowed the combine to arrange and present international guest appearances, performances by acts from the Berlin and (inter)national dance scenes and innovative theatre projects simultaneously. Lilienthal paraphrases, with his credo of displaying "a hysterical longing for reality", the fundamental cornerstones of the combine's programme. Accordingly, he has taken the offensive by including the topic of migration in the combine's repertoire, as well as offering theme weekends on the "Middle East", "Beirut" and the political transformation processes in the East. HAU works with RIMINI PROTOKOLL, a team of directors, on a permanent basis. On the stage, the team allows everyday experts to have their say rather than actors, thus producing collages consisting of striking fragments of the present. ——— HAU stands for Hebbel am Ufer (Hebbel, the German dramatist, on the riverbank) and is at once a threatening gesture and a gesture of motivation. "They are young and sweaty, their faces beaten black and blue. They looked exhausted but happy: the boxer models who were photographed for the advertising campaign for the new HAU" – wrote Eva Behrend, talking about the launch campaign. As faces of the district, the boxers represent the process of the theatre opening up to the urban topics of migration and minority on its doorstep so to speak, and not only in the HAU, but in the entire district.

www.designklicks.de *Internetseite*
Website

DESIGNKLICKS SCHAFFT NEUE PERSPEKTIVEN. Zum einen inhaltlich: als interaktives Portal für Ästhetik kann jeder mit eigenen Beiträgen ein immer neues, kollektives, ästhetisches Stimmungsbarometer erzeugen. Gleichzeitig entsteht daraus ein schier endloses Meer kreativer Momente, die mit der innovativen Art der Navigation zur neuen, kreativen Inspirationsquelle werden können. Zum anderen formal: Komplexität wird hier einfach, ein User-Manual überflüssig, alle Bedienungselemente sind aufs Äußerste reduziert, logisch nachvollziehbar und sofort verständlich. Eine sehr innovative Webanwendung mit einer richtungsweisenden Idee und einer konsequent klaren Umsetzung.

DESIGNKLICKS CREATES NEW PERSPECTIVES. For one, as far as content is concerned: as an interactive aesthetics portal individuals can produce a continuously renewing, collective, aesthetic mood barometer with their own contributions. At the same time it creates an almost infinite sea of creative moments which given the innovative type of navigation, could become a new creative source of inspiration. The other aspect is formal: Complexity becomes simple here, there is no need for a user manual, all user elements have been reduced to the extreme, are logical and immediately comprehensible. A very innovative web application with a trend-setting idea which is consistently and clearly implemented. — REINHARD BINDER

Jurybegründung

Statement of the Jury

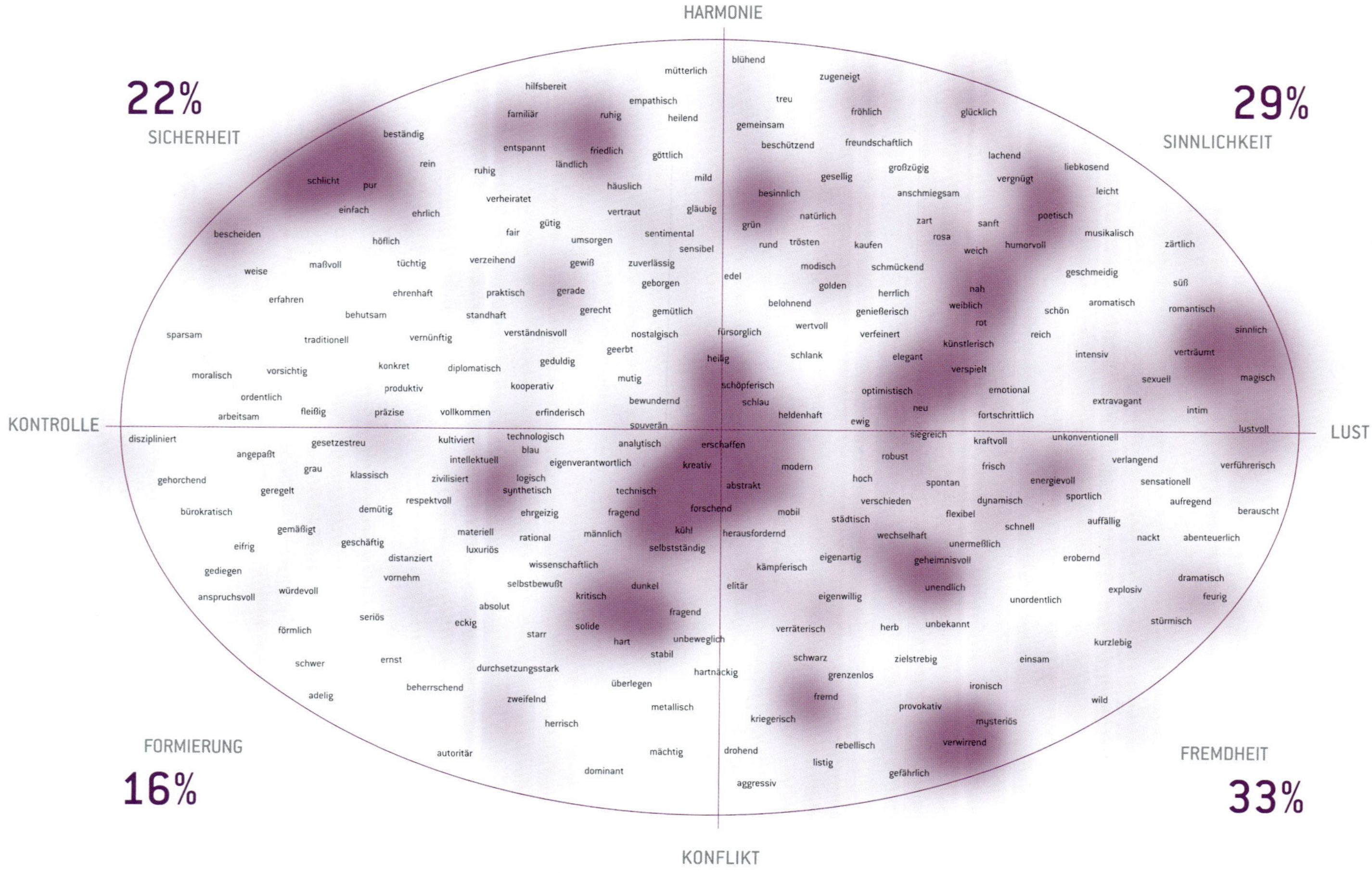

Unternehmen Company

Trendbüro Beratungsunternehmen für gesellschaftlichen Wandel — JK. PW. GmbH — Hohe Brücke 1 — 20459 Hamburg — Germany — www.trendbüro.de

Design

Deutschlandrock Design & Software Büro Stefan Landrock

DESIGNKLICKS STARTETE PASSEND ZUM THEMA „Schwarm-Intelligenz – Die Macht der smarten Mehrheit" des 10. Deutschen Trendtag im Mai 2005 als Experiment zur Suche nach einer neuen Ästhetik. Professionelle Gestalter aller Fachrichtungen – Künstler, Fotografen, Designer, Architekten etc. – sind seither aufgerufen, ihre Arbeiten, Bilder und Visionen auf *DesignKlicks.de* hochzuladen. Sie weisen ihren Werken Schlagworte zu, die den gesellschaftlichen Wertekosmos abbilden. Unterstützt durch die hoch frequentierte Infrastruktur von Spiegel-Online gelangen die Bilder zur Bewertung an die Öffentlichkeit. Jeder Nutzer wird automatisch Teil der Jury. Die Avantgarde stellt sich dem Votum der Mehrheit. Das ist eine echte Herausforderung für beide Seiten: Denn Kreation ist und bleibt ein individueller Prozess. Die Akzeptanz einer neuen Ästhetik aber ist eine kollektive Entscheidung. ——— Trendbüro nutzt diese Form der Schwarm-Intelligenz um den ästhetischen Diskurs voranzutreiben. *DesignKlicks* entwickelt sich derzeit zu einer Analyseplattform für Design und Marketing. Die Analyse der Bilder erlaubt Rückschlüsse auf das gesellschaftliche Werteempfinden.

THE START OF DESIGNKLICKS FITTED IN ADMIRABLY WITH THE THEME of the 10th German Trend Convention in May 2005, namely "Mass Intelligence – The Power of the smart Majority" as an experiment in the search for a new set of aesthetics. Since then professional designers from all specialist areas – artists, photographers, designers, architects etc. – have been called on to upload their works, images and visions on *DesignKlicks.de*. They give their works catchwords that illustrate the cosmos of values in society. Supported by the highly-frequented infrastructure of Spiegel-Online the images reach the public for evaluation. Each user automatically becomes a member of the jury. The avant-garde gives itself up to the vote of the majority. It is a genuine challenge for both sides: Because being creative is and always will be an individual process. How a new set of aesthetics is accepted, however, is a collective decision. ——— Trendbüro uses this form of mass intelligence to advance aesthetic discourse. Since then *DesignKlicks* is currently in the throes of becoming an analysis platform for design and marketing. Analysis of the images enables us to draw conclusions about the appreciation of values in society.

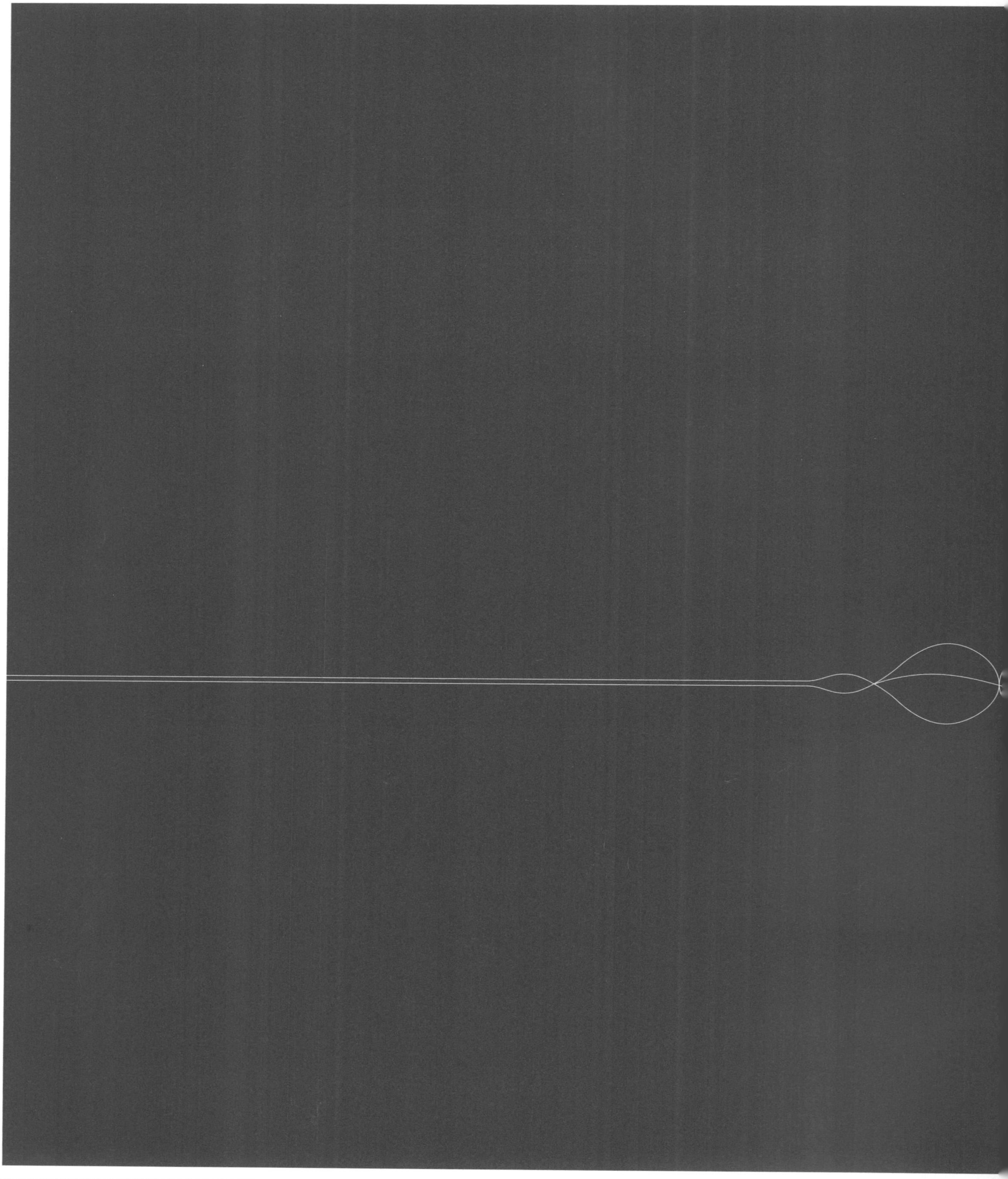

Die Nominierungen — Auswahl
The Nominees — Selection

Bereits die Silhouette macht eine Klassifizierung nach herkömmlichen Fahrzeugkategorien unmöglich: Reiselimousine, Großraumfahrzeug, Kombi oder Offroader. Die *R-Klasse* ist keines von allem und doch alles in einem. Die flach abfallende Dachpartie mündet in einem weich gerundeten, dynamischen Kuppelheck, das durch die coupéinspirierten Hecksäulen eingefasst wird. So entsteht eine einzigartige Synthese aus erkennbarer Raumbetonung und dynamischer Sportlichkeit und Eleganz. Its profile alone makes it impossible to classify this car in terms of traditional vehicle categories: Touring saloon, large-capacity vehicle, estate car or offroader, the *R-Class* is none of these and yet all of them at the same time. The roof area has a soft slant and flows into a gently rounded dynamic tailgate framed by coupé-inspired C-pillars. The result is a unique synthesis – familiar emphasis on space combined with dynamic sportiness and elegance.

Mercedes-Benz R-Klasse *PKW*

Passenger car

Unternehmen — Company DaimlerChrysler AG, Zentrale — Epplestraße 225 — 70546 Stuttgart — Germany — www.mercedes-benz.de
Design Werksdesign/Inhouse

Das neue Modell der *S-Klasse* verbindet die Anmutung souveräner Eleganz einer Limousine mit intelligenten technologischen Innovationen und einem hohen Maß an Fahrkomfort. Ihre Formensprache gründet auf den Regeln des modernen Purismus – auf der Reinheit der Formen und dem Verzicht von übertriebenem Zierrat. Das verleiht der Luxuslimousine die souveräne Ruhe und Kraft eines starken, selbstbewussten Charakters. Mit dieser Linienführung ist die *S-Klasse* auch stilistisch ein Vorbild. The new *S-Class* combines the supreme elegance of a saloon with intelligent technological innovations and advanced ride comfort. Its formal language is based on the principles of modern purism – cleanness of line and rejection of any excessive decoration. It is this that gives the luxury saloon its effortlessly superior, strong and self-confident character. With these emphatic lines, the *S-Class* represents a new design concept and forges a new Mercedes style that looks towards the future.

Mercedes-Benz S-Klasse S350L

PKW

Passenger car

Unternehmen — Company DaimlerChrysler AG, Zentrale — Epplestraße 225 — 70546 Stuttgart — Germany — www.mercedes-benz.de

Design Werksdesign / Inhouse

Mit ihrem einzigartigen Konzept als viertüriges Coupé verbindet die neue *CLS-Klasse* den Charme eines Coupés mit der Funktionalität einer Limousine und bestätigt die Rolle der Marke als Trendsetter für neue Automobil-Ideen. Der *CLS* zählt zu den emotionalsten Fahrzeugen von Mercedes-Benz. Die Ausgangsfrage für die Designentwicklung lautete: „Wie emotional kann ein Mercedes sein, ohne seine klassischen Grundwerte wie Premiumqualität, Sicherheit und Werterhalt zu vernachlässigen?" With its unique four-door coupé concept, the new *CLS-Class* combines the charm of a coupé with the functionality of a saloon and confirms the brand's role as a trendsetter in automotive thinking. The *CLS* is one of Mercedes-Benz's most emotional vehicles. The starting point for the development of the design was: "How emotional can a Mercedes be without neglecting its traditional values such as maximum quality, safety and value retention?"

Mercedes-Benz CLS-Klasse – Viertüriges Coupé
Mercedes-Benz CLS-Class – 4-doors-Coupé

PKW

Passenger car

Unternehmen — Company DaimlerChrysler AG, Zentrale — Epplestraße 225 — 70546 Stuttgart — Germany — www.mercedes-benz.de
Design Werksdesign / Inhouse — Werksdesign Sindelfingen

Der Erfolg des Roadster-Konzeptes geht auf eine gelungene Synthese aus klassischem Design und modernster Technik zurück. Trotz der klassischen Formensprache, die vom *Wiesmann*-Roadster ausgehend weiterentwickelt wurde, entstand ein komplett neues Fahrzeug. Die Basisdaten lieferte der Wiesmann Renn GT, der als Versuchsträger unter extremen Bedingungen während des 24-Stunden-Rennens auf dem Nürburgring im Jahre 2002 getestet wurde. The success of the Roadster concept is based on a successful synthesis of classic design and state-of-the-art engineering. This encouraged Friedhelm and Martin Wiesmann to develop a new, similarly fascinating, sports car on the basis of the Roadster, the *Wiesmann GT*. Basic data were gained from the Wiesmann Racing GT, which was tested as a experimental vehicle under extreme conditions during the 24-hour race at the Nürburgring in 2002.

Wiesmann GT *PKW*

Passenger car

Unternehmen — Company Wiesmann Manufaktur der Individualisten — An der Lehmkuhle 87 — 48249 Dülmen — Germany — www.wiesmann.com

Design Werksdesign/Inhouse — Martin Wiesmann

Der *Renault Modus* verdankt sein charakteristisches Aussehen den fließenden Linien und den ungewohnt großen Fensterflächen, die schon auf den ersten Blick den vorhandenen Raum deutlich unterstreichen. Denn nach dem Prinzip „außen kompakt und innen geräumig“ wird er den Gegebenheiten engen urbanen Raums sowie den dadurch entstehenden Bedürfnissen des modernen Menschen gerecht und überzeugt in der Stadt sowie auf langen Strecken. The *Renault Modus* owes its characteristic appearance to the flowing lines and exceptionally large expanses of glass that clearly underscore its spaciousness at very first glance. Because operating on the principle of "compact exterior, spacious interior", it responds to the facts of limited urban space and the resulting needs of modern people, excelling both in the city and on long trips.

Renault Modus *PKW*

Passenger car

Unternehmen — Company Renault Deutschland — Renault Nissan-Straße 6-10 — 50321 Brühl — Germany — www.renault.de
Design Renault SAS — www.renault.com

Die Plakatserie SAAB visualisiert mit einer strengen und zugleich spielerischen grafischen Gestaltung die *Saab*-Design-Entwicklung an Hand des ersten Automodels. Die vertikal und horizontal gewählte Typogestaltung, im Kontext mit der sich von Plakat zu Plakat vergrößernden Bildgestaltung in Form eines einzigen Automodells, verdichtet die Aussage zu einer visuell spürbaren Symbiose und vermittelt dem Betrachter den Erfolg des Produktes von den Anfängen bis zur Gegenwart. The SAAB poster series uses a stringent and yet playful graphic look to visualize the trend in *Saab* design taking the example of the very first *Saab*. The choice of vertical and horizontal typography in the framework of the image, which grows bigger from one poster to the next in the form of a car model, ensures that the message condenses to form a visually tangible symbiosis – conveying to the viewer the success of the product since its beginnings right through to the present.

SAAB *Plakatserie für PKW SAAB*

Poster series for passenger car SAAB

Unternehmen — Company Hanns Schmid Grafikdesign — Bruggerstrasse 69 — 5400 Baden — Switzerland — www.schmidgrafikdesign.ch

Design Werksdesign / Inhouse — Hanns Schmid

Der *Sun Ti* von *Knaus* zeichnet sich durch automobiles Design im Exterieur aus. Die Front ist durch das einmalige Panorama-Glasdach geprägt. Der Innenraum besticht durch ein Raummöbelkonzept, das sich zur Mitte hin öffnet und somit für ein maximales Raumgefühl sorgt. Zusätzlich besticht der *Sun Ti* durch neue Funktionslösungen, wie das in der Fronthutze integrierte Flachfernsehkonzept. Der *Sun Ti* zeigt ein homogenes, durchgängiges Beleuchtungskonzept mit hochwertiger Spotbeleuchtung in allen Wohnbereichen sowie einer integrierten Beleuchtung in der Fronthutze. The *Knaus Sun Ti* stands out for its exterior design. The front features a unique glass panorama roof. The interior boasts a space/furniture concept that opens towards the centre to ensure maximum felt space. In addition, the *Sun Ti* sports various new functional solutions, such as the flatscreen TV integrated into the front hood. The *Sun Ti* presents a homogeneous, end-to-end illumination with high-grade spotlights in all living sections as well as integrated light in the forward hood.

Knaus Sun Ti 650 MF *Reisemobil*

Motorhome

Unternehmen — Company KNAUS TABBERT GROUP GMBH — Helmut-Knaus-Straße 1 — 94118 Jandelsbrunn — Germany — www.knaus.de
Design DESIGNquadrat GbR — Alexander Christ — www.designquadrat.de

Der HYMER *Van* ist ein alltagstaugliches, komplett ausgestattetes Reisemobil, das sich durch seine kompakten Abmessungen und eine komfortable wie funktionale Komplett-Ausstattung auszeichnet. Das Reisemobil bietet zwei Personen vielfältige Möglichkeiten für individuelle Städtereisen, sportive Kurzurlaube oder für längerfristige Urlaubsreisen. Der HYMER *Van* präsentiert sich in einem modernen, harmonischen Design, bei welchem speziell der Fender mit einem leicht angeschrägten Übergang vom Fahrerhaus zum Wohnraum eine extra Portion Dynamik in das Gesamtstyling einbringt. The HYMER *Van* is a total equipped motor home suited for every day, which stands out because of its compact measurements and a comfortable and functional equipment. It is ideal for two people and offers them various possibilities e.g. for city trips, quick getaways or longer vacations. The HYMER *Van* has a modern and harmonic design. Especially the fender, with a slightly sloping transition for the drivers cabin to the body, tops off the total styling of the motor home by bringing in this dynamic feature.

HYMER Van *Reisemobil* *Motorhome*

Unternehmen — Company HYMER AG — Holzstraße 19 — 88339 Bad Waldsee — Germany — www.hymer.com
Design Hymer IDC — Prof. Johann Tomforde — www.hymer-idc.de // Werksdesign / Inhouse — Rainer Buck

Der *Eriba Feeling* basiert auf der Idee, den klassisch erfolgreichen Touring-Hubdach-Wagen mit seinen fahrtechnischen Vorteilen an modernste Wohnraumanforderungen anzupassen. Farblich auf das Außendesign abgestimmt ist nicht nur der Hubdach-Zeltstoff (orange oder rot), sondern auch die Polsterstoffe und Übergardinen. Einen besonderen Pfiff im Innenbereich verleihen die mittig geschwungenen Möbelklappen, die hinter der Acrylglas-Trennfuge eine Ambiente-Beleuchtung beherbergen. The *Eriba Feeling* is based on the idea, to adapt the successful classic touring elevating roof caravan with its driving-technical advantages to latest living space needs. Related and matched to the coloured exterior design is not only the pop-up-roof fabric, but also the upholstery and curtains. The rolling furniture hatches, which accommodate a hidden lightning, give an extra kick to the inside.

Eriba-Feeling *Wohnwagen*

Caravan

Unternehmen — Company HYMER AG — Holzstraße 19 — 88339 Bad Waldsee — Germany — www.hymer.com
Design Werksdesign/Inhouse — Rainer Buck

Die Babyliegeschale ist für Kinder bis ca. 18 Monate geeignet. Der 5-Punktgurt sorgt für einen sicheren Halt, die hohen und dick gepolsterten Seitenwangen bieten beim Seitenaufprall maximalen Schutz. Ein einzigartiges Schutz-Konzept verhindert beim Frontalaufprall einen Überschlag der Schale. Durch ein Luftzirkulationssystem wird die Rückenpartie des Kindes belüftet. Die innovative Einhandbedienung der Tragebügelverstellung und der Travel-System-Adaptoren machen die Babyliegeschale außergewöhnlich bedienerfreundlich. The baby seat is designed for children aged up to about 18 months. The 5-point belt ensures that the child is securely harnessed in the seat. The high, well-upholstered side wings provide maximum protection against side impact. The unique stabiliser concept prevents the seat from toppling in the event of a frontal collision. There is an air circulation system to keep the child's back ventilated. The innovative, single-handed adjustment mechanism for the carrying handle and the travel system adaptors make the baby seat exceptionally user-friendly.

ION *Babyliegeschale*
Baby seat

Unternehmen — Company CONCORD GmbH — Industriestraße 25 — 95346 Stadtsteinach — Germany — www.concord.de

Design FOURHEAD Productdesign — www.fourhead.de

Tests der Fachpresse haben weltweit bestätigt, dass die P99-D unangefochten das hellste Xenon Outdoor Lichtsystem auf dem Markt ist. Das grell-weiße Licht der Gasentladungslampen ist vergleichbar mit 120 Watt Halogenstrahlern bei nur einem Bruchteil des Stromverbrauchs. Der Wirkungsgrad der Supernova Xenon Systeme ist besser als bei jeder LED. Durch Mikroprozessoren in Lampe und Batterie und dank der ausgezeichneten Wärmeableitung durch die exakt geformten, CNC gefrästen Kühlkörper sind alle Supernova-Gasentladungslampen extrem langlebig. Multiple tests by specialist magazines have shown that the P99-D is unchallenged as the brightest battery-run Xenon outdoor light on the market. The bright-white light of the twinned discharge units is comparable to that of a 120 watt halogen bulb, at only a fraction of the energy consumption. The light yield of Supernova Xenon systems is also better than that of any LED system. Due to the microprocessors in the lamp and battery, and thanks to exactly shaped, CNC machined cooling fins, Supernova H.I.D. spotlights are extremely long lasting.

SUPERNOVA P99-D SUPERNOVA TAIL LIGHT

Xenon Fahrradbeleuchtungssystem mit LED-Rücklicht
H.I.D. Xenon bicylce lighting system with LED tail light

Unternehmen — Company SUPERNOVA LIGHTING SYSTEMS — Industriestraße 11 — 79194 Freiburg — Germany — www.supernova-lights.com
Design Marcus Wallmeyer — www.supernova-lights.com

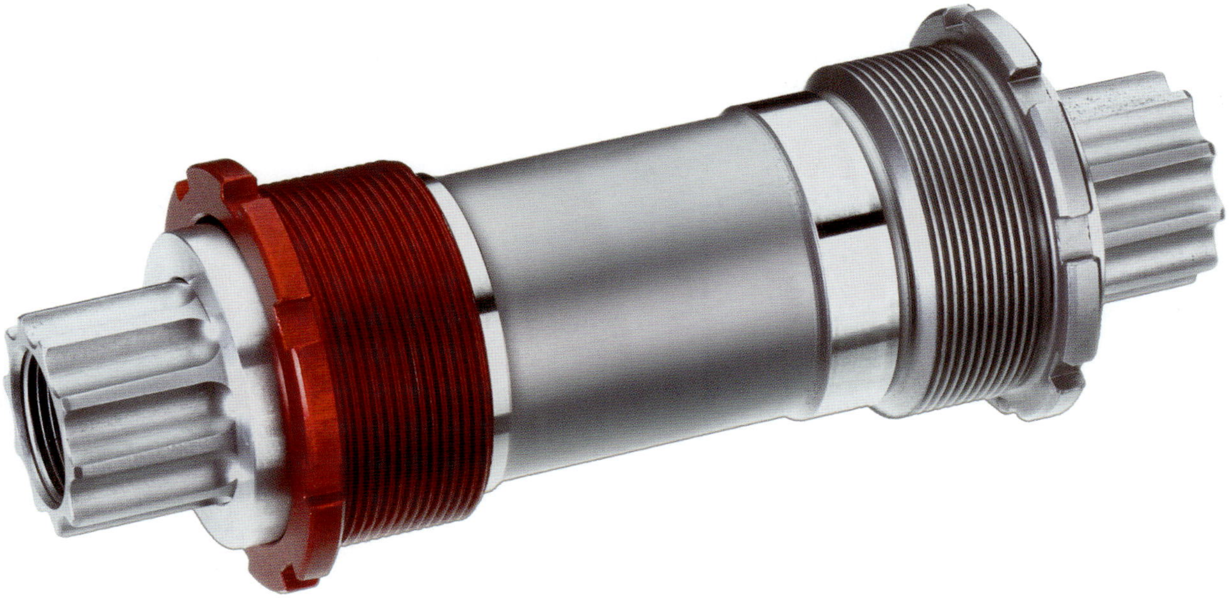

Vor 20 Jahren etablierte SKF die ersten eigenen Fahrradtretlager auf dem Weltmarkt. Schnell entwickelten sie sich zur Standardkomponente in Qualitätsfahrrädern. Nach eingehendem Studium der Anforderungen und Wünsche der Radfahrer von heute wurde klar, dass es für diese wichtige Fahrradkomponente mittlerweile viele Verbesserungspotenziale hinsichtlich Zuverlässigkeit und Leistung gibt. Deshalb hat SKF eine neue Generation von Tretlagern entwickelt – eine Generation, die neue Maßstäbe setzt. 20 years ago, SKF invented the first bottom bracket unit for bicycles. This has now become a standard feature in almost all high-quality bikes. Upon taking a close look at today's bikers' wishes and requirements, SKF found plenty of room for improving the reliability, durability and all-round performance of this most vital part of any bicycle. The new SKF bottom bracket unit sets new standards in strength and precision.

SKF *Tretlagereinheit für Fahrräder*

Bottom bracket unit for bicycles

Unternehmen — Company SKF GmbH — Gunnar-Wester-Straße 12 — 97421 Schweinfurt — Germany — www.skf.de
Design I.D.M. Müsing — Bernd Müsing // Werksdesign/Inhouse — SKF GmbH

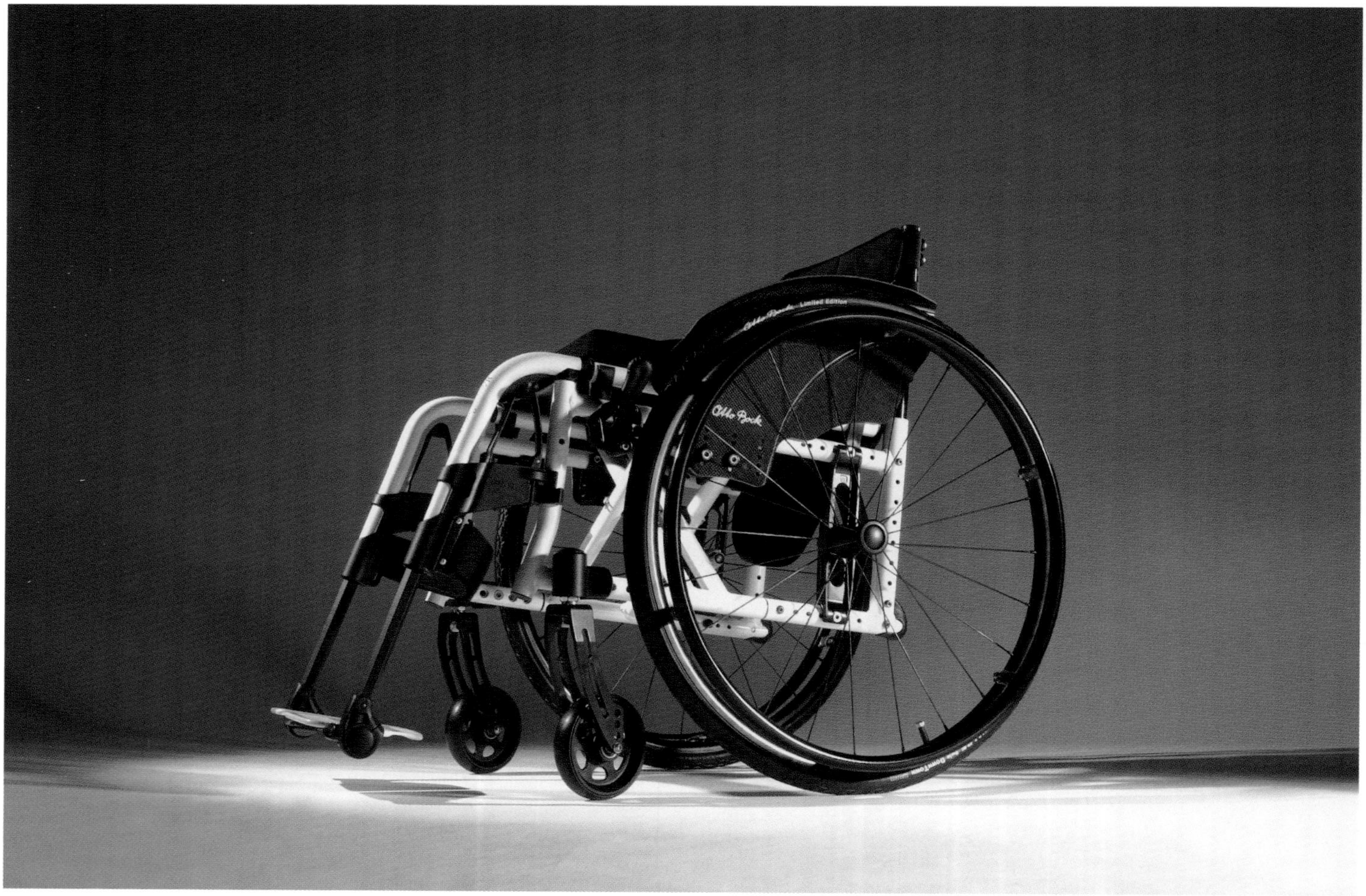

Hochkomplexe Anforderungen sind die große Stärke des Aktivrollstuhls *Avantgarde VR light*. Seine hochwertige Verarbeitung und perfekte Einstellbarkeit garantieren optimale Anpassung an den Nutzer. Bei all diesen Vorteilen ist das Gewicht dieses hochbelastbaren Aktivrollstuhls durch konsequente Reduktion an entscheidenden Punkten optimiert worden. The great strength of the *Avantgarde VR light* wheelchair for active use is an extraordinary high degree of functionality and adaptability to the most diverse and even highly complex requirements. High-quality workmanship and perfect adjustability ensure optimal adaptation to the user. Despite of all these advantages, the weight of this heavy-duty wheelchair for active use has been optimized through consistent reduction at decisive points.

Avantgarde VR Light *Manueller Rollstuhl*
Manual wheelchair

Unternehmen — Company Otto Bock HealthCare GmbH — Max-Näder-Straße 15 — 37115 Duderstadt — Germany — www.ottobock.de
Design Werksdesign/Inhouse — Achim Eiser

Der FX ONE hat einen starken Charakter, vor allem im Profil. Die geschwungenen Formen sind erst durch ein 3D-Biegeverfahren möglich (MEYRA ist eines der wenigen europäischen Unternehmen, die diese Technologie einsetzen). Die notwendigen Funktionen fügen sich durch zwölf Aluminium Schmiedeteile in die weichen Formen des Rahmens ein und verbinden Funktionalität und Belastbarkeit harmonisch mit dem Design. The FX ONE has a strong character, especially featuring by its profile. A 3D bending process enables the rolling forms (MEYRA is one of the few European companies applying this technology). The necessary functions are integrated in the soft forms of the frame by 12 aluminium forgings, harmonically combining functionality and endurance with the design. The manufacture of the FX ONE in Germany incorporates all technological novelties and is sold to the whole world.

FX ONE *Starrrahmen-Aluminium-Aktivrollstuhl*

Aluminium rigid frame everydayuse wheelchair

Unternehmen — Company MEYRA Vertriebsgesellschaft mbH — Meyraring 2 — 32689 Kalletal — Germany — www.meyra.de

Design REALO Design — Christof Meier — www.realo-design.de

Im Rahmen des integralen Küstenzonen Management-Projektes West Zeeuws-Vlaanderen werden momentan diverse Maßnahmen zur Aufwertung der gesamten Küstenzone ergriffen. Das Büro *Crepain-Binst Architekten* hat für das Dorf Breskens einen Pfad entworfen, entlang dessen besondere Elemente einen Platz bekommen. Zwei dieser besonderen Elemente wurden durch BURO LUBBERS entworfen. Ein Aussichtssteg im Hafen des Dorfes neben der Bootsrampe und eine Panoramaterrasse in den Dünen von Breskens. Beide Elemente sind von ihrer Formensprache und Materialverwendung her miteinander verbunden und passen zum Charakter des Hafens und der Küste. Within the framework of the project "Integral Coast Management West Zeeuws-Vlaanderen", several developments are taking place to upgrade the coastline of Zeeland, a province in the south of The Netherlands. For the small city of Breskens, *Crepain Binst Architects* designed a pedestrian route, linking several objects of interest. BURO LUBBERS designed two of these objects, the lookout point near the boat slip and a panoramic terrace in the dunes. Both objects are similar in form language and materialisation and harmonise with the rugged character of the harbour and the coast.

Aussichtssteg & Panoramaterrasse Breskens

Aussichtspunkte
Viewpoints

Unternehmen — Company BURO LUBBERS — Oude Dieze 17 — 5200 BN 's-Hertogenbosch — Netherlands — www.burolubbers.nl // Gemeinde Sluis, Projektleiter Tiny Maenhout // Aquavia Sas van Gent

Design BURO LUBBERS landschapsarchitectuur & stedelijk ontwerp — Peter Lubbers, Marian de Vries, Caroline de Feijter, Roy Wouters

Der *Mega-Fußball* ist ein Seilspielgerät. Zur optimalen Ausnutzung des Geräterahmens in Form eines Ikosaeders wurde eine neuartige Raumnetzstruktur entwickelt: Eine aus 12 Fünf- und 20 Sechsecken bestehende Ballnetzstruktur. Die Außenstruktur eines klassischen Fußballs. Im Fall des *Mega-Fußballs* wurden die Fünfecke des innersten Raumnetzballs mit Hilfe von Gummimembranen ausgefacht, so dass der Eindruck entsteht, dass der Ball im Netz schwebt. The *Mega-football* is a rope play climbing structure. The space frame with the shape of an icosahedron is filled by a new developed space net structure. The ball net structure is made out of 12 pentagons and 20 hexagons. The shape of a classical football. In the case of the *Mega-football* the pentagons of the inner netball structure are filled out with a black rubber membrane, so that the impression originates that the ball floats in the net.

Mega-Fussball *Kletter-Seilspielgerät*

Rope Play Climbing Structure

Unternehmen — Company Berliner Seilfabrik GmbH & Co. — Lengeder Straße 4 — 13407 Berlin — Germany — www.berliner-seilfabrik.com

Design Werksdesign/Inhouse — Karl Köhler

Die frei geschwungene und ausschließlich von oben gehaltene Segelfläche verleiht dem STRUCTURELAB S1 seine unverwechselbare Form. Entwickelt wurde das Schirmsystem mit dem Ziel, ein einzigartiges Verschattungs- und Gestaltungselement für Tag und Nacht zu kreieren. Es gelang eine überzeugende Synthese aus der Eleganz eines gespannten Segels mit der Flexibilität und Funktionalität eines Schirmes. The distinctive shape of the STRUCTURELAB S1 is defined by its curved textile membrane, which is held up at the top only. Development of the system aimed at creating a unique look plus a shade element – resulting in an impressive synthesis of the elegance of a taut sail and the flexibility and functionality of a mobile sunshade.

Structurelab S1 Schirmsystem
Structurelab S1 sunshade system

Verschattungssystem
Sunshadesystem

Unternehmen — Company Bikatec Metall-und Textilverarbeitungs GmbH — Ten-Brink-Straße 14a — 78269 Volkertshausen — Germany — www.bikatec.de
Design StructureLab GmbH — Jürgen Schubert, Jürgen Gendriesch, Alexander Prang — www.structurelab.com

byNo. #4 ist eine schön brennende Fackel aus Edelstahl in schlichtem Design und mit außergewöhnlicher Funktionalität. Sie ist mit einem verborgenen Docht konstruiert, der bei der Anwendung keine Justierung erfordert, und hinterlässt bei der Bedienung keine schmutzigen Hände. Die Brenndauer der Fackel reicht über den ganzen Abend, bevor Lampenöl nachgefüllt werden muss. Nie war es einfacher, im Außenbereich das ganze Jahr über für eine behagliche Atmosphäre zu sorgen. *byNo. #4* is a beautiful stainless steel torch with a simple look and extraordinary functionality. It includes a covered wick that does not require adjusting when the torch is lit and it is very clean to operate. The torch burns for an entire evening before it needs refilling with lamp oil. Creating a pleasant outdoor atmosphere all the year round could not be simpler.

byNo. #4 *Gartenfackel*
Garden torch

Unternehmen — Company Blindkilde — Gunnar Clausens Vej 58 — DK-8260 Viby J — Denmark — www.byno.dk
Design DOK 54 — Jacob M. Lund, Thomas Brinch-Møller — www.dok54.dk

Für die *Designlinie Landmark* ist eine zeitlose Gestaltung für den Einsatz in verschiedenen urbanen Zusammenhängen charakteristisch. Sie zeichnet sich durch die gelungene Verbindung von Präsenz und Zurückhaltung aus. In die rechte Seitenwand und Teile der Dachkonstruktion der Buswartehalle ist ein auf LED-Technologie basierendes Beleuchtungssystem integriert. Dies sorgt bei Dunkelheit für angenehme Lichtstimmung und wechselnde Farben. A characteristic of the *Landmark line of designs* is its timeless concept for use in different urban contexts. Its central feature is a successful combination of outward appearance and reticence. A lighting system, based on LED technology, is integrated into the right-hand side wall and into parts of the roof section of the bus stop. This provides agreeable lighting and changing colours during the night.

Designlinie Landmark
Design landmark

Stadtmöbelfamilie

Street furniture

Unternehmen — Company Wall AG — Friedrichstraße 118 — 10117 Berlin — Germany — www.wall.de

Design Hadi Teherani AG — Hadi Teherani — www.haditeherani.de

Die *fibreStone* Mehrschichtplatte aus Glasfaserbeton ist eine exklusive Weltneuheit, die dauerhaft mit einem hochwertigen und preiswerten Betonsockel (Stärke je nach statischen Erfordernissen) verbunden ist. Ein eigens dafür entwickelter Spezialmörtel bewirkt den fest haftenden Verbund von Betonsockel und der 13 mm starken Glasfaserbetonplatte. Die Mehrschichtplatte wird derzeit in den drei fixen Formaten 30 / 120 / 5, 60 / 120 / 5 und 60 / 60 / 5 cm angeboten. Die Anfertigungen sind hinsichtlich der Größe, Farbe und Oberfläche frei wählbar. The *fibreStone* composite board is an exclusive world first, bonded durably with a high-grade and cost-effective concrete base (strength depends on load-bearing properties). A purpose-developed mortar guarantees a firm bond between the concrete base and the 13 mm-thick fibre-glass-reinforced concrete board. The latter is currently available in three set sizes: 30 / 120 / 5, 60 / 120 / 5 and 60 / 60 / 5 cm. Special formats are available on request.

fibreStone *Mehrschichtplatten*

Multilayer boards

Unternehmen — Company Rieder Faserbeton Elemente GmbH — Am Glasberg 1 — 83059 Kolbermoor — Germany — www.rieder.cc

Design Werksdesign / Inhouse

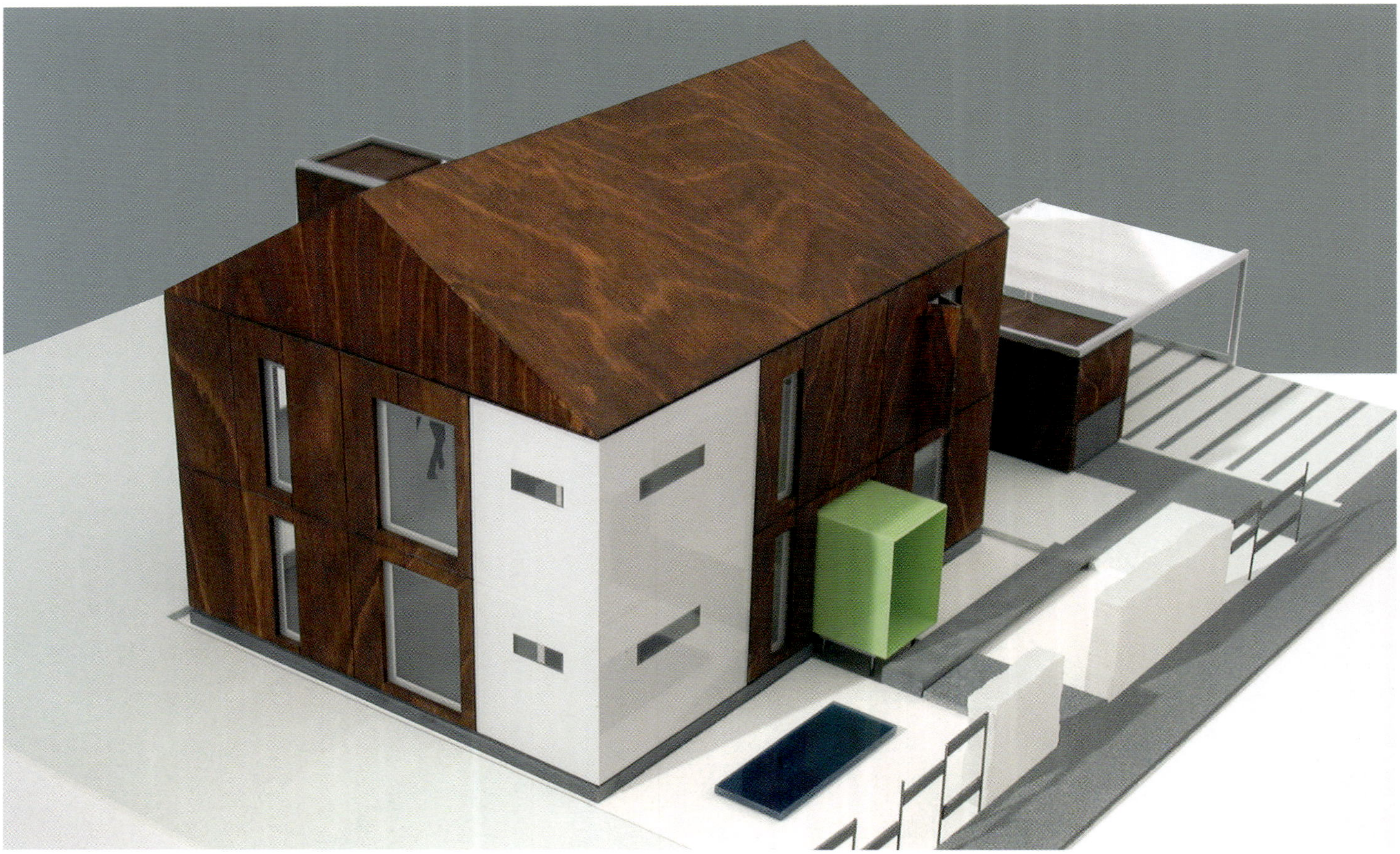

Ein wesentliches Merkmal des *Editionshaus 478* ist sein durchgängig variables Konzept und sein modulares Gestaltungsprinzip. Es werden ausschließlich ökologisch unbedenkliche Baustoffe verwendet, die ein gesundes Wohnen gewährleisten. Gestalterische Grundidee ist eine kubische Gebäudeform, die einen hohen Wiedererkennungswert hat. Gebäudeteile wie Loggien und Balkone werden nicht addiert, sondern aus dem Körper herausgeschnitten. Individuelle Fassaden- und Grundrissvariationen erlauben vielfältige Gestaltungsmöglichkeiten. A key feature of *Editionshaus 478* is its completely variable concept and its modular design system. It exclusively uses ecologically friendly building materials that guarantee healthy living. The underlying design idea: a cube-shaped building that has an unmistakable look to it. Building sections, such as loggias and balconies, are not added on, but cut out of the building's carcass. The design opportunities are immense given Individual façades and footprint versions.

Editionshaus 478

Holzfertighaus
Timber houses

Unternehmen — Company Bau-Fritz GmbH & Co.KG, seit 1896 — Alpenstraße 25 — 87746 Erkheim — Germany — www.baufritz.de
Design Gatermann & Schossig — Elmar Schossig — www.gatermann-schossig.de

Die Aluminium-Tür der *Designer Collection* besticht durch ihre zeitlose Form. Ziel war es, den Begriff von Wertigkeit, Exklusivität und Signifikanz eines Eingangsbereichs neu zu definieren. Gefertigt wird die Haustür mit beidseitig flügelüberdeckender Türfüllung. Akzente setzen Sonderverglasung mit LED Beleuchtung sowie Zebrano Echtholz. Verdeckt liegende Bänder sowie ein Stoßgriff in Edelstahl komplettieren die Komposition edler Materialien. The aluminium entrance door is distinguished through its ageless form. The aim was to redefine the terms of quality, exclusiveness and significance of house entrances. The *Groke* door is produced with aluminium overlay panels on both sides. Accents set the special glazing with LED lighting and zebrano wood. Cover-lying hinges and a push handle of high-grade steel complete the fine combination of materials.

Groke Designer Collection Charactere 100144

Hauseingangstür aus Aluminium
Aluminium entrance door

Unternehmen — Company Groke Türen und Tore GmbH — Wikingerstraße 10 — 76189 Karlsruhe — Germany — www.groke.de
Design Frank König

Mit diesem neuartigen, patentierten Schiebewandsystem können Transparenz, große Glasflächen, mobile Raumtrennung und ausgezeichnete Schalldämmung in die Architektur integriert werden. Nach Wunsch können die Gläser satiniert, emailliert, geätzt, auch strukturiert und selbstverständlich in verschiedenen Farbtönen gewählt werden. Innen liegende, elektrisch zu betätigende Jalousien für den Sichtschutz sind bei Bedarf ebenfalls erhältlich. This innovative and patented sliding wall system allows glazed surfaces, mobile room partitioning and excellent architecture. If desired, the glass may be frosted, enamelled, naturally there are various colours to choose from. Visual protection are also possible on request.

H & T SGS *Schalldämmende Glasschiebewand*

Sound-absorbing all-glass wall

Unternehmen — Company H & T Raumdesign AG — Falt-und Schiebewände — Rohrerstrasse 20 — 5001 Aarau — Switzerland — www.schiebewaende.com
Design Werksdesign / Inhouse — Thomas Zulauf, Christian Horlacher

ORNILUX SB1 ist ein Isolierglas, das in allen Fenstern und Fassaden wie ein konventionelles Wärmedämm-Isolierglas eingesetzt werden kann. Das Max-Planck-Institut für Ornithologie hat im Auftrag der Glaswerke Arnold an der Vogelwarte Radolfzell eine umfangreiche Untersuchung zum Erkennen von Glasscheiben durch Vögel durchgeführt. Gemäß der durchgeführten Tests kann ORNILUX SB1 zur Einschränkung von Vogelschlag bereits an Architekten bzw. Bauträger empfohlen werden. ORNILUX SB1 is a specially developed impact-resistant insulated glass that can be used in all window or façade installations, just like conventional insulated glass. The Max Planck Institute for Ornithology supported the development of ORNILUX by conducting extensive field testing in the Radolfzell Bird Sanctuary to investigate a bird's ability to recognise glass sheeting. Based on these test results, ORNILUX SB1 can now be recommended to architects or builders to minimize the effects.

ORNILUX SB1 *Isolierglas gegen Vogelschlag*

Insulation glass against bird strike

Unternehmen — Company Glaswerke Arnold — Alfred-Klingele-Straße 15 — 73630 Remshalden — Germany — www.glaswerke-arnold.de

Design Werksdesign / Inhouse — Christian Irmscher

Herausgefordert durch den Wunsch nach Individualität und dem Bedürfnis, besonders dem Wohnbereich eine persönliche Note zu geben, haben die *Schmitz-Werke* eine neue Balkon- und Terrassenmarkise entworfen. Die *markilux® 6000* sorgt durch verschiedenste Ausführungsarten für den individuellen Charakter. Drei Stilwelten stehen zur Verfügung: Club, Studio und Lounge passen sich den verschieden Hausstilen an – variable Ausstattungsmerkmale in Farbe und Design geben unterschiedliche Akzente. Challenged by clients' desire to individualize and give their living space a personal touch, the *Schmitz-Werke* have developed a new balcony and patio awning. The *markilux® 6000* creates an individual character through its versatile style options. The style lines Club, Studio and Lounge individually match different housing styles – various options for colour and design allow to place unique accents.

markilux® 6000 *Gelenkarm-Markise* *Folding-arm awning*

Unternehmen — Company markilux® — Schmitz-Werke GmbH + Co. KG — Hansestraße 87 — 48282 Emsdetten — Germany — www.markilux.com
Design kramerDesign — Andreas Kramer — www.kramer-produkt-design.de // Werksdesign / Inhouse — Sven Kröner, Luise Schniggendiller, Martin Dieckmann, Holger Honrath

DORMA TEGO® *Glas* ist eine designorientierte Systemlösung einer serienmäßig gefertigten automatisierten Ganzglas-Drehflügeltür. Durch ihre beeindruckende optische Wirkung unterstützen die einzelnen technischen Bausteine in der Transparenz die hohe ästhetische Qualität von TEGO® *Glas*. Ein besonderes Highlight ist die effektvoll farblich beleuchtbare Bediensäule, die das Gesamtbild unübertroffen elegant erscheinen lässt. The DORMA TEGO™ *Glass* is a design-aligned, fully glazed system solution providing a series-manufactured, automated swing door of the utmost elegance. The impressive visuals of the individual technical components enhance the outstanding aesthetic qualities of TEGO™ *Glass* in all its enticing transparency. And the operating column with its strikingly attractive coloured lighting provides a fitting culmination, imparting an overall effect of unequalled appeal.

TEGO® Glas *Ganzglas-Türsystemlösung*

All-glass door system solution

Unternehmen — Company DORMA GmbH + Co. KG — Breckerfelder Straße 42–48 — 58256 Ennepetal — Germany — www.dorma.com

Design Kaluza Design — Büro für Industrial Design — Georg Kaluza — www.kaluza-design.de // Werksdesign / Inhouse — Michael Glanz

Die neuen *Schüco Designgriffe* überzeugen durch eine klare, zeitlose Formensprache. Das einheitliche Erscheinungsbild bezieht mit Fenstern, Türen und Schiebesystemen alle Anwendungsbereiche ein. Die – wahlweise abschließbaren – Öffnungselemente lassen sich sowohl mechanisch als auch elektrisch komfortabel bedienen. Die Ansteuerung der Fensterperipherie (z.B. der Oberlichter) funktioniert per integrierter Funk-Fernbedienung. Für die variantenreiche Oberflächengestaltung stehen Edelstahl und RAL-Farben zur Verfügung.

The new *Schüco design handles* combine elegant, timeless design with state-of-the-art technology. The identical appearance relates to all purposes: windows, doors, sliding systems. The mechanical and electric opening units are easy to operate. A remote control to operate window periphery (e.g. toplights) is integrated, a locking for opening units is optionally available. Last but not least there is a comprehensive range of surface finishes: stainless steel, RAL, EV1 and stainless steel-look.

Schüco Griff-Designlinie
The new Schüco design handles

Griffe
Handles

Unternehmen — Company Schüco International KG — Karolinenstraße 1-15 — 33609 Bielefeld — Germany — www.schueco.de
Design Teams Design GmbH — www.teamsdesign.com // Werksdesign/Inhouse — Antje Bönisch, Konrad Kaiser

Schüco TipTronic ist das erste komplett mechatronische Drehkipp-Fenster, das komfortable und intuitive Bedienung in Echtzeit mit der gesamten Welt der Gebäudeautomation verbindet. Die Merkmale: Intuitive Bedienung am Fenster über neues Griffkonzept, Kippfunktion über zwei Tasten, Drehfunktion wie bei herkömmlichen Fenstern, Gruppenschaltung von bis zu 30 Fenstern, Bedienung über Taster oder BUS-Schnittstellen sowie Sicherheit über integrierten Klemmschutz und Verschlussüberwachung, WK2. *Schüco TipTronic* is the first fully mechatronic turn/tilt window to combine convenient and intuitive operation with building management systems in real time. The Features: Intuitive window operation using a new handle design, tilt function using two buttons, turn function as for conventional windows, up to 30 windows can be grouped together and are operated using standard keypads or BUS interfaces, safety via integrated anti-finger-trap feature and integrated closing monitor, WK2.

Schüco TipTronic *Mechatronische Beschlagsgeneration*

The mechatronic generation of fittings

Unternehmen — Company Schüco International KG — Karolinenstraße 1-15 — 33609 Bielefeld — Germany — www.schueco.de

Design Teams Design GmbH — www.teamsdesign.com // Werksdesign/Inhouse — Konrad Kaiser, Antje Bönisch

Der kompakte Pelletofen *Bellina* passt perfekt vor allem in kleinere Wohnungen. Durch die klaren ausgewogenen Formen und die farbigen, austauschbaren Seitenteile passt er sich perfekt jedem Wohnstil an. Für noch mehr Bedienkomfort sorgt die Steuerung mit der sämtliche Funktionen über eine übersichtliche Menüführung einfach und sicher bedient werden können. Integrierte Anschlüsse für Staubsauger und eingebaute Bürsten erleichtern die Reinigung. Durch den hohen Wirkungsgrad von über 90 % und die minimalen Emissionen überzeugt der Ofen auch ökologisch. The compact pellet stove *Bellina* fits perfectly into the smallest flat or studio. Its clear, balanced lines and the coloured, removable sides go with virtually any type of furnishing taste. Menu control makes operating the stove child play; all functions are accessible at a touch of your finger. Integrated connections for vacuum cleaners and built-in brushes facilitate cleaning. With its high-energy efficiency (over 90 %) and minimal emissions the *Bellina* is also a very eco-friendly stove.

Bellina 5 kW *Pelletofen*
Pellet stove

Unternehmen — Company Calimax GmbH – Bundesstraße 102 – 6830 Rankweil – Austria – www.calimax.com
Design Werksdesign / Inhouse

Der Kaminofen HWAM 3320 mit patentierter Verbrennungsautomatik wurde für den täglichen Gebrauch entwickelt. Eine neue und moderne Luftspülung mit erwärmter Luft hält die Glasscheibe frei von Ruß. Die Tür ist mit einer eleganten Hebe-/Senkfunktion versehen, die beim Öffnen vor dem Kaminofen leicht nach oben gleitet. Der HWAM 3320 ist mit einer neuen, zum Patent angemeldeten Ascheentnahme ausgestattet. Der Kaminofen kann entweder auf einem Metallsockel stehen oder an der Wand montiert werden. The stove HWAM 3320 with patented combustion automatics was designed to simplify the daily use. A new, sophisticated pane rinsing system using heated air keeps the glass pane clear of soot. The door is provided with an elegant lift/lower function and slides effortlessly up in front of the stove. The HWAM 3320 is equipped with a new patent pending ash removal system. The stove can be wall-mounted or supplied with a metal plinth.

HWAM 3320 *Kaminofen*

Stove

Unternehmen — Company HWAM Heat Design AS — Nydamsvej 53 — 8362 Hoerning — Denmark — www.hwam.com

Design STRAND + HVASS — Niels Hvass, Christina Strand — www.strand-hvass.com

Geradlinige Struktur mit höchsten formalen Ansprüchen – so zeigen sich die quadratischen Einbau-Downlights der Baureihe ATHENIK. 16 farbstabilisierte LEDs mit einer 3W Einzelleistung und einer fotometrischen Gesamtleistung, die einer 18W-Kompakt-Leuchtstofflampe entspricht, erzeugen eine surreale, beleuchtete Hightech-Umgebung. Einer der wichtigsten Aspekte bei der Umsetzung von LED-Konzepten ist das Wärmemanagement. Hier setzt TRILUX mit einem speziellen Kühlkörperdesign neue Maßstäbe. The square recessed downlights of the ATHENIK series reveal a straight-lined structure designed for the highest formal demands. 16 colour-stabilised power-LEDs with an individual performance of 3W and a total photometric performance equalling that of an 18W compactfluorescent lamp create a surreal, high-tech illuminated environment. One of the critical aspects in the implementation of LED concepts is the thermal management. TRILUX has set a new standard with a specially designed cooling-body.

Athenik LED

Leuchte
Luminaire

Unternehmen — Company TRILUX-Lenze GmbH + Co KG — Heidestraße 4 — 59759 Arnsberg — Germany — www.trilux.de
Design Werksdesign/Inhouse

Der *sensIQ* ist ein Infrarot-Bewegungssensor der Referenzklasse mit hoher Erfassungspräzision. Er bietet einen Erfassungswinkel von 300° und – per Einstellregler am Gerät – erstmals eine intuitive, metergenaue Reichweiteneinstellung in drei Richtungen. Es werden Decken- und Wandmontage durch einen innovativen Drehmechanismus in einem Gerät vereint. Zusätzlich bietet der *sensIQ* unter anderem Unterkriechschutz, Rückfeldüberwachung, einen modularen Aufbau sowie Fernsteuerbarkeit. *sensIQ* is a reference-class infra red motion sensor with precision detection. The product provides 300° sensing range and intuitive, unique user guidance with clearly defined scales for adjusting reach in three directions. The design combines a ceiling and wall mounting capability by using an innovative turning mechanism. It also comes with a sneak-by guard, a back-field monitor, a remote control and many other innovative features.

sensIQ *Infrarot-Bewegungssensor*
Infrared motion sensor

Unternehmen — Company STEINEL — Vertrieb GmbH — Dieselstraße 80-84 — 33442 Herzebrock-Clarholz — Germany — www.steinel.de

Design Kurz Kurz Design — Dorian Kurz — www.kurz-kurz-design.de

Stella – eine innovative Operationsleuchte die einzigartig Halogen- und Gasentladungstechnologie kombiniert. Großflächiges, schattenkontrolliertes Licht mit zusätzlicher, kontrastreicher Tiefenausleuchtung leuchtet das Wundfeld perfekt aus. Die Steuerung des Lichts erfolgt via sterilem Handgriff oder über ein separates Bedienfeld direkt durch den Chirurgen. *Stella* besticht durch ihre ergonomische Form, ihre leichte Beweglichkeit und beste Hygieneeigenschaften. *Stella* – an innovative surgical light that uniquely combines halogen and gas-discharging technology. Broad, shadow-controlled light with additional depth illumination and good contrast illuminates perfectly the surgical field. The control of the light takes place directly by the surgeon via a sterilised handle or via a separate control panel. *Stella* convinces with ergonomic shape, easy handling, and best hygiene characteristics.

Stella *Operationsleuchte*
Surgical light

Unternehmen — Company Dräger Medical AG & Co. KG — Moislinger Allee 53–55 — 23542 Lübeck — Germany — www.draeger-medical.com
Design Barskidesign — Olaf Barski — www.barskidesign.com

Leuchte *Josephine* ist eine moderne Neuinterpretation des Archetyps der Pendelleuchte, die kristalline und organische Elemente zu einer harmonischen Einheit in sich verschmelzen lässt. Sie verbindet die sonst nur Glas zugesprochenen Eigenschaften wie Transparenz und Lichtbrechung mit den Vorzügen von Kunststoff wie Elastizität, Leichtigkeit und individueller Einfärbbarkeit. Ihre Struktur erscheint multielementar, sie wird jedoch in einem Stück produziert. In a modern reinterpretation of the archetypal pendant, *Josephine* blends crystalline and organic elements in a harmonious whole. The luminaire combines qualities usually attributed only to glass – such as transparency and refractivity – with the advantages of plastic: elasticity, lightness, and the capacity to take on a whole rainbow of colours in varying shades. The structure appears comprised of multiple elements, but is produced as a single piece.

Josephine *Pendelleuchte* *Pendant*

Unternehmen — Company Koziol - ideas for friends GmbH — Werner-von-Siemens-Straße 90 — 64711 Erbach — Germany — www.koziol.de
Design Werksdesign / Inhouse — Jürgen Diehl

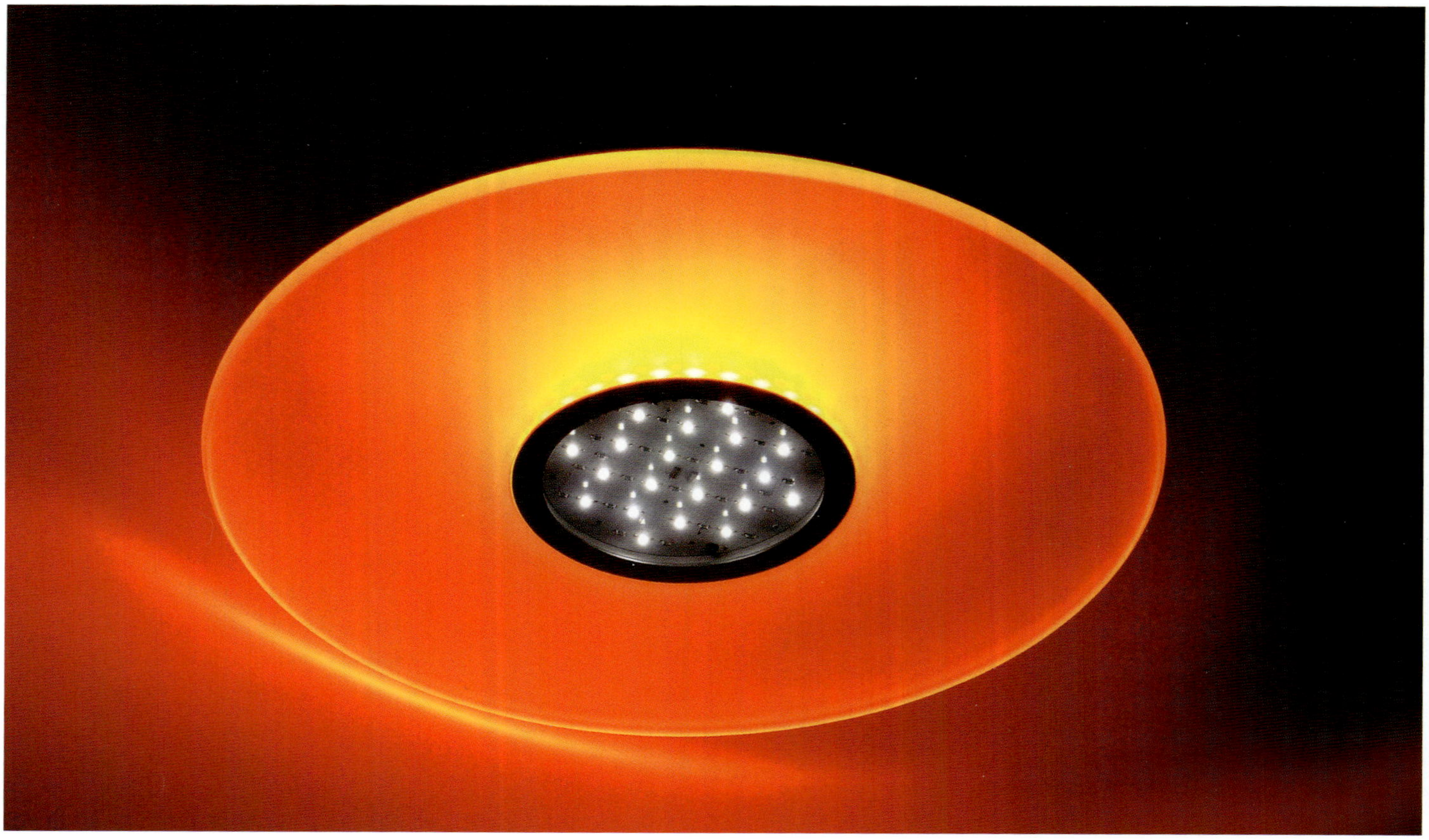

Mit ihrer neutralen Kreisform passt UFO perfekt in jedes Ambiente: Konferenzräume, Empfangsbereiche, Foyers, Bistros oder Wohnzimmer. Die Leuchte ist höhenverstellbar, dadurch ist sie sowohl als Decken- wie auch als Hängeleuchte einsetzbar. Zusätzlich zu einer Lichtsequenz durch alle Farben können jederzeit die integrierten, dimmbaren weißen Hochleistungs-LEDs zugeschaltet werden. Mit Hilfe einer Fernbedienung lässt sich bequem eine spezielle Farbe oder eine automatische Farb-Abfolge wählen. UFO, with is neutral circular shape, blends perfectly into any ambience: conference rooms, reception areas, foyers, bistros and living rooms. The luminaire is adjustable in height, making it suitable as a ceiling light or as a pendant. In addition to a sequential display of all the colours, the integrated and dimmable high-power LEDs can be switched in at any time. A specific colour or an automatic colour sequence can be conveniently selected by remote control.

12.6003 – UFO *Hängeleuchte*

Pendant

Unternehmen — Company Lightwave B. Flury -LED Products — Schlattweg 558 — 5704 Egliswil — Switzerland — www.lightwave-led.com
Design Werksdesign / Inhouse — Beat Flury, Steven Knobel

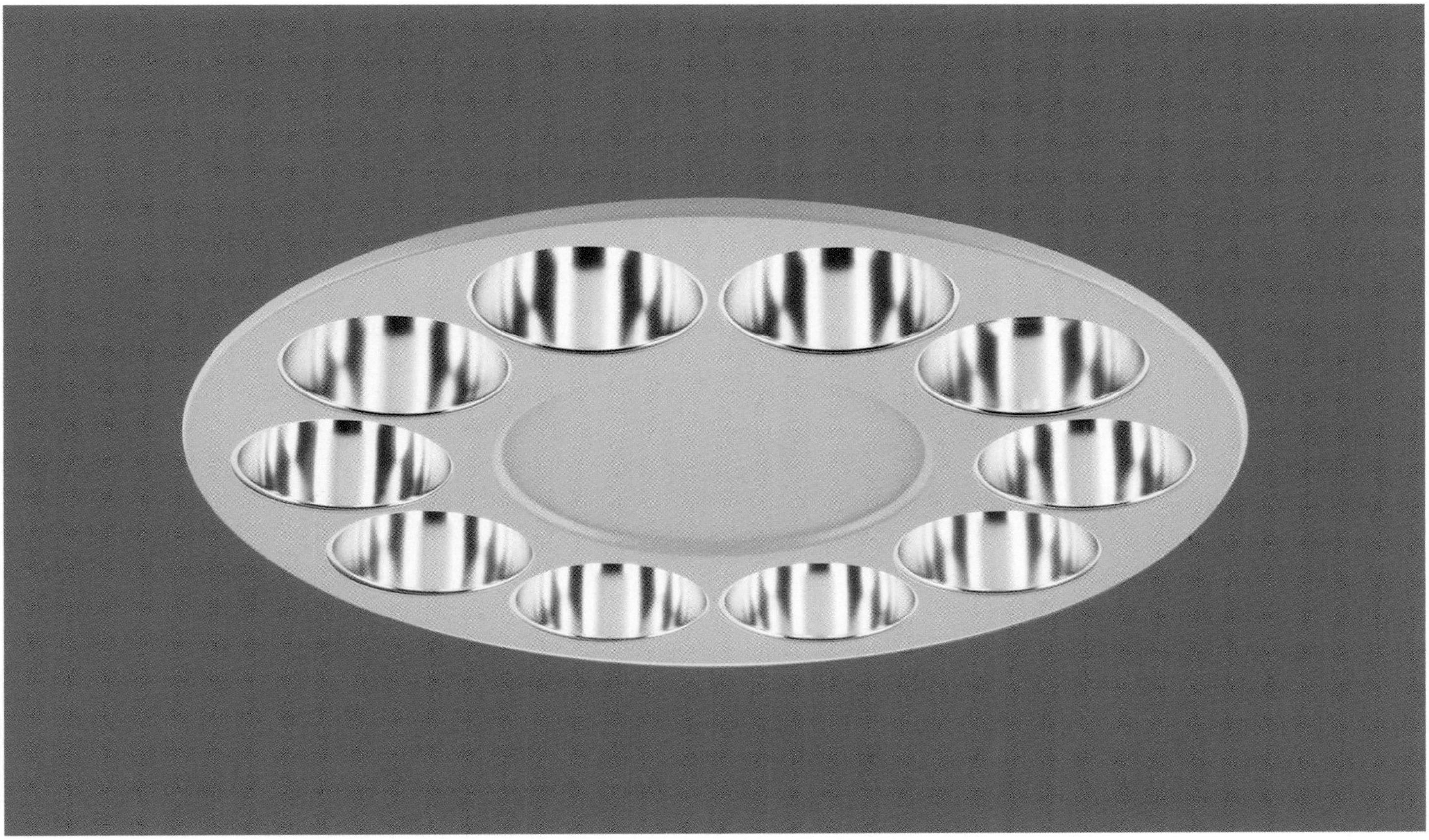

Die kompakte Leuchte für energiesparende, ringförmige T-16 Leuchtstofflampen stellt eine integrative Systemlösung dar und lässt durch die niedrige Einbauhöhe Freiraum für elektrische Versorgungs- bzw. Klimakanäle. Für die Entblendung sorgen zehn Primärreflektoren sowie eine zusätzliche konzentrische Ringblende aus blendungsreduzierendem Aluminium MIROSILVER®. Die Leuchtdichtebegrenzung entspricht mit rundum ≤ 1000 cd / m^2 der DIN EN 12464. The compact luminaire represents an integrated system solution for low energy, circular T-16 fluorescent tubes and its low installation height permits sufficient free space for the additional installation of electrical supply systems or air conditioning ducts. The use of ten primary reflectors and an additional concentric reflector in glare-reduced louver material MIROSILVER® ensures all-round luminance limitation of L ≤ 1000 cd / m^2 in accordance with DIN EN 12464.

421-388 FC10 *Deckeneinbauleuchte*

Recessed ceiling luminaire

Unternehmen — Company Ludwig Leuchten KG — Frühlingstraße 15 — 86415 Mering — Germany — www.ludwig-leuchten.de

Design Bartenbach Lichtlabor GmbH — www.bartenbach.com

CustomLight ist ein neuartiges, wandlungsfähiges Beleuchtungssystem, das technische und dekorative Beleuchtung verbindet. Durch Einsatz und Kombination verschiedener produktspezifischer Elemente werden Räume je nach Wunsch individuell gestaltet; auch nach der Endmontage des Systems stehen alle Möglichkeiten offen. Darüber hinaus sind Einsatzort und Erscheinungsbild der Leuchten austauschbar und erfüllen gleichzeitig höchste beleuchtungstechnische Anforderungen. *CustomLight* is a new technical-decorative lighting concept characterised for its personalisation. *CustomLight* is more than a luminaire, it is a versatile system prepared in order to personalise spaces by selecting and combining the different elements from a specific product range. Even after the initial installation, customers shall be able to select, combine and change the lighting, location and appearance of the lights, obtaining at the same time the highest luminotechnic performance required.

CustomLight *Beleuchtungssystem*
Lighting system

Unternehmen — Company Troll Lighting — Passeig de la Ribera 109 — 08420 Canovelles — Spain — www.troll.es
Design Werksdesign / Inhouse — Troll Lighting Team

LightPoints sind die einzigartige Möglichkeit, Leuchtdioden im Glas frei schweben und leuchten zu lassen. Eine transparente stromleitende Grundglasscheibe wird mit LEDs – ebenfalls in verschiedenen Farben – bestückt und im PVB-Folienverbund mit einer Deckglasscheibe zusammengefügt. Verschiedene Größen mit unterschiedlicher Anzahl von LEDs, auf Wunsch auch einzeln ansteuer- und programmierbar, sind möglich. Das PVB Folienverbundglas erfüllt die gesetzlichen, bautechnischen Anforderungen nach EN 12 600. *LightPoints* offer LEDs the unique possibility to float freely inside glass, while emitting light. The PVB flat glass laminate contains a transparent conductive base glass with LEDs – available in different colours – as well as a cover glass. The glass units are available in different sizes with various numbers of LEDs. Alternatively they can be programmed to customers' requirements. The PVB laminated glass also satisfies the building regulations that apply to EN 12 600.

LightPoints *LEDs im PVB-Flachglas-Folienverbund*
LEDs in flatglas laminate

Unternehmen — Company lif GmbH — SCHOTT AG — Etzwiesenstraße 33 — 72108 Rottenburg — Germany — www.lif-germany.de

Design Werksdesign/Inhouse — Christoph Döppner, Rolf A.O. Schneider, Bernd Albrecht, Andreas Nickut

Im Wellness-, Beauty- oder Hairstylingbereich entspannen Kunden durch atmosphärisch angenehme Lichterlebnisse. Wechselnde Farbwellen mit Sternenhimmel verwöhnen während der Anwendung. Dem behandelnden Friseur dient ein zuschaltbares neutrales Arbeitslicht. Ein Blendschutz bewahrt dabei die entspannende Lichtstimmung für den Kunden. Das sanft geschwungene satinierte Lichtsegel mit seiner filigranen Edelstahlabhängung ist durch zwei unauffällige Zugschalter leicht zu bedienen. Clients in the wellness, beauty and hairstyling area relax in an atmosphere evoked by a soothing light experience. Changing colour waves with a starry sky spoil the client during his treatment. The hairdresser working with the client, has at his disposal an additional working lamp. A screen ensures that the relaxing light effect is maintained for the customer. The gently curved satin-finished light sail with its delicate suspension in stainless steel is easy to operate by two pull switches.

WaveLight *Lichtsegel*
Light sail

Unternehmen — Company OLYMP GmbH & Co. KG — Hohenheimer Straße 91–97 — 70184 Stuttgart — Germany — www.olymp.de
Design pan it's design — Oliver Panitz — www.pan-its-design.de

Lumex Line Light ist die Synthese einer konsequenten Entwicklung aus lateral ausstrahlender Glasfaseroptik und der Lichttechnologie randgespeister Lichtflächen. Wie in einem Lichtleiter wird das Licht durch die gesamte Länge des Acryl-Bands transportiert und mit Hilfe einer Gitter-Technologie an jeder Stelle mit derselben Lichtdichte freigesetzt. Die gesamte Lichttechnik ist in ein stranggepresstes Aluminiumgehäuse eingebettet. Die Leuchtenoberfläche und die Abdeckplatte sind aus widerstandsfähigem mattiertem Polycarbonat gefertigt. *Lumex Line Light* combines a natural development of lateral emission fibre optics with the technology of edge-lit light surfaces. As in an optical fibre, the light is carried along the entire length of the acrylic strip and, using grid technology, it is released with the same light density at each point along its length. All the light-emitting technology is embedded in an extruded aluminium sheath. The illuminated surface and cover-plate are made of hard-wearing, matt polycarbonate.

Line Light *LED-Leuchte*
LED light

Unternehmen — Company Lumex Lichttechnik — Bonholzstraße 6 — 71111 Waldenbuch — Germany — www.leuchtrahmen.de
Design Werksdesign/Inhouse — Gerhard Müller

AQUADESIGN ist spritzwassergeschützt nach IP 44, witterungsbeständig, kratzunempfindlich und UV-stabil. Ein spezieller Adapter ermöglicht die Verwendung aller elektrotechnischer Steuerungsoberflächen aus dem Merten Produktprogramm „System M". Von der Schalterwippe über Innenbewegungsmelder bis hin zu Kommunikationseinsätzen. Die Rahmen sind als rastende Standardausführung oder als verschraubte Variante erhältlich. In dieser Ausführung bietet das Verdübeln in der Wand erhöhten Diebstahlschutz.

AQUADESIGN meets the IP 44 standard. Moreover it is weather-resistant, scratch-proof and UV-stable. Special adapters ensure that AQUADESIGN switches can also be used with all electrotechnical control components of the Merten product range "System M". This ranges from switch rockers and indoor movement detectors through to communication inserts. The frames are available as snap-in or screw-in version. Dowelled into the wall, the screw-in version provides increased protection against theft.

AQUADESIGN *Unterputz-Spezialprogramm – spritzwassergeschützt, IP44*

Special flush-mounted range – splash-proof, IP44

Unternehmen — Company Merten GmbH & Co. KG — Fritz-Kotz-Straße 8 — 51674 Wiehl — Germany — www.merten.de

Design Werksdesign/Inhouse

Dem hochwertigen, bruchfesten Thermoplast sind Silber-Ionen zugefügt. Diese stellen die antibakterielle Wirkung sicher und beseitigen Bakterien innerhalb von 24 Stunden vollständig. Im Gegensatz zu Antibiotika sind Silber-Ionen in der Lage, die Entwicklung resistenter Mikroorganismen zu verhindern. Dieser natürliche Prozess belastet weder Mensch noch Umwelt. Die antibakterielle Wirkung bleibt über die gesamte Lebensdauer erhalten. Silver ions are added to the high-quality, shockproof thermoplastic switch material. These guarantee an anti-bacterial effect and completely remove bacteria within 24 hours. Unlike antibiotics or disinfectants, silver ions are able to prevent the development of resistant microorganisms. This natural process has no negative effects on either humans or the environment. The antibacterial action remains effective for the products whole life.

M-SMART Aktivweiß
M-SMART Active White

Antibakterieller Schalter

Anti-bacterial switch

Unternehmen — Company Merten GmbH & Co. KG — Fritz-Kotz-Straße 8 — 51674 Wiehl — Germany — www.merten.de

Design Werksdesign/Inhouse

Unverwüstlich ist der dickwandige Alu-Druckguss-Metallrahmen, der zusammen mit einer isolierstoffgefütterten Metallabdeckung in der Wand verankert wird. Auch die Metallwippe ist für höchste Belastungen ausgelegt und hält dadurch zuverlässig harten Stößen, Angriffen mit Werkzeugen oder offenem Feuer stand. Ein Plombierungssystem verhindert den Zugriff auf die Halteschrauben. Diebstahl oder Beschädigung ist nur durch Aufbohren möglich. Simply indestructible: the heavy-walled, coated metal frame comes in a 1-gang, 2-gang or 3-gang version. The metal frame is used to anchor an insulated, equally thick-walled and coated cast aluminium metal cover to the wall. This cover, in conjunction with 4 mm wall fixing screws, ensures that the switch stays right where you put it.

Panzer Anti-Vandalism

Unterputzspezialprogramm

Special flush-mounted switch range

Unternehmen — Company Merten GmbH & Co. KG — Fritz-Kotz-Straße 8 — 51674 Wiehl — Germany — www.merten.de
Design Werksdesign/Inhouse

Die neuen Türgriffschalter der TGM-Baureihe erlauben die Integration von Bedien- und Anzeigefunktionen in den Türgriff von Maschinen. Damit kann auf ein separates Bedien-Paneel, zum Beispiel für den Not-Aus-Taster oder für Anzeige-LEDs, verzichtet werden. Der Bediener hat also, im wahrsten Sinne des Wortes, alles im Griff. Das spart Kosten und optimiert die Ergonomie. Der modulare Aufbau der Baureihe erlaubt die Konfiguration des Türgriffs nach individuellen Anforderungen des Kunden. The new door handle switches TGM allow the integration of operating and indicating functions of machines. Thus there is no need for a separate operating panel e.g. for emergency-stop push-button or indicating LEDs. The operator has in the true sense of the word everything under control. This is cost-efficient and optimises the ergonomics. The modular design of this series allows for the configuration of the door handle according to the requirements of the customer.

TGM *Türgriffschalter*

Door handle switch

Unternehmen — Company steute Schaltgeräte GmbH & Co. KG — Brückenstraße 19 — 32584 Löhne — Germany — www.steute.de

Design Margit Herzog Design — Margit Herzog — www.margitherzog-design.de

Ausgangspunkt war die Idee, das Kommunikationssystem *Siedle-Steel* für den Eingangsbereich (mit zahlreichen Funktionen wie Sprechen, Sehen, Licht, Informieren, Postempfangen usw.) auch als Orientierungssystem im Innenbereich fortzuführen. Auf diese Weise kann dem Architekten und Planer ein ganzheitliches System mit einem für den Außen- und Innenbereich einheitlichen Erscheinungsbild an die Hand gegeben werden. It all started with the idea of continuing the *Siedle-Steel* communication system for the entryway area (featuring the whole range of speech, vision, light, information, post reception functions etc.) as a system of orientation for the interior – so providing architects and planners with an integral system based on a cohesive appearance for inside and outside.

Siedle-Steel *Orientierungssystem* *Orientation system*

Unternehmen — Company S. Siedle & Söhne Telefon-und Telegrafenwerke OHG — Bregstraße 1 — 78120 Furtwangen — Germany — www.siedle.de
Design Werksdesign/Inhouse — S. Siedle & Söhne Telefon-und Telegrafenwerke OHG — Eberhard Meurer

Zodiaq® präsentiert sich als hochwertiges, dekoratives Oberflächenmaterial, das die praktischen Vorteile und die Ästhetik von natürlichem Quarz mit einer attraktiven Farbpalette sowie gleich bleibenden Strukturen und Farbtönen verbindet. *Zodiaq*® besitzt eine unverwechselbare, luxuriöse Anmutung, die sich in einer Vielzahl horizontaler und vertikaler Oberflächenanwendungen im privaten und öffentlichen Bereich einsetzen lässt. *Zodiaq*® is a premium decorative surfacing material that combines the strength and beauty of natural quartz with a palette of attractive colours, and consistency of structures and tones. *Zodiaq*® has a distinctive, luxurious aesthetic that lends itself to a variety of horizontal and vertical surfacing applications in both residential and commercial environment.

Zodiaq® Concrete Block *Technischer Stein*

Engineered stone

Unternehmen — Company DuPont de Nemours (Deutschland) GmbH — Du Pont-Straße 1 — 61352 Bad Homburg — Germany — www.zodiaq.com

Design Werksdesign / Inhouse

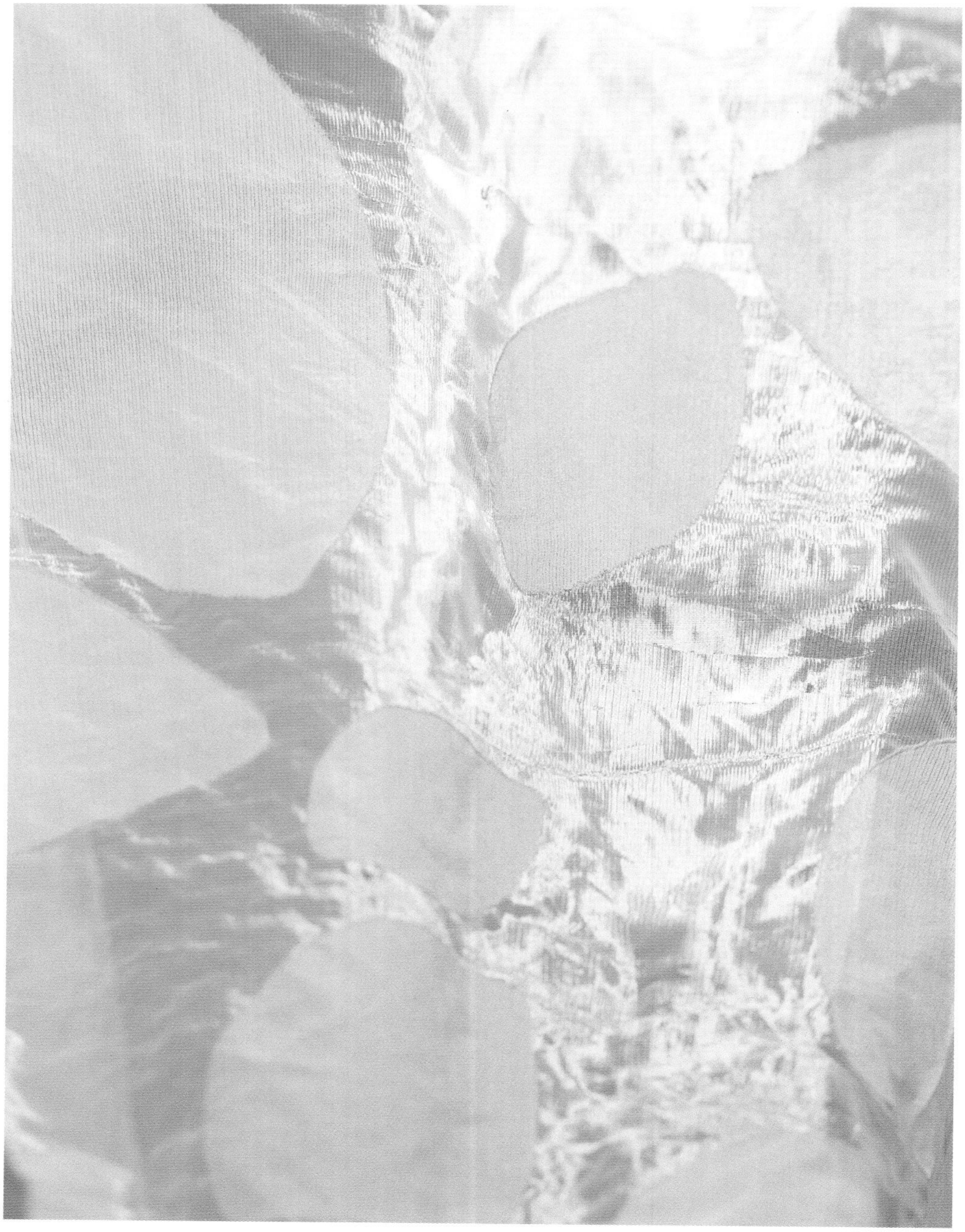

SPIRIT, ein Coup-à-fils, ist aus feinstem Stahl gewebt und zeichnet sich durch die Kombination des verwendeten Stahlmaterials in Verbindung mit der eingesetzten Scherli-Webtechnik aus. Großzügig angelegte, abstrakte Blätter scheinen im Raum zu schweben. Die außergewöhnliche Materialität entsteht aus dem Kontrast von matten Flächen und dem speziell metallischen Glanz, der an eine Wasseroberfläche erinnern lässt.
Material: 20 % Metall, 80 % Polyester Trevira CS
Maße: 300 cm hoch

SPIRIT, a coup-à-fils, is woven with finest steel. In combination with a clip-cord weave this steel makes a very distinguishing mark. Generous abstract leaves appear to float in space. This exceptional material generates a contrast between its matt areas and the special metallic shine which resembles the surface of water.
Material: 20 % Metal, 80 % Polyester Trevira CS
Measurements: 300 cm high

SPIRIT *(Dekorations-)Stoff*

(Interior decoration) Fabric

Unternehmen — Company Kinnasand GmbH — Danziger Straße 6 — 26655 Westerstede — Germany — www.kinnasand.com
Design Werksdesign/Inhouse — Kinnasand Design Team

Die Naturteppich-Kollektion GEYSIRA wird aus 100 % isländischer Schurwolle hergestellt. Die Qualität dieser Wolle ist einzigartig, ebenso ihre natürliche Farbigkeit. In GEYSIRA verbinden sich Natur und Design: die Wolle wird im 45 °C heißen Wasserdampf der Geysire gereinigt. Die Farbgebung entsteht ausschließlich aus den natürlichen Wollfarben der Schafe, so entsteht ein Unikat – made by OSCHWALD, designed by nature. GEYSIRA garantiert natürliches Wohlbefinden im Arbeits- und im Wohnbereich.

The natural carpet collection GEYSIRA is made from 100 % Icelandic wool. The quality of this wool is unique, as is the natural richness of its colours. GEYSIRA combines nature and design: the wool is washed at 45 °C using natural steam from local geysers. The colours are determined solely by the natural colour of the sheep's wool, thus making each carpet unique – made by OSCHWALD, designed by nature. GEYSIRA guarantees a natural sense of wellbeing, both at work and in the home.

Geysira *Teppichboden-Kollektion*
Floor carpet collection

Unternehmen — Company OSCHWALD Boden aus Natur GmbH — Rudolf-Blassingstraße 7 — 79183 Waldkirch — Germany — www.oschwaldkirch.de

Design Werksdesign / Inhouse

Das Garn in Form einer gestrickten Kordel verleiht diesem handgetufteten Teppich sein besonders lässiges Auftreten. Über einem Fond aus dichter kurzer Schurwolle fällt leger der sehr lange Pol aus Strickgarn. Dieses Garn aus bester Neuseelandschurwolle wird in einem sehr aufwändigen Verfahren hergestellt und anschließend so bearbeitet, dass es als geschnittener Flor eingesetzt werden kann. Die Strickstruktur bewirkt, dass das Garn einen sehr weichen Fall hat und somit eine unvergleichliche Optik entsteht. The yarn in the form of a knitted cord conveys this hand-tufted carpet is especially casual look. Over a ground of dense, short pure new wool falls the very long pile from knitting yarn to give a casual appearance. This yarn from best New Zealand wools is produced in a special production technique and treated afterwards in such a way that it can be used as a cut pile. Due to the knitting structure the yarn has a very soft drape and an incomparable look.

LEM 7205/520 *Teppich-Kollektion LANA L III*
Rug collection LANA L III

Unternehmen — Company JAB TEPPICHE HEINZ ANSTOETZ KG — Dammheider Straße 67 — 32052 Herford — Germany — www.jab.de
Design Werksdesign / Inhouse

LASA ist ein weicher Teppich aus gefilztem Merinogarn. Diese neue Qualität überzeugt durch ihre einmalig kuschelige Oberfläche. Erstmalig wurde ein besonders dickes Kernmantelgarn aus feinster neuseeländischer Merinowolle hergestellt. Durch eine ausgefeilte Technik hat das Garn eine sehr klare Schnittfläche, die eine dunkle Garnseele von der farbigen Ummantelung abgrenzt. Den Untergrund des Teppichs bildet ein dichter Flor aus feiner Schurwolle. LASA is a soft carpet made from a felted merino yarn. This new quality captivates with its cosy surface. For the first time an especially thick core yarn has been produced from the finest New Zealand merino wool. By using cutting-edge techniques the yarn has a very clear-cut surface which closes of the dark yarn core from the colourful sheath. The ground of the carpet is a dense pile from fine pure new wool.

LASA 7204 / 311 *Teppich-Kollektion LANA L III*
Rug collection LANA L III

Unternehmen — Company JAB TEPPICHE HEINZ ANSTOETZ KG — Dammheider Straße 67 — 32052 Herford — Germany — www.jab.de
Design Werksdesign / Inhouse

Dieser handgetuftete Teppich erhält seinen einzigartigen Charakter durch ein mehrfarbiges Streifengarn. Dieses Schurwollgarn bildet in der Teppichkonstruktion die hohe weiche Oberfläche. Den Fond bildet ein dichter kurzer Flor, der farblich passend durch den offenen verspielten Pol der oberen Fläche hervorblitzt. Der Teppich LOA wird in fünf ausgefallenen Farbstellungen hergestellt. Durch die Verwendung bester Wollen aus Neuseeland ist LOA ein aktuelles und wohnliches Schmuckstück in der modernen Wohnlandschaft.

This hand-tufted carpet achieves its unique character through the use of a multicolour stripe yarn. This yarn form pure new wool forms the high, soft surface of this carpet construction, whilst a dense, short pile in matching colours in the ground shows through the open, playful pile of the surface. The carpet LOA is produced in five outstanding colourways. By using best New Zealand wools LOA is an outstanding and comfortable gem for modern interiors.

LOA 7200/760 *Teppich-Kollektion LANA L III*

Rug collection LANA L III

Unternehmen — Company JAB TEPPICHE HEINZ ANSTOETZ KG — Dammheider Straße 67 — 32052 Herford — Germany — www.jab.de

Design Werksdesign/Inhouse

TAURUS wird in der von ANKER patentierten 40-Riet-Technik produziert und erhält durch dichte Schlingen eine puristische Optik. Die Oberfläche ist durch dreidimensionale Linienstrukturen dezent modelliert. Gestalterische Qualität und strapazierfähige Funktionalität bilden eine Einheit. Die eingesetzten Solution-Dyed-Fasern sind lichtecht und resistent gegen chlorhaltige und desinfizierende Mittel. TAURUS is produced using the 40 reed manufacturing technique patented by ANKER and gets its purist design through dense loops. The surface texture is created by the three dimensional structure of the lines. Creative quality and sturdy functional form a unity in design and texture. The solution dyed fibres are light-fast and proof against hyper-choric and disinfecting cleaning agents.

TAURUS *Objekt-Teppichboden*

Contract carpet

Unternehmen — Company ANKER-TEPPICHBODEN — Gebr. Schoeller GmbH + Co. KG — Zollhausstraße 112 — 52353 Düren — Germany — www.anker-teppichboden.de

Design Werksdesign/Inhouse — Ines Binder

Dyckerhoff TERRAPLAN ist ein monolithischer Betonboden ganz ohne Beschichtung. Mit seiner perfekten Optik sowie seiner dichten und diffusionsoffenen Oberfläche vereint er die Vorzüge eines Terrazzobodens mit hoher Wirtschaftlichkeit. In der mit Weißzement auch farbig ausführbaren Betonrezeptur bestimmen die individuell kombinierbaren Gesteinskörnungen nach der finalen Oberflächenbearbeitung maßgeblich das Erscheinungsbild der jeweiligen Boden-Unikate. *Dyckerhoff* TERRAPLAN is a monolithic concrete floor with no coating. Thanks to its perfect visual appearance and its dense surface, that enables diffusion, it combines the advantages of terrazzo flooring with great cost efficiency. The concrete mix hinges on white cement and can thus also be supplied in coloured versions, whereby the individual stone grain sizes can be combined individually and, after final surface treatment primarily define the look of the respective unique flooring.

Dyckerhoff TERRAPLAN *Betonboden*

Concrete floor

Unternehmen — Company Dyckerhoff AG — Biebricher Straße 69 — 65203 Wiesbaden — Germany — www.dyckerhoff.com

Design Werksdesign / Inhouse — Andre Conrady, Dr. Reinhard Winzer

Mit großen Plattenmaßen und einem dadurch geringen Fugenanteil bildet *Cristalite granite* von Schock sehr ruhige Flächen – sowohl an der Wand als auch auf dem Boden. Durch den hohen Quarzanteil erreicht das Material aus Quarz und Acryl eine außergewöhnliche Härte und Langlebigkeit bei nur 6 mm Aufbauhöhe. *Cristalite granite* bietet viele Gestaltungsmöglichkeiten mit einer Vielfalt aufeinander abgestimmter Farbtöne, Strukturen und Texturen. Thanks to extra-large sheets and fewer seams, Schock *Cristalite granite* creates extremely smooth, calm surfaces – for walls or floors. The high quartz content gives the quartz / acrylic material remarkable hardness and hard-wearing durability despite an added surface height of just 6 mm. *Cristalite granite* offers a wealth of creative options with a wide variety of coordinated colours, patterns and textures.

Cristalite granite *Wand- und Bodenbelag*

Flooring and wall cladding

Unternehmen — Company Schock GmbH — Gmünder Straße 65 — 73614 Schorndorf — Germany — www.schock.de

Design Werksdesign / Inhouse — Dr. Gudrun Janke, Martin Kirschbaum

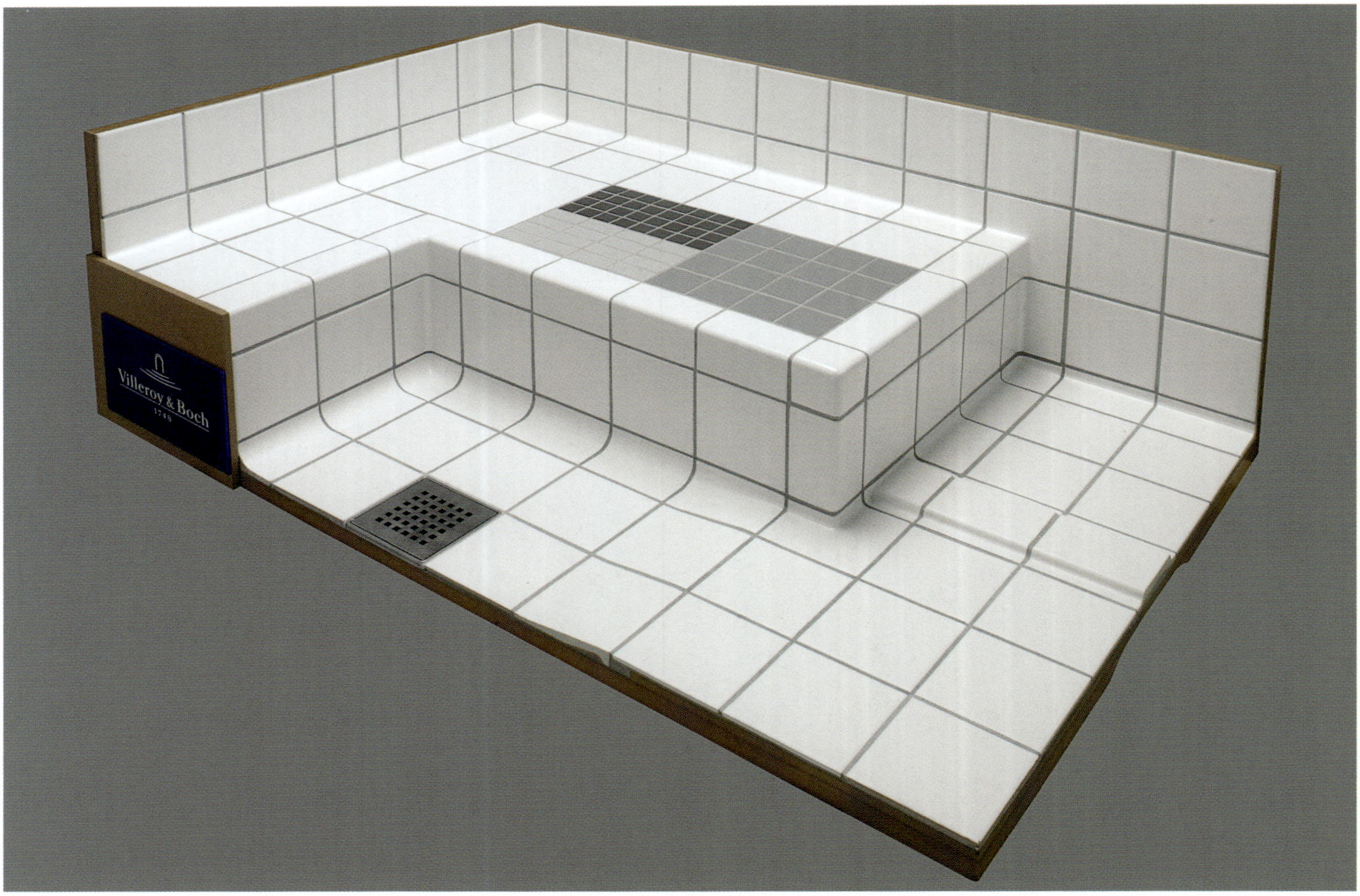

Ein keramisches Fliesen-Programm, das die Ausbildung weicher Übergänge zwischen Flächen ermöglicht. Außer keramischen Duschformationen können Nischen, Ablagen, Sitze und ganze Landschaften gebildet werden. Beim Einsatz der Schwellen-Formteile wird nur ein minimaler Höhenaufbau benötigt. Alle Duschtassen sind problemlos mit Rollstühlen zu befahren und leisten damit einen wichtigen Beitrag zum barrierefreien und rollstuhlgerechten Bauen. A ceramic tile programme which enables a soft transition between surfaces. Ceramic shower formations as well as niches, shelves, seats and whole landscapes can be created. The underlying structure requires only a minimal height with these threshold moulded parts. All shower trays can be accessed with wheelchairs making an important contribution to barrier-free and wheelchair-friendly building.

Pro Architectura New-Sinus

Fliesenserie mit Formfliesensystemen
Tile collection with pre-shaped tile systems

Unternehmen — Company V & B Fliesen GmbH — Rotensteinerweg — 66663 Merzig — Germany — www.villeroy-boch.com
Design Architekturbüro Frank Huster — Frank Huster — www.frankhuster.de

Mosa Terra XXL ist die bekannte Kollektion Terra Maestricht in neuen Dimensionen: neue XXL Formate von 90 x 90 cm, 60 x 120 cm und 30 x 120 cm. Unglasiertes Feinsteinzeug, quadratisch oder rechteckig, in einer steinharten und verschleißfesten Qualität. Die Fliesen werden zu Keramikflächen mit minimalen Fugen (≥ 2 mm). *Terra XXL* ist darüber hinaus sehr pflegeleicht, frostbeständig, hat einen Antirutschwert von R10 und ist chemikalienresistent.

Mosa Terra XXL is the well-known *Terra* Maestricht series taken to its extreme: new XXL sizes of 90 x 90 cm, 60 x 120 cm and 30 x 120 cm. Unglazed porcelain ceramic tiles, square or rectangular, of tough, absolutely wear-resistant quality. The tiles become expanses of ceramic with minimal grouting (≥ 2 mm). *Terra XXL* is very scratch and wear resistant, easy to maintain, frost resistant, anti-slip R-10 and resistant to chemicals.

Terra XXL *Fliesen-Kollektion*

Collection of tiles

Unternehmen — Company Royal Mosa — Meerssenerweg 358 — 6201 BA Maastricht — Netherlands — www.mosa.nl
Design Werksdesign/Inhouse

Kubisch, geradlinig und minimalistisch in der Form, ist die großzügige Acryl-Badewanne mit einer oder zwei Rückenschrägen ausgestattet oder auch ganz ohne Rückenschräge zu haben. Ein umlaufender transluzenter Wassergraben erlaubt via Fernsteuerung individuelle Farb-Spiele. Statt eines Überlaufs fangen wasserbeständige, gradlinige Holzeinsätze das überfließende Wasser im außen liegenden Graben ein und bringen eine wohnliche Optik ins Wasserparadies.
The spacious acrylic bath with one, two or without backrest is cubist, understated and minimalist. A translucent water channel around the rim allows individual interplays of colour via remote control. Instead of an overflow, water-resistant, straight wooden inserts – placed on the rim of the channel – absorb the overflowing water in the surrounding channel and add a homely touch to the watery paradise.

Starck X *Badewanne freistehend mit Holzaufleger für Wannengraben*
Bathtub freestanding with wooden inlays for tub trench

Unternehmen — Company Duravit AG — Werderstraße 36 — 78132 Hornberg — Germany — www.duravit.de
Design Philippe Starck — www.philippe-starck.com

Die Badkollektion *Bellevue* verbindet und gestaltet Räume. Ästhetisch ansprechende Möbel und Accessoires schaffen fließende Übergänge vom Bad über den Ankleideraum bis hinein ins Schlafzimmer. Als Bestandteil der Kollektion verkörpert ein großer Waschtisch auf der Wandkonsole mit 1.000 mm Breite eindringlich die Extravaganz und Großzügigkeit der Serie. Für die Badkollektion *Bellevue* wurde die klassische Kombination von reinweißer Keramik und ausdrucksstarkem Holz neu interpretiert. The *Bellevue* bathroom collection connects and designs rooms. Aesthetically appealing furniture and accessories create a flowing transition from the bathroom and the change room to the sleeping room. A big washstand on the wall bracket, 1,000 mm in width, most strikingly shows the extravagance and liberality of the collection. For *Bellevue* the classical combination of pure white ceramic and expressive wood has been interpreted anew.

Bellevue *Badkollektion*
Bathroom collection

Unternehmen — Company Villeroy & Boch AG — Saaruferstraße — 66693 Mettlach — Germany — www.villeroy-boch.com
Design .molldesign Reiner Moll & Partner — Daniel Mader — www.molldesign.de

Die Eckwanne *Paiova* (zu deutsch: Paar) verändert die Weltanschauung. Statt sich länger gegenüber zu sitzen, tauchen die Badenden nebeneinander ein. Durch bequeme Nackenstützen genießt man zu zweit am Fußende der Wanne zum Beispiel das Heimkino auf dem dazugehörigen Rollcontainer. Eine Stufe in drei verschiedenen Größen erleichtert den Einstieg. Komfortable Abmessungen bieten Badegenuss, die Formensprache ermöglicht auch die optimale Integration in kleinere Bäder. *Paiova* corner bath (in English: pair) changes the philosophy of life. Instead of sitting face to face, *Paiova* invites you to lie back and relax side by side. With your head reclining on the comfortable neck rests, enjoy a home movie together on the matching *Paiova* mobile storage. A raised step in three different sizes facilitates getting in and out of the bath. Enjoying bathtime bliss in luxurious dimensions, its design also allows optimal integration into smaller bathrooms.

Paiova *Badewanne mit Möbelverkleidung und Rollcontainer*

Bathtub with furniture casing and mobile storage

Unternehmen — Company Duravit AG — Werderstraße 36 — 78132 Hornberg — Germany — www.duravit.de

Design EOOS Design GmbH — Martin Bergmann, Gernot Bohmann, Harald Gründl — www.eoos.com

Die Badewanne als zentrales Produkt der Entspannung wird gestalterisch maßgeblich geprägt durch das technisch notwendige Overflow-System. VISIGN M3 fokussiert dieses Thema und konzentriert technische Qualität in gestalterischer Modernität. Als integratives Element und klassisches Zukaufprodukt ergänzt VISIGN M3 selbstverständlich die zeitgemäß schlichte Architektur- und Designsprache, ohne seinen technischen Charakter zu verleugnen. Der ergonomische Griffbereich ist selbst für „seifige" Situationen bestens vorbereitet und nebenbei leicht zu reinigen.

As central product of relaxation the bathtub is in its design significantly stamped by the technical necessary overflow-system. VISIGN M3 focuses this topic and concentrates technical quality in formative modernity. As integrative element and classical accessory product, VISIGN M3 complements naturally those today's plain architecture and design language without disowning its technical character. The ergonomic handle is at best prepared even for "slippery situations", and besides, it is easy to clean.

VISIGN M3 *Überlaufgarnitur*

Overflow system

Unternehmen — Company VIEGA — Ennester Weg 9 — 57439 Attendorn — Germany — www.viega.de

Design ARTEFAKT industriekultur — Achim Pohl, Thomas Fiegl — www.artefakt.de

Aus der fast 400 mm breiten Armatur ergießt sich das Wasser als breiter Schwall in die Waschschüssel. Ihr stammähnlicher, schlanker Korpus wächst in einer leichten Kurve empor. Dabei schafft das Miteinander organischer und kantiger Formen den Bezug zu Schlichtheit und Schönheit alles Natürlichen. Im Sinne der Philosophie Jean-Marie Massauds verleiht ihr dies die Anmutung einer enttechnisierten Skulptur. From the almost 400 mm wide mixer water pours in a wide stream into the basin. Its stem-like, slender corpus sprouts in a slight curve. With its balanced blend of organic and sharp-edged forms, the mixer reclaims the simplicity and beauty of the natural world. That gives it the grace of a de-technified sculpture, which is quite in the sense of Jean-Marie Massauds philosophy.

Axor Massaud *Einhebel-Waschtischmischer*

Single-lever washbasin mixer

Unternehmen — Company Hansgrohe AG — Auestraße 5–9 — 77761 Schiltach — Germany — www.hansgrohe.com

Design Jean Marie Massaud — www.massaud.com

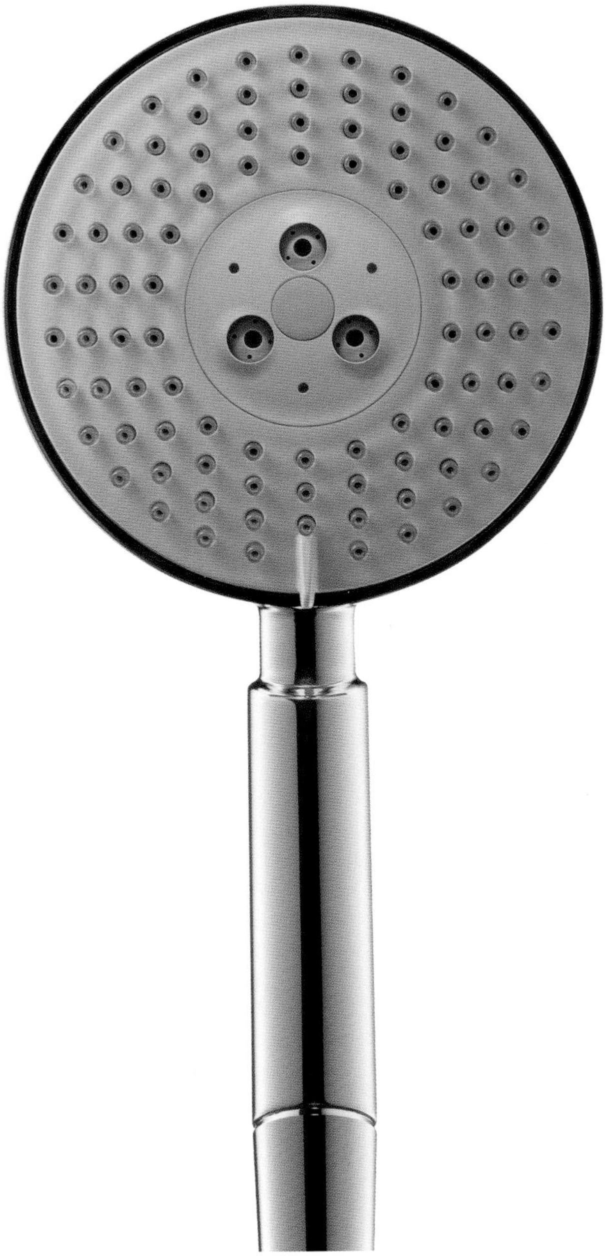

Formal überzeugt *Raindance* durch ihren unprätentiösen Auftritt. Die auffallend flache Gestalt und kompromisslos gradlinige Formen gehen einher mit ergonomischer Handhabung und ausgeprägter Funktionalität. Die integrierte AIR-Technologie lässt alltägliche Duschroutine zum wahren Erlebnis werden: Wie ein Mantel umhüllt das Luftblasen-Wassergemisch aus großen, sanften Tropfen den Körper des Duschenden. In shape, the *Raindance* hand shower appeals with its simple unpretentiousness. Its amazingly flat form and clear design language are teamed with superlative functionality. The typical short Raindance handle, its ergonomic form and its lightweight all guarantee extremely satisfactory handling. Furthermore, the AIR technology turns the day-to-day shower routine into a true experience to remember.

Raindance S 120 AIR 3jet *Handbrause*
Hand shower

Unternehmen — Company Hansgrohe AG — Auestraße 5-9 — 77761 Schiltach — Germany — www.hansgrohe.com
Design Phoenix Design — www.phoenixdesign.de

PLUSMODO heißt die neue Küchenarchitektur, die der renommierte spanische Designer Jorge Pensi für *Poggenpohl* entworfen hat. Charakteristisch für PLUSMODO ist die neuartige Inszenierung von offenen und geschlossenen Bereichen. Ausziehbare Trays positioniert Pensi direkt unter der 13 Zentimeter dicken Arbeitsplatte. Dies erzeugt eine große Transparenz sowie eine starke Betonung der Horizontalen. Verstärkt wird dieser Effekt durch das Beleuchtungskonzept.
PLUSMODO is the name of the new kitchen architecture created for *Poggenpohl* by the renowned Spanish designer Jorge Pensi. PLUSMODO is characterised by the novel presentation of open and closed functional areas. Pensi places pull-out trays directly below the 13 cm thick worktop. This creates a new transparency, as well as strongly emphasizing the horizontals. This effect is heightened by the atmospheric lighting concept.

PLUS**MODO** *Küchenarchitektur*
Kitchen architecture

Unternehmen — Company Poggenpohl Möbelwerke GmbH — Poggenpohlstraße 1 — 32051 Herford — Germany — www.poggenpohl.de
Design Jorge Pensi Diseño S.L. — Jorge Pensi // Werksdesign / Inhouse — Poggenpohl Möbelwerke GmbH

Die funktional wie ästhetisch gestalteten *Jumbo*-Unterschränke werden in der Küchenzeile oder am Abschluss der Küche eingeplant. Mit großzügigen Dimensionen und einem übersichtlich angeordneten Fach für Flaschen schaffen sie einen zusätzlichen Stauraum. Die Fronten sind in zehn Hochglanzausführungen sowie in Edelstahl oder Makassar furniert erhältlich. The functionally and aesthetically designed *Jumbo* floor cupboards are integrated into the kitchenette or added to the array upon completion of construction. With generous dimensions and a clearly arranged storage space for bottles, they provide additional storage capacities. The fronts are available in ten high-gloss colours, stainless steel and veneered Makassar.

Jumbo *Küchenmöbel*

Kitchen furniture

Unternehmen — Company Holzprojekte Härtenberger HKH GmbH — Hauptstraße 39 — 94336 Hunderdorf — Germany — www.haertenberger.de

Design ALNO - Produktdesign — www.alno.de // Werksdesign / Inhouse — Stephan Härtenberger

Einzigartige technische Funktionen sorgen bei *Modia®* für eine perfekte Körperabstimmung: 100%ige Liegefläche durch bis über den Außenholm reichenden Leisten im Kopf- und Fußbereich, stufenlos einstellbare Mittelzone, Rutschstopp der Matratze durch hochgezogene Kappen und Strukturoberfläche, Memory-Einstellung der Kappen mit individueller Härtegradeinstellung und Sperrfunktion sowie eine Federfunktion der Leisten bis zu den Randbereichen des Bettes. *Modia®* features unique technical functions to ensure perfect body support: 100 % sleeping surface guaranteed with slats extending over the outer frame in the head and foot sections, continuously adjustable middle section, surface structure and raised end caps prevent mattress from sliding, memorised adjustment of the caps with personalised firmness settings and locking function. The slat spring functionality ranges out to the edges.

Modia®

Lattenrost

High performance frame

Unternehmen — Company SCHLARAFFIA — RECTICEL-SCHLAFKOMFORT GmbH — Schlaraffiastraße 1-10 — 44867 Bochum — Germany — www.schlaraffia.de
Design Werksdesign/Inhouse — Giacomo Lanza, Andreas Schrade, Pierre Bourgeois, Markus Veutgen

Einzigartige technische Funktionen sorgen bei *Flexmont* für eine perfekte Körperabstimmung. Der sehr flache Motor eignet sich optimal für den Einsatz in dem ebenfalls sehr flachen Bett. Eine weiche Schulterkomfortzone sorgt für verbesserte Wirbelsäulenunterstützung. Durch den niedrigen Federweg ergibt sich eine Mittelzonenverstärkung, der Rutschstopp entsteht durch die hochgezogenen Kappenenden sowie die strukturierte Oberfläche. Die Federfunktion der Leisten reicht bis zu den Randbereichen des Bettes. *Flexmont* (features unique technical functions) to ensure perfect body support: Ultra-flat motor – also suitable for use in flat beds. A soft shoulder comfort zone for improved vertebra column support. Middle section reinforcement with low spring travel. Mattress prevented from sliding with raised end caps and surface structure. The slat spring functionality ranges out to the edges.

Flexmont *Motorrahmen* *Motor frame*

Unternehmen — Company SCHLARAFFIA — RECTICEL-SCHLAFKOMFORT GmbH — Schlaraffiastraße 1-10 — 44867 Bochum — Germany — www.schlaraffia.de
Design Werksdesign / Inhouse — Giacomo Lanza, Andreas Schrade, Pierre Bourgeois, Markus Veutgen

Die *Phi-ton 3D-Matratze* ist aus dem funktionellen 3D-Abstandsgewirke gefertigt. Im 3D-Textil sind Ober- und Untergewebe durch Millionen Polyester-Fäden miteinander verbunden. Sie sorgen mit dem richtigen Verhältnis von Stabilität und Elastizität für den nötigen Komfort. Darüber hinaus sind die Fäden der Garant dafür, dass die *Phi-ton 3D-Matratze* auch in punkto Luft- und Feuchtigkeitsdurchlässigkeit sowie Hygiene und Allergikerfreundlichkeit unübertroffen ist.
The *Phi-ton 3D-mattress* is made of the highly functional material 3D braided fabric. In this textile upper and bottom layer are connected through millions of polyester fibres. They give the body great comfort with the right balance between stability and elasticity. And they make the *Phi-ton 3D-mattress* unsurpassed in the air and moisture regulation as well as hygiene. Notorious causers of allergies do not feel at home.

3D-Matratze
3D-mattress

Schlaf- und Komfortprodukt
Sleeping and comfort product

Unternehmen — Company Phi-ton B.V. — Zijpendaalseweg 51A — 6814 CD Arnhem — Netherlands — www.phi-ton.com
Design Werksdesign/Inhouse — Fried-Jan Unger

Ohne die optischen Beschränkungen eines Unterrahmens und ohne Gefährdung durch freie bewegliche Teile, bietet das *Phi-ton*-Bett Spielraum für Individualität. Vier der fünf Segmente des *Phi-ton*-Bettes lassen sich stufenlos einzeln verstellen. Das Bett passt sich der Geometrie des menschlichen Körpers nahtlos an. Nur das mittlere Segment des Bettes ruht auf den schlanken Füßen. Die gibt es in vier Varianten, alle so konstruiert, dass die schwebende Optik des Bettes optimal zur Geltung kommt. Without the visual restrictions of an underframe and the risks of free moving parts, the *Phi-ton* bed always offers room for individuality. Four of the five segments of the *Phi-ton* bed are independently, continuously adjustable. The bed shapes itself flawlessly to the geometry of the human body. Only the middle segment of the bed rests on the slender feet. Available in four different versions, all constructed so that the floating appearance of the bed is at its best.

Phi-ton *Bett*

Bed

Unternehmen — Company Phi-ton BV — Zypendaalseweg 51A — 6014 Arnhem — Netherlands — www.phi-ton.com

Design Werksdesign/Inhouse — Fried-Jan Unger

Ein Bett, dessen Rahmen aus Stahlrohr und Polyurethan gefertigt ist. Das Kopfteil besteht aus Polyurethan-Schaum mit Stahleinsatz. Der abnehmbare Bezug des Bettes besteht aus Stoff oder Leder. Sechs Breiten stehen zur Verfügung: 154, 164, 174, 184, 194 und 214 cm. In der Version mit festem Kopfteil beträgt dessen Höhe 49 cm und die Länge des Bettes 226 cm. Bei der Version mit elektrisch höhenverstellbarem Kopfteil lässt sich die Höhe des Kopfteils von 76 bis zu max. 90 cm variieren, die Länge ist von 235 bis zu 240 cm einstellbar. A bed with frame in steel tubing and polyurethane. Headboard in polyurethane foam with steel inserts. Fully upholstered in fabric or in leather: always removable. Six widths are available: 154, 164, 174, 184, 194 and 214 cm. In the version with fixed headboard the headboard height is 49 cm and the length of the bed 222 cm. In the Version with electrically controlled height-adjustable headboard the headboard height varies from 76 to max. 90 cm and the length from 235 to 240 cm.

Eleen *Bett* *Bed*

Unternehmen — Company MDF Italia — Via Morimondo 5/7 — 20143 Milano — Italy — www.mdfitalia.it

Design Soda Designers — Nada Nasrallah, Christian Horner — www.soda.at

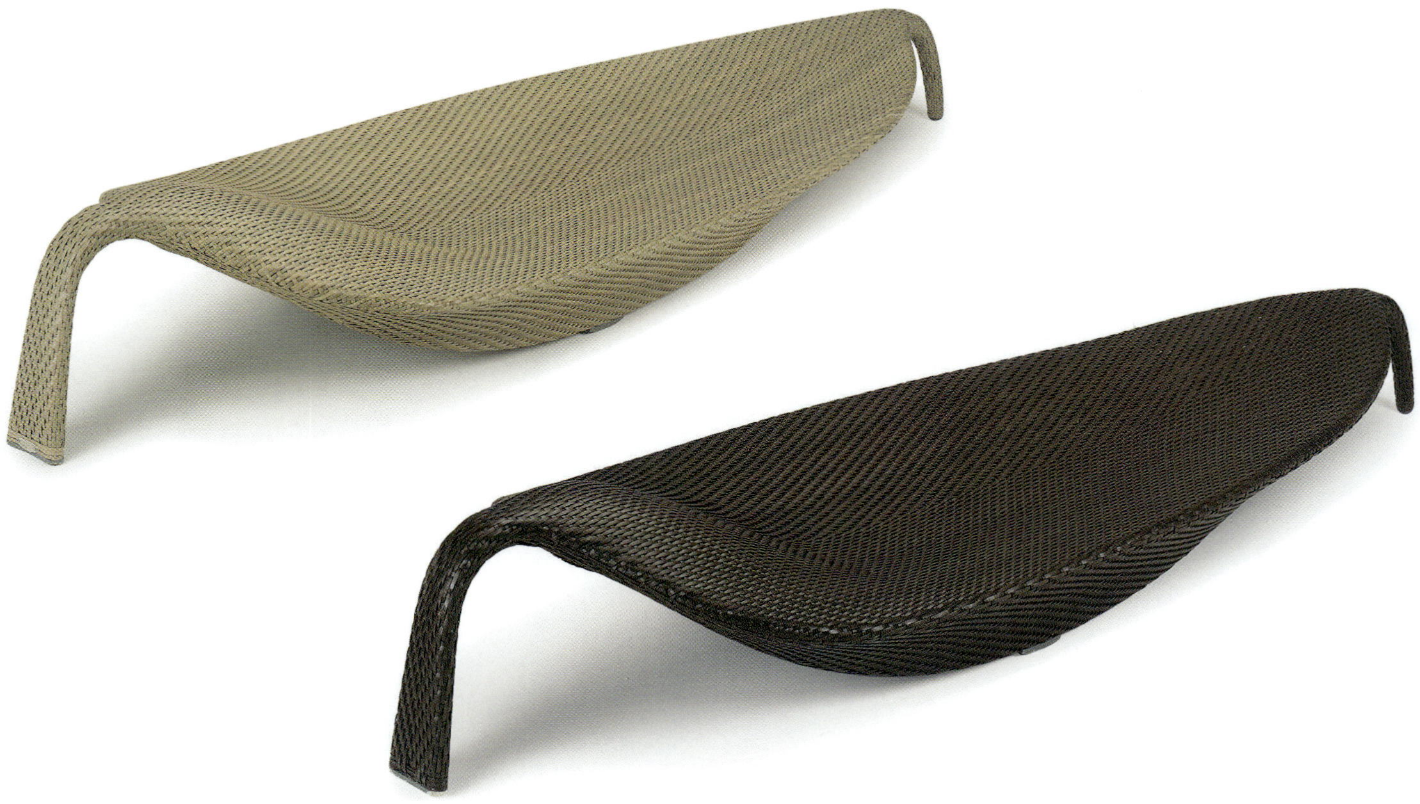

Die anmutige Form elegant geschwungener Blätter stand Pate für diese einzigartige Liege. Zwei unterschiedliche Größen – als Einzel- oder als Doppelbett – sind erhältlich. Die einem Blatt nachempfundene natürliche Form mit ihrer leichten, längsseits verlaufenden Teilung in der Mitte lädt mit einer sanften Mulde zum Verweilen ein. Ein bequemes Kissen aus atmungsaktivem Schaumgummi ist in die Oberseite der Strandliege integriert. The graceful shape of elegantly curved leaves is the spin off for this unique beach chair. Two sizes – single or double bed – are available. A leaf's natural shape with its light division running down its length creates a gentle depression for relaxed lounging. A comfortable cushion made of breathable foam rubber has been integrated into the top of the beach chair.

LEAF *Gartenmöbel*
Outdoor furniture

Unternehmen — Company DEDON GmbH — Zeppelinstraße 22 — 21337 Lüneburg — Germany — www.dedon.de
Design Frank Ligthart

System kvaro ist ein Baukastensystem, dessen multifunktionaler Aluminiumdruckgussknoten Korpusflächen sowie einzelne Module horizontal und vertikal miteinander verbindet; als Kantenschutz garantiert er zudem Langlebigkeit. Zusatzelemente werden in die Knoten eingeschraubt. Da die Verbindungen leicht lösbar sind, lassen sich die Elemente immer neu kombinieren und räumlichen Veränderungen anpassen. *System kvaro* is a modular construction system where the multi-functional aluminium pressure-casted joining knot ties up corpus surfaces as well as single modules horizontally and vertically. Additionally, the edge protection ensures longevity. Supplementary elements are screwed into the joining knots. As the joining knots can easily be dissolved the elements can be combined again and again and thus rearranged in every room.

system kvaro *Kastenmöbelsystem*

Furniture system

Unternehmen — Company Flötotto GmbH — Senner Straße 32 — 33335 Gütersloh — Germany — www.floetotto.de

Design KOLBERGdesign — Justus Kolberg — www.kolbergdesign.com

MYponte entstand aus dem Anspruch, neben ästhetischen Objektmöbeln auch attraktive und technisch innovative Möbel für den Wohnbereich zu schaffen. Das Highboard aus dem System *MYponte* nimmt die Linienführung der dazugehörigen Tische auf und bietet durch die Ausstattung mit Glaswinkelschiebetüren mit Selbsteinzug bzw. einer radargesteuerten Elektromotorik vielfältige Gestaltungsmöglichkeiten. Durch seine Flexibilität ist das System im Bürobereich wie auch im gehobenen Wohnsegment einsetzbar.

MYponte arose from the wish to offer not only aesthetically pleasing mass-purpose furniture, but also attractive and technically innovative furniture for residential purposes. The *MYponte* system highboard reflects the lines of the respective table and thanks to the self-closing glass angled sliding doors and a radar-controlled electronic motor affords any number of design opportunities. Due to its flexibility the system can be used in office environments as well as in sophisticated living areas.

System: MYponte *Sideboard*

Sideboard

Unternehmen — Company Rainer Euskirchen Büro Design GmbH — Jöllenbecker Straße 314 — 33613 Bielefeld — Germany — www.myponte.com / www.euskirchen-moebel.de

Design Bic Design, Dirk Togtmeyer — www.bic-design.de // Werksdesign / Inhouse

Ein System mit Wandpaneelen, Nischen und Behältern mit Schubkästen, das durch den bedeutenden Abstand von der Wand gekennzeichnet ist. Durch die freie Kombinierbarkeit der verschiedenen Elemente können innovative Konfigurationen entstehen, die als Bücherregal, als Platz für schöne Objekte oder auch als Unterbringung für einen Plasma- oder LCD-Fernseher sowie weitere moderne Geräte verwendet werden können. A wall system which incorporates panels, recessed compartments and drawer units, characterised by a significant distance from the wall. By freely combining these various elements, you can create highly-innovative configurations, which allow to store books, objects, plasma or LCD TVs and any other modern media equipment.

INSIDE ART INSIDE LINE

Regalsystem

Shelving system

Unternehmen — Company MDF Italia — Via Morimondom 5/7 — 20143 Milano — Italy — www.mdfitalia.it

Design Piergiorgio Cazzaniga

RONDO STAND wurde als Aktions- und Ruhemöbel konzipiert. Hier können Katzen ihren Spieltrieb ausleben, auf den speziellen Kratzhüllen ihre Krallen wetzen, schlafen und klettern. Durch die drei verschiedenen Ausführungen – Filz, Korb oder Leder – ist RONDO in jedes anspruchsvolle Wohnambiente integrierbar. Der Standfuß aus Edelstahl ist so entwickelt, dass auch bei größeren Katzen keine Gefahr durch Umwerfen besteht.
RONDO STAND was designed as furniture for both action and rest. Cats can fulfil their desire to play, sharpen their claws on the special scratch covers, sleep and climb. Thanks to the three different designs in felt, wicker or leather, it can be integrated into any sophisticated living environment. The standing base made from stainless steel is designed so that there is no danger of it tipping over, even when used by larger cats.

RONDO STAND Filz *Katzenmöbel*

Cat interior

Unternehmen — Company cat-interiors.de — Bahnhofstraße 5 — 96215 Lichtenfels — Germany — www.cat-interiors.de
Design Meyer Design — Otto Meyer, Christel Meyer — www.meyer-design.net

Die *Rough&Ready*-Bank ist widerstandsfähig und robust. Die gebogene Rückenlehne bietet überraschend guten Komfort. Die Bank wird aus 100 % recyceltem Holz hergestellt. Die dicken Balken werden aus 20 Jahre alten Hafenpfosten aus Basralocus-Holz gesägt. Das gut konservierte, charaktervolle Holz altert auf schöne Art und Weise. Mittels patentiertem „Streetlock comb system“ werden die Balken mit den Beinen verbunden: wartungsfrei, dauerhaft und vandalismusresistent. The *Rough&Ready* bench is tough and robust. A curved backrest realises surprisingly good comfort. The bench is sustainable since it's made of recycled wood. The thick beams were sawn from 20 year old harbour posts of Basralocus. The untreated wood is well preserved, has gained character and greys beautifully. The beams are fixed to the legs by means of the patented "Streetlock comb system": maintenance free, durable, vandal-proof.

Rough & Ready 6

Bank, R&R-6-2340 aus der R&R-5/6 Serie
Bench, R&R-6-2340 from the R&R-5/6 series

Unternehmen — Company Streetlife — Oude Singel 144 — 2312 RG Leiden — Netherlands — www.streetlife.nl
Design Werksdesign/Inhouse — Peter Krouwel, Arjan van der Vegte

BIG APPLE, der Maßholztisch aus Apfelbaum, ist im Grundmaß 280 cm, auf 420 cm Gesamtlänge ausziehbar und dabei freitragend. Die beiden zusätzlichen Einlegeplatten werden, wenn sie nicht benötigt werden, einfach im Tisch untergebracht. Die Zargenhöhe ist so niedrig, dass selbst Armlehnstühle unter den Tisch passen. Durch seine reduzierte Formgebung und seine klare, schlichte Gestaltung lässt sich der Tisch im Privatbereich ebenso wie im Office-Bereich einsetzten. Die Holzrichtung und die Bauweise der Füße passen sich dem Apfelholz an. BIG APPLE, a customized wooden table made of apple wood has a basic length of 280 cm and can be extended to 420 cm, unsupported. The two additional feed boards can simply be stored in the table if they are not needed. The height of the frame is so low that even chairs with armrests fit under the table. With its minimalist shape and its clear, simple design the table is equally comfortable in the home and in the office. Both the grain of the wood and the leg structure have been shaped as befits the apple wood.

BIG APPLE *Esstisch*

Dining table

Unternehmen — Company Schneider „Der Ideenschreiner" GmbH — Porschestraße 6 — 70736 Fellbach — Germany — www.der-ideenschreiner.de

Design Werksdesign / Inhouse

Die Stühle der Modellreihe *Orplid* unterstützen mit ihrer flexiblen Holzlamellenlehne ein dynamisches und ermüdungsfreies Sitzen. Entsprechend dem Aufbau der menschlichen Wirbelsäule, sind Holzlamellen mit Wollfilzstücken kombiniert. So kann die Lehne den Bewegungen des Rückens folgen und stützt diesen gleichzeitig optimal ab. Zusätzlicher Komfort ergibt sich aus den nur kleinen Kontaktflächen zum Körper. So ist eine freie Luftzirkulation gewährleistet und Stauwärme wird wirksam verhindert. Chairs of the *Orplid* series have a flexible wood-rib back that supports a dynamic and relaxing seating posture. Modelled on the human spine, the back of the chair interleafs wooden ribs with pieces of dense wool felt. This allows the back of the chair to follow the movements of the human body while simultaneously offering optimal support. The *Orplid* chair further enhances comfortable sitting through its relatively small areas of contact with the human body, which allows air to circulate freely.

Orplid *Stuhl* *Chair*

Unternehmen — Company Schreinerei Schleiter/Holzgestaltung — Hopfenmühle 1 — 95233 Helmbrechts — Germany — www.andreas-schleiter.com
Design Werksdesign/Inhouse — Andreas Schleiter

Die Stuhlserie ist ein Konzentrat aus 30 Jahren Stuhlentwicklung. Edle Massivhölzer in innovativer Längsholzverarbeitung sorgen durch ein geringes Gewicht (ab 3,5 kg), weiche Kantenausformung, Einhandbedienbarkeit sowie Stapelbarkeit für eine hohe Gebrauchstauglichkeit. Dank genauestens dem menschlichen Körper folgenden Formen überzeugen die Stühle durch ihren hohen Komfort. The range of chairs is a conclusion of 30 years of chair development. Best quality solid wood, worked innovatively with the grain with greater utility through lightweight and softly formed edges. The chairs can easily be lifted with one hand, are stackable, convincingly comfortable and formed exactly to suit to the human body.

Stuhl „Caresse der Rückenschmeichler"
Chair "Caress your back"

Stuhlserie

Range of chairs

Unternehmen — Company Klaus Bartels — Echtholzdesign — Jägerstraße 14 — 95349 Thurnau — Germany — www.echtholzdesign.de

Design Werksdesign/Inhouse — Klaus Bartels

This chair ist aus Hightech-3D-Fasern gefertigt. Die Form des Stuhls erfordert lediglich eine 5 mm starke Sitzschale. Daher ist der Stuhl relativ leicht. Und er verbraucht nur einen geringen Anteil natürlicher Ressourcen. *This chair* ist sehr hoch stapelbar. Er ist gemäß der strengsten Brandschutzbestimmungen verknüpfbar. *This chair* kostet lediglich 140 Euro. Die zurückhaltende Formgebung erlaubt seine Nutzung in unterschiedlichsten Umgebungen. *This chair* is made with high tech 3D finer. The shape is made in a way the shell only needs to be 5 mm thick. Therefore the chair is relatively light. And it uses relatively a little amount of natural resources. *This chair* is very high stackable. *This chair* is connectable according the heaviest regulations of the fire department. *This chair* only costs 140 Euro. The modest shape allows the chair to fit in different surroundings.

This chair *Konferenz-Stuhl, stapelbar*

Contract chair, stackable and connectable

Unternehmen — Company Lensvelt office furniture — Minervum 7003 — 4817 ZL Breda — Netherlands — www.lensvelt.nl

Design Richard Hutten Studio bv — Richard Hutten — www.richardhutten.com

Die Stuhlreihe GLOOH umfasst Freischwinger, Stuhl, Thekenstuhl und Barhocker; alle in Holzausführung oder gepolstert oder mit Polstersitz und Holzrücken. Die ergonomisch geformten Sitzschalen sind mit einer für diese Anwendung weltweit einmaligen, innovativen Verbindungstechnik gefertigt, die eine perfekte Passung zum Gestell ermöglicht. Zum optionalen Zubehör gehören unter anderem Reihenverbinder und Griffloch.

The range of GLOOH consists a cantilever chair, side chair, counter stool and bar stool; all with wooden seat moulding or upholstered or with upholstered seat and wooden back. The ergonomically formed seat mouldings are produced with an innovative world wide unique new technique which offers a perfect fit between seat moulding and steel frame. Optional available: e.g. row connections and grip hole.

GLOOH *Stuhlserie*
Chair series

Unternehmen — Company KFF Design — Bahnhofstraße 27 — 32657 Lemgo — Germany — www.kff.de
Design Werksdesign / Inhouse — Hansjörg Helweg, Rolf Helweg

Das neu entwickelte Prinzip des Einklappens der Tischbeine ist von einer bisher nicht gekannten Perfektion: Ein Druck auf das breite Tastenelement genügt und die Tischbeine klappen seitlich versetzt auf die Tischplatte. Weitere Funktionen unterstreichen die Attraktivität und Vielseitigkeit der Tische: Verkettbarkeit, Tischgestelle einsparende Einhängeplatten sowie Winkelverkettungen. The new designed folding mechanism is from up until today an unknown perfection. To unlock the legs simply push on the wide locking profile and the legs slide with a lateral movement when turned down. Additional functions such as table-to-table connectivity and linking tops emphasise the versatility of the table.

system 24 *Klapptisch*
Folding table

Unternehmen — Company Hiller Objektmöbel GmbH & Co. KG — Kippenheimer Straße 6 — 77971 Kippenheim — Germany — www.hiller-moebel.de
Design Korb+Korb Architektur-Innenarchitektur-Design — Daniel Korb — www.korb-korb.ch

Der Stuhl ST. PETER wurde für die „Jugend-Kultur-Kirche St.Peter“ in Frankfurt am Main entwickelt. Die formale Qualität des Stuhls wird durch Zurückhaltung und Leichtigkeit bestimmt. Funktionale Qualitäten bieten die angenehm federnde Sitzschale für hohen Sitzkomfort und das speziell geformte Kufengestell, das radiale und lineare Reihungen ermöglicht. Das Kufengestell aus Rundrohr sichert Stabilität bei geringem Gewicht und ermöglicht eine Stapelung in hoher Anzahl. The chair ST. PETER is the result of a competition by the German Designer Club and was developed for the “St.Peter Jugend-Kultur” church in Frankfurt. The chair's superior quality comes from its discrete look and light weight. Functional qualities include the springy shell that provides surprising comfort and the frame, which makes it possible to position chairs in radial or linear rows. The runner frame is made of circular shaped tubes, which ensure stability. This chair is light and can be stacked high.

ST. PETER Modell 1080 *Stapelstuhl*

Stacking chair

Unternehmen — Company Brunner GmbH — Stühle | Tische | Objektmöbel — Im Salmenkopf 10 — 77866 Rheinau — Germany — www.brunner-group.com

Design f/p design gmbh — www.fp-design-gmbh.com

Minimierter Materialeinsatz, maximale Transparenz, niedriges Gewicht und einfachste Handhabung sind hier in einer schlüssigen, genial einfachen Gestaltung integriert. Der besondere Clou ist die elegante Ausbildung des Sitzrahmens nach dem „Schienen-Kufen-Prinzip". Das ermöglicht ein äußerst kompaktes, sicheres und formschlüssiges Stapeln und schont durch den definierten Abstand die Sitzbespannung. Minimum material input, maximum transparency, low weight and easy handling have been integrated into a coherent and ingeniously simple design. The ingenious feature is the elegant design of the seat frame that incorporates interlocking rails: self-engaging, stackable seat frame allows neat, extremely compact storage, while protecting the membrane seat and backrest.

Programm 230 Aline *Mehrzweck Stuhl*

Multi-purpose chair

Unternehmen — Company Wilkhahn — Wilkening + Hahne GmbH+Co. KG — Fritz-Hahne-Straße 8 — 31848 Bad Münder — Germany — www.wilkhahn.com
Design Störiko product design GmbH — Andreas Störiko

Form follows science … der neue Bürostuhl, der Design und Wissenschaft kombiniert und optimiert. Durch neue, wissenschaftlich untersuchte ergonomische Vorteile wie Rücken- und Lumbalunterstützung sorgt *Zody* für eine optimale Sitzhaltung mit 3-Punkt-Synchronmechanik, Gewichts- und Höheneinstellung, asymmetrischer PAL Back Lumbalunterstützung, Sitztiefeneinstellung, Rückenlehnen-Torsion, Back Stop und Sitzneigeverstellung sowie 4D-Armlehnen. Shape follows science … this new office chair combines and optimises design and science. *Zody* incorporates new scientific ergonomic findings, such as the back and lumbar support, three-point synchronous mechanism, weight and seat height adjustment, asymmetrical patented PAL back system, seat length adjustment, backrest tension control and back stop adjustment, forward tilt as well as 4D armrests for an ideal sitting position.

Zody_System 89 *Bürodrehstuhl-Serie*

Office chair series

Unternehmen — Company Haworth GmbH — Am Deisterbahnhof 6 — 31848 Bad Münder — Germany — www.haworth-europe.com

Design ITO Design Team — www.ito-design.com

Vergessen Sie den Arbeitsplatz, wie Sie ihn bisher kannten. Stellen Sie sich eine Arbeitsumgebung vor, die auf Zusammenarbeit basiert. Eine motivierende, inspirierende Arbeitsumgebung, in der nicht der individuelle, sondern der Gruppenarbeitsplatz im Fokus steht. Eine klare Struktur von Arbeitsplatten und Gestellen, die den Charakter des Raumes so belässt, wie er ist. Großzügige Dimensionen, keine überflüssigen Details. Mit zweckmäßigem Zubehör, Kabelrinnen und maximaler Bewegungsfreiheit. Forget the work place as you currently know it. Picture the working environment, based on team spirit. Here, the team's workplace is the focal point instead of the individual workplace. It motivates and inspires. A clear structure of tabletops and supports, which accepts space as it is. Substantial dimensions without unnecessary details. By adding functional accessories and storage space for cables, MI offers maximum freedom of movement.

MI *Büromöbelsystem*
Office furniture system

Unternehmen — Company Kembo — Koningsschot 55 — 3905 PR Veenendaal — Netherlands — www.kembo.com
Design Werksdesign / Inhouse — Erik Munnikhof

AL *Conferencing* ist ein Gesamtpaket, das Möbel, Raum, Technologie und Ästhetik vereint. Die klare Gestaltung des Mobiliars unterstützt die Übersichtlichkeit und einfache Bedienbarkeit der Technik. Der Konferenztisch steht für Klarheit in der Kommunikation, für die Reduktion auf das Wesentliche, AL steht für den sichtbaren Werkstoff Aluminium – ein Symbol für Leichtigkeit, Hightech und Mobilität. Die gesamte Medien-Infrastruktur ist optisch zurückhaltend in das Konferenzmobiliar integriert. Über einen Medienbus kommunizieren alle Geräte – vom Laptop bis zum Flatscreen, vom Projektor bis zur ausfahrbaren Leinwand. Gesteuert wird zentral über eine eigens konzipierte Infrarot-Fernbedienung. Das System ist für alle Gerätetypen kompatibel. Eine intelligente Lösung für Präsentation, Besprechung und Konferenz. AL *Conferencing* is a total package, which unites furniture, area, technology and aesthetics. The AL *Conference* Table stands for clarity in communication, for reduction to the essentials. AL stands for aluminium as a visible material – a symbol for lightness, hi-tech, and mobility. The entire media infrastructure is integrated into the conference furniture. All devices communicate via a media bus – from notebook to flatscreen, from projector to extendable screen. All this is controlled via a specially designed infra red remote control. The system is compatible for all types of devices. An intelligent solution for presentations, meetings and conferences.

AL Conferencing *Konferenz-Programm*

Conference programme

Unternehmen — Company Bene AG — Schwarzwiesenstraße 3 — 3340 Waidhofen/Ybbs — Austria — http://bene.com

Design Werksdesign/Inhouse — Kai Stania

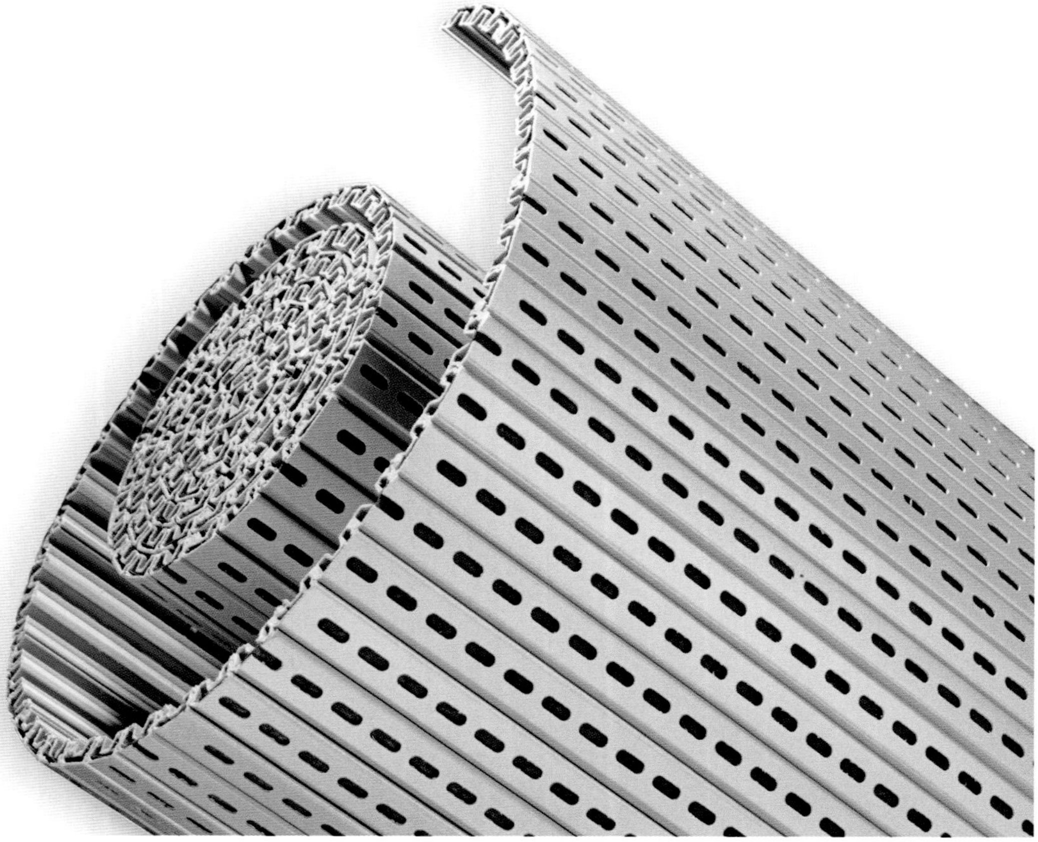

RAUVOLET *acoustic-line* ist der erste Rollladenschrank zur Schallabsorption und damit Verbesserung der Raumakustik, der mit einer individuell gestaltbaren Optik ausgestattet ist. Dies ermöglichen eine Polypropylenoberfläche mit Lochstanzung und ein spezielles, eingezogenes Vlies. So wird der Nachhall in Großraumbüros deutlich reduziert, was die Zertifizierung nach DIN ISO 354 bestätigt. Der Werkstoff PP wird ökologischen Ansprüchen gerecht, denn er ist recycelbar, wärme- und lösungsmittelbeständig. RAUVOLET *acoustic-line* is the first roller shutter cupboard for sound absorption and therefore for improvement in room acoustics whose appearance can be uniquely tailored. This is made possible by a polypropylene surface with punched holes and a special concealed sound absorbing non-woven fabric. In this way, echo is considerably reduced in open plan offices and this is confirmed by the certification to DIN ISO 354. The material PP meets ecological requirements as it can be recycled and is heat and solvent resistant.

RAUVOLET acoustic-line *Schrankrollladensystem*

Roller shutter system

Unternehmen — Company REHAU AG + Co. — Otto-Hahn-Straße 2 — 95111 Rehau — Germany — www.REHAU.com

Design Werksdesign/Inhouse — Stefan Bräuer, Klaus Grießhammer, Klaus Schoberth, Carsten Neufert

Die Einsatzmöglichkeiten der Möbelrolle SMILES reichen von Möbeln im Kinderzimmer, auf Krankenstationen bis hin zu Displays und Verkaufshilfen. Mit verschiedenen Gesichtern und unterschiedlichen Farben bieten die Rollen variantenreiche Gestaltungsmöglichkeiten. Der Raddurchmesser der Rollen beträgt 50 mm. Sie sind mit und ohne Feststeller verfügbar und können mit allen am Markt bekannten Befestigungen ausgerüstet werden. The possible uses of the furniture castor SMILES range from furniture in children's rooms, to hospital wards to displays and sales aids. With a variety of faces and an assortment of colours, the castors open up completely new design possibilities. The castors have a wheel diameter of 50 mm. They are available with or without a brake, and can be fitted with all the well-known attachments available on the market.

SMILES Serie 5520 *Möbelrolle*

Castor for furniture

Unternehmen — Company Bruandet S.A. — 13, rue de Chamois — 39700 La Barre — France — www.bruandet.com // TENTE-ROLLEN GmbH — Herrlinghausen 75 — 42929 Wermelskirchen — Germany — www.tente.com

Design Emmanuel Dietrich — www.emmanueldietrich.com // Werksdesign / Inhouse — Christian Marchand

Eine neuartige Sanitärlösung, die Stabilität, Funktionalität und Design bis ins Detail bietet. Türen und Wände bestehen aus 10 mm starkem ESG-Glas (11 Farben). Die innovative Tragkonstruktion ist von außen nicht sichtbar. Fußelemente sowie stabilisierender Fries sind durch das hierfür entwickelte T-Verbindungssystem zurückgelegt und verschwinden somit aus dem Blickfeld. Die glattflächige Front, das reduzierte Fugenbild, die verdeckten Befestigungselemente in Kombination mit den Materialien Glas und Aluminium unterstreichen den optischen Eindruck. An innovative solution offering stability, functionality and design in every detail. Doors and walls are made of safety glass (11 colours). Supporting legs as well as the stabilising frieze are recessed due to a special T-connection system, therefore disappearing out of sight. This feature accounts for the suspended character of NOXX. The combination of the materials glass and aluminium and the smooth surface with the specially developed joints underline the extraordinary impression of this system.

NOXX *Trennwandsystem*

Cubicle system

Unternehmen — Company KEMMLIT-Bauelemente GmbH — Sanitäreinrichtungen für höchste Ansprüche — Maltschachstraße 37 — 72144 Dusslingen — Germany — www.kemmlit.de

Design wöhr mieslinger architekten — Wolfram Wöhr, Jörg Mieslinger // Werksdesign/Inhouse — Dr. Marc Kemmler, Kurt Binder

Das Theken- und Präsentationssystem *Delta* umfasst eine Produktfamilie, die auf einem intelligenten Strangprofil basiert. Das Strangprofil ermöglicht die Aufnahme diverser Präsentationselemente. *Delta* kann individuell und schnell in Farbe, Material sowie Funktion an unterschiedlichste Anforderungen angepasst werden. Die Produktfamilie besteht aus den Varianten Infotheke, Bartheke, Präsentationscounter, Sideboard und Sitzbank. The *Delta* counter and presentation system represents a family of products, which can be quickly adapted in colour, material and function to the individual requirements of our clients by means of an intelligent extrusion profile and interior construction. This allows the incorporation of a wide variety of presentation elements. The *Delta* product family comprises of an info counter, bar counter, presentation counter, sideboard and bench.

Delta *Theken- und Präsentationssystem*

Counter and presentation system

Unternehmen — Company Expotechnik — Heinz Soschinski GmbH — Aarstraße 176 — 65232 Taunusstein — Germany — www.expotechnik.com

Design NOSE AG Design Intelligence — Ruedi August Müller, Christian Harbeke — www.nose.ch // Werksdesign/Inhouse — Heinz H. Soschinski

Die *tesa*-Roller verfügen über ein innovatives, ergonomisches Design, durch das schnelles und präzises Kleben und Korrigieren kinderleicht wird. Für jeden Einsatzzweck gibt es den richtigen Roller: rot steht für dauerhaftes Kleben, gelb bedeutet wieder ablösbares Kleben und weiß schließlich heißt Korrigieren. Durch ihre unterschiedlichen Farben sind die Roller klar voneinander zu unterscheiden. Alle Roller sind in zwei Varianten erhältlich: als handliche Einwegroller oder als Nachfüllroller mit praktischem 2-Klick-Nachfüllsystem. The *tesa* rollers have been ergonomically designed so that precise and fast gluing and correcting becomes child's play. tesa offers the right roller for every kind of use: red means permanent adhesion, yellow stands for non-permanent, re-positionable adhesion and white at last for correction. The different rollers and their applications can be told apart by their contrasting colours. All rollers are available in two versions: in handy disposable form or refillable with practical 2-click system.

tesa Korrektur- und Kleberoller „Roll & Relax"
tesa Correction and Adhesive Rollers "Roll & Relax"

Korrektur- und Kleberoller
Correction and adhesive rollers

Unternehmen — Company tesa AG — Quickbornstraße 24 — 20253 Hamburg — Germany — www.tesa.de
Design Koop Industrial Design — Norbert Koop — www.koopdesign.de

Der Schreiblernfüller, der mitwächst. Für Schreibanfänger mit kleiner Hand ist ein kurzer und leichter Schaft beigefügt. Wenn die Kinderhand weiter gewachsen ist, kann der kurze Schaft gegen den langen ausgetauscht werden. Der Füller wurde zusammen mit Pädagogen entwickelt und in ausgiebigen Schultests geprüft. Er besitzt ein ergonomisch geformtes, elastisches Griffstück. Es wird automatisch richtig angefasst. So wird das Schreibenlernen sinnvoll unterstützt. Die Kappe regt zum Spielen an und fördert zusätzlich unverkrampftes Lernen. The pen which is growing up with the user. For beginners with small hands, there is a short and lightweight barrel. When the child's hand is growing the short barrel could be replaced by a longer one. The Fountain pen has been created and developed together with teachers and has been extremely tested in schools. The grip is ergonomically shaped and made of elastic material. So the pen is automatically held in a right way. To learn writing is useful supported. The cap initialises to play with it and all together helps to learn easier and faster.

Base Kid *Schreiblernfüller*

Fountain pen for beginners

Unternehmen — Company Schneider GmbH — Martin-Heinrich-Klaproth-Straße 28 — 38855 Wernigerode — Germany — www.dachslschneider.de

Design Atelier für Gestaltung — Andreas Papenfuss, Christina Albert, Josef Landmann, John Mitsching — www.atelierpapenfuss.de

Beim Textmarker *Aqua-Lighter®* kann durch eine weltweit einzigartige Technik das Farbdepot mehrfach mit Wasser aufgefüllt werden. Das Nachfüllen ist bequem und das Handling umweltfreundlich und kinderleicht. Durch sein kompaktes Format passt der Textmarker in nahezu jedes Federmäppchen. Der *Aqua-Lighter®* ist in den vier Leuchtfarben Gelb, Pink, Grün und Orange erhältlich. The *Aqua-Lighter®* text marker has an ink tank that can be refilled several times with water thanks to technology that is unique world-wide. Refilling is very easy as is handling, which is also ecofriendly. With its compact size, the text market fits in almost any pen box. The *Aqua-Lighter®* is available in luminous yellow, pink, green and orange.

Aqua-Lighter® *Textmarker*

Highlighter

Unternehmen — Company SENATOR Schreibgeräte GmbH & Co. KGaA — Bahnhofstraße 57 — 64401 Gross-Bieberau — Germany — www.senator-pen.com
Design Werksdesign/Inhouse — Georg Keil

Die Malstifte des patentgeschützten Systems sind weltweit die einzigen Fasermaler, die mit Leitungswasser nachfüllbar sind. Es stehen drei unterschiedliche Malspitzen zur Verfügung. Mit dem AQUA-WONDER® *Breit* mit dicker Malspitze lassen sich große Flächen schnell und kinderleicht ausmalen. Konturen, Linienmuster und Details sind mühelos mit dem AQUA-WONDER® *Fein* zu realisieren. Der AQUA-WONDER® *Brush* mit Pinselspitze schließlich eignet sich bestens zum Malen im Kunstunterricht. Dank ihrer Wasserlöslichkeit lassen sich die Farben miteinander vermalen, und in Kombination mit einem wassergetränkten Haarpinsel sind selbst Aquarelleffekte realisierbar. The colour pens in the patent-protected system are the world's only fiber pens that can be refilled from a tap. There are three different tips available. The AQUA-WONDER® *broad tip* enables large surfaces to be painted quickly and easily. Contours, line patterns and details can then be realised using the AQUA-WONDER® *fine tip*. And the AQUA-WONDER® *brush tip is* ideal for painting in art class. Thanks to their water soluble ink, the colours can be mixed when painting, and in combination with a water-drenched brush the effect is similar to that of a water colour painting.

AQUA-WONDER® *Fasermaler*

Fibre pen

Unternehmen — Company SENATOR Schreibgeräte GmbH & Co. KGaA — Bahnhofstraße 57 — 64401 Gross-Bieberau — Germany — www.senator-pen.com

Design Werksdesign / Inhouse — Georg Keil

Beim Druckkugelschreiber *New Spring Chrome* – „Made in Germany", setzt eine im Gehäuse integrierte Spirale neue Akzente in dieser Produktgattung. Die Drückerfunktion wird durch dieses Gestaltungsprinzip zusätzlich deutlich visualisiert und verleiht dem Kugelschreiber, auch durch seine glänzenden Metallteile, eine hochwertige Anmutung. Der *New Spring Chrome* ist mit einer blauen Kunststoff-Super-Großraummine G2 ausgestattet. Erhältlich ist er in den Farben Schwarz, Anthrazit, Dunkelblau und in Dunkelrot-metallic. The spring ballpoint *New Spring Chrome* – "Made in Germany" creates a new look in this product genre – with the spiral integrated into the barrel. The spring function is additionally visualized strongly by this design, lending the ballpoint a high-grade feel, something emphasised by its gleaming metal parts. *New Spring Chrome* comes with a blue plastic super-large capacity G2 refill. It is available in black, anthracite, dark blue and metallic dark red.

New Spring Chrome *Kugelschreiber*

Ballpoint pen

Unternehmen — Company SENATOR Schreibgeräte GmbH & Co. KGaA — Bahnhofstraße 57 — 64401 Groß-Bieberau — Germany — www.senator-pen.com
Design Rommel und Schoen Design — Volkmar Rommel — www.rommelundschoendesign.de

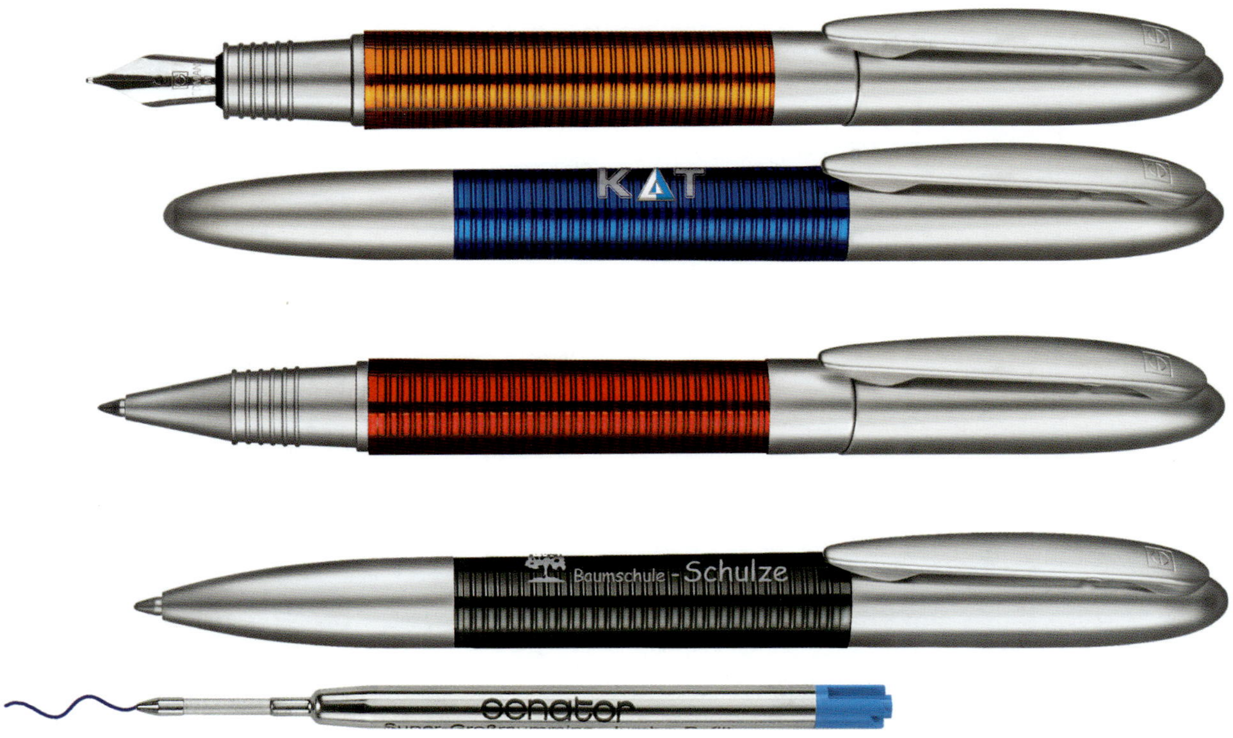

Der gekonnte Mix aus unterschiedlichen Materialien, leuchtenden Farben und die aufwändige Verarbeitung machen die neue Schreibgeräte-Serie SOLARIS zum Blickfang. Das Design besticht durch hochwertig veredeltes mattes Metall, kombiniert mit einer innovativen Struktur unter schimmerndem, transparentem Kunststoff in Blau, Bernstein, Rot und Rauchgrau. Drei Schreibsysteme stehen zur Auswahl. Ein Patronenfüllhalter, ein Rollerball und ein Kugelschreiber mit hochwertiger Drehmechanik. Alle Schreibgeräte haben stabile, matt verchromte Metallclips, die die eleganten Konturen betonen. The successful mix of different materials, plus luminescent colours that capture the light and high-quality workmanship make the new SOLARIS series of writing utensils genuine eye catchers. The impressive design is based on high-grade, refined, matte metal combined with an innovative form under shimmering, transparent plastic in shades of blue, amber, red and anthracite. Three writing systems are available: a cartridge pen, a roller pen and a ballpoint pen with quality twist mechanism. All pens come with matte, chromed metal clips emphasising the utensil's elegant contours.

SOLARIS *Metall-Schreibgeräte*

Metall writing instruments

Unternehmen — Company SENATOR Schreibgeräte GmbH & Co. KGaA — Bahnhofstraße 57 — 64401 Gross-Bieberau — Germany — www.senator-pen.com

Design Hariolf Briggl

Upside Down ist durch Karabinerhaken und Umhängeband mit Sicherheitsclip jederzeit griff und schreibbereit. Er ist in den acht matt transparenten Farben Blau, Rot, Grün, Pink, Gelb, Schwarz, Weiß und Orange mit matt weißem Innenteil sowie farblich passendem Umhängeband erhältlich. Alle Varianten gibt es auch ohne Umhängeband. Der *Upside Down* ist mit einer handelsüblichen Kunststoff Super-Großraummine ausgestattet. *Upside Down* is at hand ready for use at any time – thanks to the snap hook and lanyard. It comes in eight transparent satin colours: blue, red, green, pink, yellow, black, white, and orange, with a satin white inside section and a matching-tone lanyard. All versions are also available without the lanyard. *Upside Down* comes with the customary plastic super-thick lead.

Upside Down *Drehkugelschreiber*

Ballpoint pen

Unternehmen — Company SENATOR Schreibgeräte GmbH & Co. KGaA — Bahnhofstraße 57 — 64401 Gross-Bieberau — Germany — www.senator-pen.com
Design Werksdesign/Inhouse — Hans-Jürgen Beyer

Beim Drehkugelschreiber SENATOR*@tract* „Made in Germany" sorgt ein Dreh für das Ausfahren der Mine, ein weiterer Dreh versetzt die Mine wieder in Ruhestellung und fährt gleichzeitig einen Halte-Clip aus. Dieser „Safetyclip" liegt während der Schreibstellung flach am Schaft an und stört nicht beim Schreiben. Spitze, Clip und Ring bestehen aus Metall, das Gehäuse aus hochwertigem ABS. Der Schreiber ist in verschiedenen Farben erhältlich und kann sowohl mit einer Großraummine als auch mit einer nachfüllbaren Gelmine bestückt werden. The special mechanism SENATOR*@tract* "Made in Germany" enables the pen to extract or retract by a simple twist and has a retractable security clip. This "safety clip" lies flat along the shaft and does not disturb the writing process. The tip, clip and ring are made of metal, the body of high-grade ABS. The pen is available in various colours and can be fitted with either a large ink cartridge or a refillable gel cartridge.

@tract *Kugelschreiber*
Ballpoint pen

Unternehmen — Company SENATOR Schreibgeräte GmbH & Co. KGaA — Bahnhofstraße 57 — 64401 Groß-Bieberau — Germany — www.senator-pen

Design Rommel und Schoen Design — Volkmar Rommel — www.rommelundschoendesign.de

Die gelaserte, schwarzbraune Krokostruktur aus Edelharz ist kombiniert mit glänzendem Champartien und einem gefederten Clip. Den Bleistift zeichnet eine stufenlose Drehmechanik, die robuste, 1,4 mm starke Polymermine, die Härte B für konstanten, weichen Schreibfluss sowie der extra große, austauschbare Radiergummi unter der Kappe aus. Der Kugelschreiber ist mit einer austauschbaren, dokumentenechten Großraummine ausgestattet. The laser-engraved crocodile-like surface made of brownish-black precious resin is combined with chrome-plated and polished metal parts and a spring-loaded clip. Pencil: smooth twist mechanism, tough 1.4 mm polymer lead in line, width B for a constant flow of writing and extra-large replaceable eraser under the cap. Ballpoint pen: large-capacity refill, suitable for documents.

e-motion Drehkugelschreiber und Drehbleistift aus Edelharz e-motion Twist ballpoint pen and twist mechanical pencil made in precious resin

Schreibgerät

Writing instrument

Unternehmen — Company A. W. Faber-Castell Vertrieb GmbH — Nürnberger Straße 2 — 90546 Stein — Germany — www.faber-castell.de

Design Industrial Design Team — Heinrich Stukenkemper

Ein elf Zentimeter langer Taschenkugelschreiber mit Wegrollsicherung am Stiftende. Die Mine (Kugelschreibermine D1) fährt durch eine kurze und schnelle Bewegung des Handgelenks aus, durch die gleiche Bewegung wird sie wieder versenkt. Front- und Endstück sind matt verchromt, der Schaft besteht aus Naturkautschuk mit Längsrillen. An 11cm-long pocket ballpoint pen with a non-slip end. The tip (ballpoint D1 tip) is extended by a short swift movement of the wrist, and returns to its place with the same action. The front and end sections are stain chromed, and the barrel is made of natural rubber with length-wise ribbing.

P'3140 *Kugelschreiber*

Ballpoint pen

Unternehmen — Company A. W. Faber-Castell Vertrieb GmbH — Nürnberger Straße 2 — 90546 Stein — Germany — www.faber-castell.de

Design Porsche Design GmbH — Roland Heiler — www.porsche-design.com

Durch Verpressen eines Goldringes, in den zuvor eine tiefe Linie gefräst wurde, bildet sich eine gleichmäßig verlaufende Wellenlinie im Ring. Außen mit einer feinen Bewegung, innen deutlich fühlbar als plastische Erhebung. Die Gestaltung von *Sinus* entsteht durch die wohldosierte Materialverformung in Verbindung mit hochlegiertem Gold. Das Ergebnis ist die klare Form mit einer Zeichnung die nur durch den Einsatz von Technik so präzise entstehen kann. A deep line cut into a golden ring changes by compressing to a continuously running wave. Outside presenting a fine movement, inside clearly felt as a soft rise. The design of the *Sinus* ring is based on the careful forming of the material in combination with the precious metal. The result is the clear design with a pattern, which only can be that precise by the technique well-used.

Sinus *Schmuckringe*
Rings

Unternehmen — Company Gebr. Niessing GmbH & Co. — Schmuckmanufaktur — Butenwall 117 — 48691 Vreden — Germany — www.niessing.com
Design Werksdesign / Inhouse — Susanne Winckler, Timo Küchler

Dacapo ist ein neu entwickeltes Schmucksystem, bei dem Schmuckelemente mit einer Schnur aus Edelmetall verbunden werden. Die Innovation liegt in der Verwendung von Klettmaterial im Innenleben einzelner Farbscheiben. Die Scheiben bestehen aus hochwertigem Gold, Platin und Perlmutt. Das Schmuckelement fungiert gleichzeitig als Verschluss. Alles ist jederzeit veränderbar und lädt zum kreativen Spiel mit Material und Form ein.

Dacapo is a jewellery system combining jewellery elements with a coil made of fine metal. The innovation about this system is the use of a Velcro fastener on the inside of individual colour discs. The discs consist of high-quality gold, platinum and mother pearl. The jewellery element is ornament and clasp at the same time. At any time changes are possible and induce to creative games with material and form.

Dacapo *Schmucksystem*
Jewelry system

Unternehmen — Company Gebr. Niessing GmbH & Co. — Schmuckmanufaktur — Butenwall 117 — 48691 Vreden — Germany — www.niessing.com

Design Werksdesign / Inhouse — Susanne Winckler

NEOLOG richtet sich an anders Denkende und soll bestehende Klischees in Frage stellen. Sie befreit sich von der kulturellen und geschichtlichen Verankerung der Ziffern- und Zeigermodelle. Die Zeit wird in Mengen dargestellt, ohne abstrakt zu werden. Die Uhrzeit ist in Stunden, Zehner- und Einer-Minuten unterteilt. Für schnelles Ablesen sind diese in Dreiergruppen untergliedert. Keine Zeiger, keine Ziffern lediglich die Zeit. Gehäuse und Schnalle bestehen aus Edelstahl, das Armband aus Silikon. NEOLOG appeals to free thinkers and calls existing stereotypes into question. It gets rid of deepseated cultural and historical hand-based or cipher-based models by illustrating time as a quantity but without being abstract. The time is divided into hours, tens of minutes and single. The bars are organised into groups of three for faster recognition. No hands, no numbers, simply time. Case and buckle are made of stainless steel; the wristband is out of silicone.

NEOLOG A-24 II *Armbanduhr* *Wristwatch*

Unternehmen — Company SCOPETIME GmbH — Krausestraße 102 — 22049 Hamburg — Germany — www.neolog.cc
Design Werksdesign/Inhouse — Arman Emami

Ein Chronograph in puristischer Optik, reduziert auf das Wesentliche: Klare Linien, Material und Funktion. Die Zeiger sind in den Goldlegierungsfarben Gelbgold, Rotgold und Weißgold, das Zifferblatt in Silber und Weiß erhältlich. Stoppfunktionen sind von einer Sekunde bis zwölf Stunden mit Zwischenzeiten möglich. Der Chronograph ist mit einem Ronda Startech-Werk ausgestattet. Das Gehäuse ist aus antiallergischem Titan, das Armband mit Titanschließe besteht aus Kautschuk. Der Gehäusedurchmesser beträgt 37 mm, die Gehäusestärke 9 mm. Titanium Chronograph in a minimalist design, reduced to the pure lines, the material and functions. The straight metallic hands symbolise the different gold colours as yellow, red and white gold tone. Further dials available in silver and white. Stopfunctions from one second to twelve hours with lap time possible. Movement: Ronda startech. Black Rubber band with titanium buckle. Case diameter 37 mm. Case height 9 mm.

Jacob Jensen Chronograph 600 *Titan Armbanduhr*

Titanium watch

Unternehmen — Company S. Weisz Uurwerken BV — Schubertstraat 21 — 1077 GP Amsterdam — Netherlands — www.jacob-jensen.com // Filius & Haake GmbH & Co. KG — Kuckucksweg 55 — 33607 Bielefeld — Germany — www.filius-haake.de

Design Jacob Jensen Design — Timothy Jacob Jensen — www.jacob-jensen.com

Die Ferngläser *Victory 7x, 8x* und *10x42 T** FL wurden für höchste Ansprüche entwickelt. Innovation bei Optik (unter anderem wurden erstmalig fluoridhaltige Sondergläser sowie eine LotuTec Schutzschicht eingesetzt) und Mechanik ermöglichen eine äußerst kompakte, handliche und leichte Bauweise, die den ergonomischen Bedürfnissen der Nutzer besonders Rechnung trägt. Die Geräte zeichnen sich weiter durch eine bisher unerreicht hohe Auflösung und Detailerkennbarkeit, ein großes Sehfeld sowie hohe Lichtreserven für die Dämmerung aus. The binocular models *Victory 7x, 8x* and *10x42 T** FL have been designed for the highest professional requirements. Innovations in optics (e.g. for the first time fluoride containing special glasses and LotuTec protective coating) and mechanics made possible an extremely compact, handy and light design, which takes into account the specific ergonomic requirements of the users. The binoculars offer unprecedented high resolution and detail, large fields of view and substantial light reserve for dawn.

Victory 42 FL *Fernglas*
Binocular

Unternehmen — Company Carl Zeiss Sports Optics GmbH — Gloelstraße 3-5 — 35576 Wetzlar — Germany — www.zeiss.de/sportsoptics
Design zed AG — Thomas Steuri — www.zednetwork.com

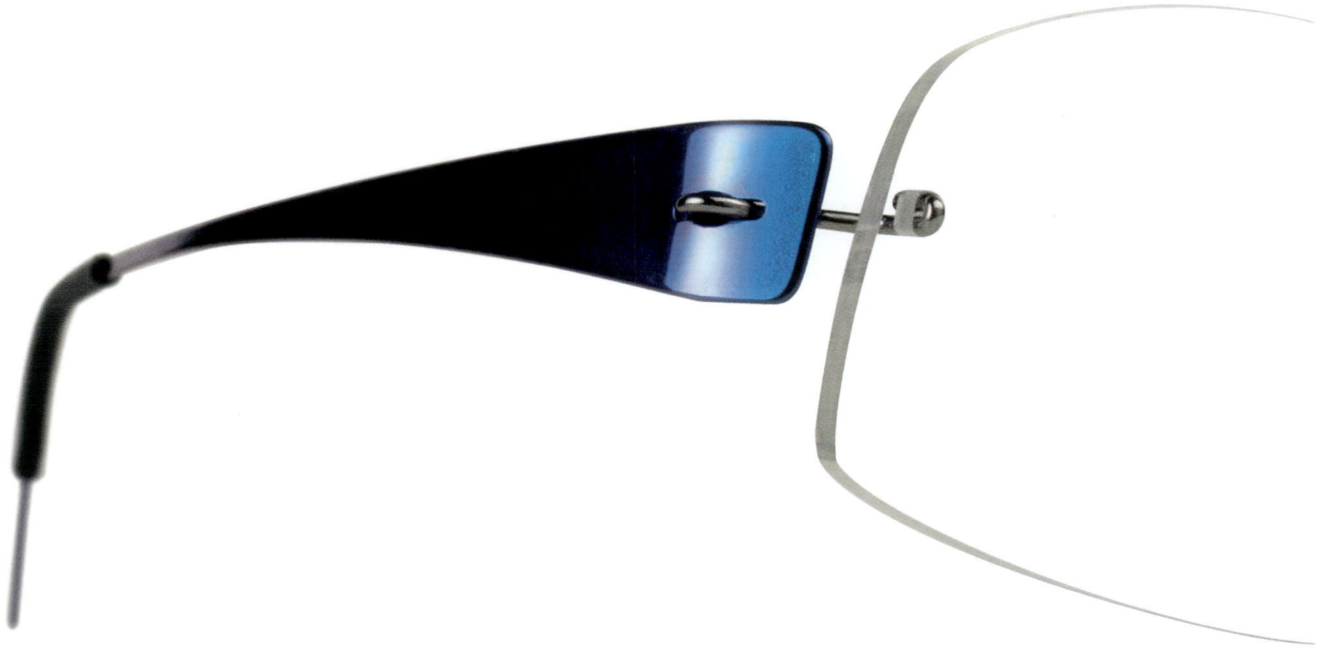

LINDBERG lanciert eine Reihe von spannenden und gewagten Bügel-Modellen für das schon erfolgreiche *Spirit*-Konzept. Die sehr minimalistische *Spirit*-Front kombiniert mit den Platten-Bügeln ermöglichen neue, fantastische Ausdrücke und erweitern die Möglichkeiten der Individualisierung der Brille. LINDBERG hat ein neues Scharnier entwickelt, das dieselbe zylindrische Form anderer LINDBERG-Scharniere hat. LINDBERG introduces a number of exciting and challenging temple models for the company's already successful *Spirit* concept. The ultra-minimalistic *Spirit* front and the various distinctive temples combine to create a fascinating style, allowing modern and discerning wearers to help create their own unique eye wear. LINDBERG has developed a new hinge that repeats the cylindrical shape which has become part of LINDBERG's visual identity.

LINDBERG 2000 Spirit Plate Temples *Brillen* *Spectacles*

Unternehmen — Company LINDBERG — Bjarkesvej 30 — 8230 Aabyhoej — Denmark — www.lindberg.com

Design Werksdesign / Inhouse — LINDBERG Designgruppe

Die Marke *logstoff.com* wurde im Jahr 2002 gegründet. Hersteller ist die Firma *Sternjakob*, bekannt durch die Marken 4YOU und SCOUT Schulranzen. Für die Betreuung der Produkte und der Marke ist die Agentur ID4 Industriedesign zuständig. *logstoff*-Taschen folgen keinen Trends, sondern sind auf Langlebigkeit ausgelegt und bieten einfache, gute Funktionen. Das Grundthema von *logstoff.com* ist Kommunikation. Jede Tasche hat ihre eigene E-Mail Adresse. Im Internet gibt es eine Kommunikationsplattform mit mittlerweile über 10.000 festen Mitgliedern. The brand *logstoff.com* was created in 2002. It is manufactured by *Sternjakob*, the company known for its popular brands 4YOU and SCOUT school bags. ID4 industriedesign is the agency that handles both the product and the brand. *logstoff* bags simply function well and are at the customers' service for a long time. *logstoff.com* is about communication. Each bag has its own e-mail address. The Internet forum has more than 10,000 steady members by now.

logstoff.com *Taschenkollektion* *Bag collection*

Unternehmen — Company Alfred Sternjakob GmbH & Co. KG — Frankenstraße 47–55 — 67227 Frankenthal — Germany — www.logstoff.com
Design ID4 Industriedesign — Axel Ricker — www.id4.de

Die Taschen-Serie ist modular aufgebaut, so dass die kleinen Modelle außen an die großen Taschen angedockt werden können, um den Stauraum zu erweitern. Darüber hinaus bieten die Taschen ein großes gepolstertes Hauptfach mit Netztasche und innen liegendem Einteiler, eine Reißverschluss-Vordertasche, gepolsterte Tragegriffe, abnehmbare, verstellbare Schultergurte mit Schulterpolstern sowie Ösen zum Andocken der kleinen Modelle. The group is composed modularly; the small models can be attached to the outside of the bigger models to enlarge packing space. Furthermore the models feature a large padded compartment with mesh pocket and divider inside, a zippered front pocket, padded carrying handles, a detachable, adjustable shoulder strap with shoulder pad and rings to attach the smaller sized models.

Spencer *Reisetaschenset*

Duffle bag with side pocket

Unternehmen — Company lemonfish GmbH — Im Rank 10 — 73655 Plüderhausen — Germany — www.lemonfish.de
Design Werksdesign/Inhouse — Bettina Burchard, Alexandra Dittrich

Die Gestaltung des *Scout 1975* lehnt sich an die ersten *Scout*-Schulranzen aus den 70er Jahren an, die mit einer neuen Leichtigkeit und Farbigkeit die damals üblichen schweren Lederschulranzen ablösten. Diese neue Tasche ist damit der Ausdruck des Lebensgefühls einer ganzen Generation und die Marke *Scout* ein Symbol für die Schulzeit der Zielgruppe der heute 30-jährigen. In Erinnerung an den „Ur-Scout" ist der *Scout 1975* mit einer eigenen Formensprache aber auch eine eigenständige Interpretation des Themas. Der *Scout 1975* wird in limitierter Stückzahl produziert.

The design of the *Scout 1975* follows the first *Scout* school satchel from the 1970s, which replaced the then conventional heavy, black leather satchels. Thus, this new bag expresses the feeling of a whole generation and the *Scout* brand is a symbol of the school days of the target group of today's thirty-year-olds. Reminiscent of the original scout, the *Scout 1975* is an independent interpretation with its own language of forms. The *Scout 1975* is produced in a limited edition.

Scout 1975

Freizeit-/Business-Taschen-Serie

Bag series

Unternehmen — Company Alfred Sternjakob GmbH & Co. KG — Frankenstraße 47–55 — 67227 Frankenthal — Germany — www.scout1975.de

Design Estragon GmbH — www.estragon.ch

Ein Ball an Stelle von Rädern: der völlig neue Bürstensauger DC15 umkurvt leicht Möbel und macht das Staubsaugen scheinbar mühelos. Herkömmliche Bürstensauger mit fest stehenden Achsen sind schwer zu steuern, besonders an beengten Stellen. Der Motor des Saugers befindet sich im Ball. Ein Teleskoprohr lässt sich einfach herausziehen, um auch in entfernte Ecken zu gelangen. Wie bei allen *Dyson* Staubsaugern leistet die *Dyson* Cyclone Technologie konstante Saugkraft. A new type of vacuum cleaner with a ball instead of wheels: so it freely zigzags around furniture. DC15 makes cleaning a lot more efficient. Traditional upright vacuums can be hard going – they don't steer. The motor is located inside the ball, lowering the centre of gravity, increasing manoeuvrability. A telescopic wand allows reaching tight corners simply by popping up the wand from the handle, extending it. Like all *Dyson* vacuum cleaners it doesn't lose suction.

DC15 The Ball *Bürstensauger*

Vacuum cleaner upright

Unternehmen — Company Dyson GmbH — Lichtstraße 43b — 50825 Köln — Germany — www.dyson.de

Design Dyson Team Forschung/Design/Entwicklung

Die *Neff* Wandesse DKG 59 mit Glasschirm beeindruckt durch ihre Leichtigkeit. Das Edelstahl-Frontprofil mit feinen Bedientasten betont die Horizontale und greift mit einem sanften Schwung das *Neff*-typische Gestaltungselement der Ellipse auf. Im Zentrum: die markante Absaugeinheit aus Edelstahl. Ein einzigartig attraktiver Look: Designqualität „Made in Germany“. Die Leistung lässt sich durch Elektroniksteuerung fein dosieren. Halogenlampen setzen die Kochkünste ins beste Licht. The *Neff* wall-mounted extractor hood DKG 59 is impressive because of its lightness. A stainless steel front profile with light touch buttons emphasises the horizontal and gently sweeps up to the typical *Neff* shape of the ellipse. The distinctive, stainless steel siphon unit lies at the centre, with its singularly attractive look: quality design „Made in Germany“. The performance can be fine-tuned through electronic control. Halogen lamps show off the cuisine in its best light.

DKG 59 *Wandesse*
Wall mounted hood

Unternehmen — Company Bosch und Siemens Hausgeräte GmbH — Markendesign Neff/Constructa — Carl-Wery-Straße 34 — 81739 München — Germany — www.neff.de
Design Werksdesign/Inhouse — Ralf Grobleben, Gerhard Nüssler

Sein markantes Design macht den *Neff* Backofen *Mega* SH 4664 A zum Blickfang in der Küche – die Formensprache von Ellipse und Rechteck als Kontrast zur Geradlinigkeit der Küchenarchitektur. Für den einzigartigen Look dieses Einbaubackofens mit 14 Betriebsarten sorgen hochwertige Details wie der Einleger des Anzeigenfeldes aus diamantgeschliffenem Facettenglas, die außergewöhnliche *NeffContRoll®* Einknopf-Bedienung, die „samtige" Oberfläche Trend Aluminium und der mitdrehende Bügelgriff SLIDE®. Distinctive design makes the *Neff* Oven *Mega* SH 4664 A catch your eye in the kitchen. The elliptical and rectangular forms contrast the straight lines of the kitchen design. High quality details like the depositor of the indicator field, made from diamond cut facet glass, ensure the distinctive look of this built-in oven with its 14 functions, along with the extraordinary *NeffControll®* single button function, the "velvety" Trend Aluminium surface and revolving handle SLIDE®.

SH 4664 A *Elektro-Einbaubackofen*

Built in electric oven

Unternehmen — Company Bosch und Siemens Hausgeräte GmbH — Markendesign Neff/Constructa — Carl-Wery-Straße 34 — 81739 München — Germany — www.neff.de

Design Werksdesign/Inhouse — Ralf Grobleben, Gerhard Nüssler

Durch die innovative *Krups* Motor-Technik läuft der leistungsstarke 1100-Watt-Motor nicht nur mit konstanter Drehzahl, sondern auch so leise, wie kaum ein anderer – egal welche Zutaten verarbeitet werden. Drei elektronisch geregelte Geschwindigkeitsstufen und drei Knetprogramme sorgen für perfekte Ergebnisse. Die integrierte Waage sorgt mittels Sensoren für präzise Wiegeergebnisse. Außerdem ist ein programmierbarer Timer enthalten, der wahlweise vorwärts oder rückwärts zählt. Owing to innovative *Krups* motor technology the powerful, 1100 Watt motor not only runs at a constant rpm but also more quietly than practically any other – regardless of the ingredients being prepared. Three electronically controlled speed levels and three kneading programmes produce perfect results. The integrated scale provides precise weight measurements by means of sensors. It also includes a programmable timer which can be set to run either forwards or backwards.

KRUPS Prep Expert / KA 8027 *Kompakt-Küchenmaschine* *Food processor*

Unternehmen — Company Groupe SEB Deutschland GmbH — Geschäftsbereich KRUPS — Herrnrainweg 5 — 63067 Offenbach — Germany — www.krups.de
Design Konstantin Grcic Industrial Design — Konstantin Grcic — www.konstantin-grcic.com

Dank der *Krups* Motor-Technik ist er das leistungsstärkste Gerät auf dem Markt. Der Motor arbeitet stets konstant und erfährt selbst beim Dauereinsatz oder beim Zufüllen der Zutaten keinen Leistungsabfall. Das strapazierfähige Alu-Druckgussgehäuse steht stabil und sicher auf der Arbeitsfläche und ist absolut unempfindlich gegen Kratzer und Fingerabdrücke. Fünf Geschwindigkeitsstufen, Ice-Crush- und Pulse-Funktion garantieren perfekte Mixergebnisse. This powerful device is on the market thanks to *Krups* drive technology. The motor always runs constantly, and even when used for long periods or filling with ingredients it experiences no loss of power. The sturdy aluminum die-cast housing stands firmly and securely on the work surface and is absolutely unsusceptible to scratches and fingerprints. Five speed settings, ice-crush and pulse function guarantee perfect mixing results.

KRUPS Prep Expert / KB 7207 *Standmixer*
Blender

Unternehmen — Company Groupe SEB Deutschland GmbH — Geschäftsbereich KRUPS — Herrnrainweg 5 — 63067 Offenbach — Germany — www.krups.de

Design Konstantin Grcic Industrial Design — Konstantin Grcic — www.konstantin-grcic.com

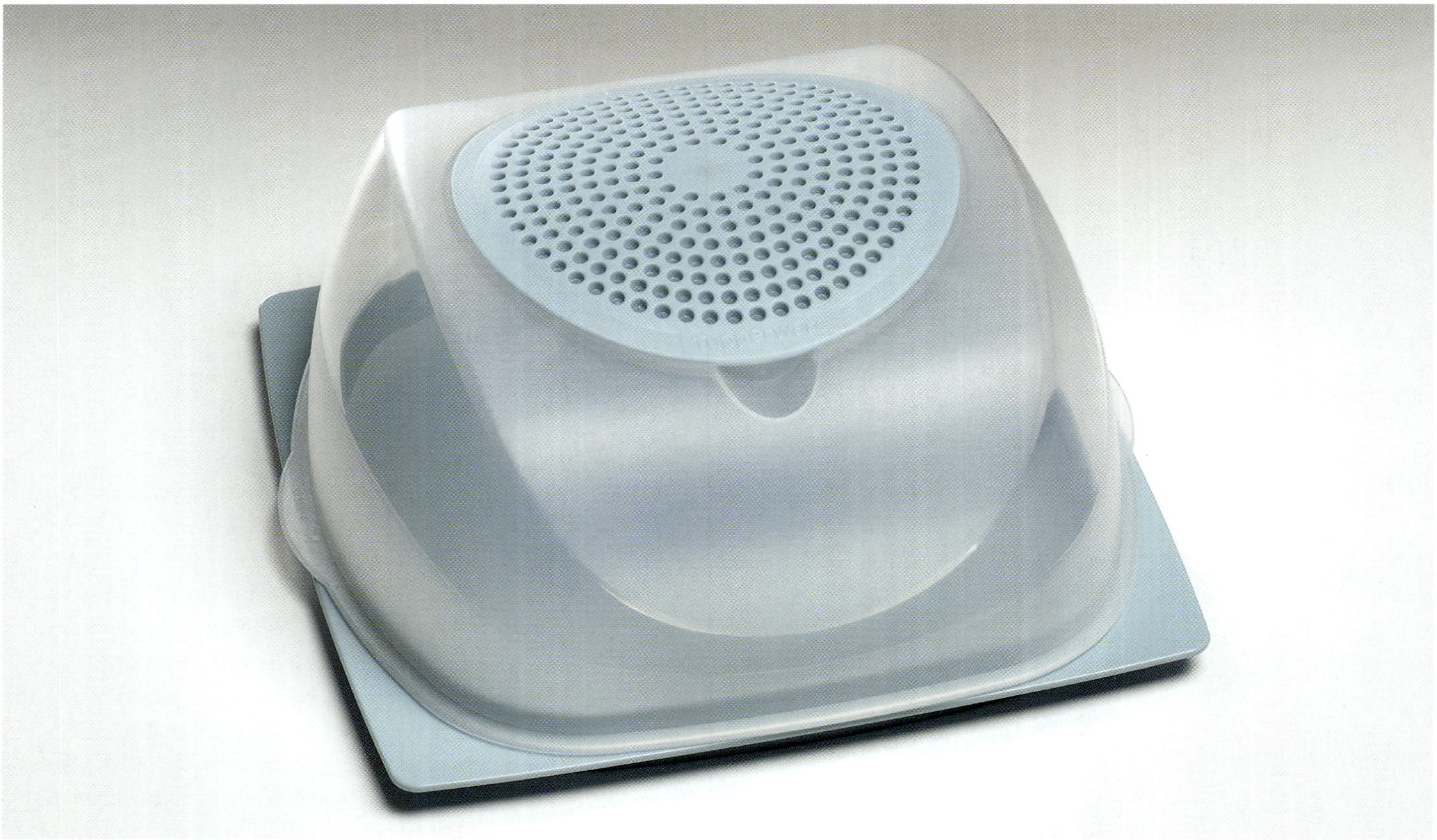

Die *CondensControl*-Behälter wurden zur optimalen Aufbewahrung von Käse im Kühlschrank entwickelt. Dank des patentierten *CondensControl*-Systems können verschiedene Käsesorten gemeinsam in einem Behälter aufbewahrt werden, ohne dass sie durch zu hohe Feuchtigkeit im Behälter vorzeitig verderben. Die Feuchtigkeit gelangt durch ein in der Haube befestigtes, atmungsaktives Foliengitter nach draußen, so dass nur eine geringe Menge Feuchtigkeit im Behälter zurückbleibt. Der Käse hat dadurch ideale Lagerbedingungen.
This *CheeSmart* server is specially designed for storing all types of cheese in the refrigerator. Due to the patent pending *CondensControl* system more than one type of cheese in the same container can be stored, without the risk of condensation. The *CondensControl* system consists of a grid with breathable surface, incorporated into a snug-fitting cover, which adds to the effectiveness of the *CondensControl* grid, by making sure that the transfer of moisture can only take place through the grid, and preventing any air from getting in or out through other openings.

Käse Max *Käsebehälter*
Cheese servers

Unternehmen — Company Tupperware Belgium N.V. — Pierre Corneliskaai 35 — 9300 Aalst — Belgium — www.tupperware.com
Design Tupperware General Services N.V. — Victor Cautereels — www.tupperware.com

Die stapelbaren Schüsseln der *Mini-Max*™-Serie lassen sich ganz einfach auf die benötigte Größe bringen, indem man die „Falten“ stufenweise aufdrückt. Ist die richtige Größe eingestellt, werden die Lebensmittel eingefüllt, mit dem Sicherheitsverschluss verschlossen und aufbewahrt. Die Behälter sind spülmaschinenbeständig und entweder mit einem Fassungsvermögen von 700 ml oder 950 ml erhältlich. Zusammengefaltet beanspruchen sie mit ihrer Höhe von nur 2 cm kaum Platz in der Geschirrspülmaschine oder im Küchenschrank. Expand the stackable bowl to the size you need, by popping out each layer one by one. Once it's set to the required size, just pour in the food, seal it, and store. *Mini-Max*™ containers are dishwasher safe and available in two handy sizes: 700 ml and 950 ml. Being only 2 cm high in the collapsed position, they take minimum space in the dishwasher or afterwards in the cupboard.

Mini-Max™ *Zusammenfaltbare Lebensmittel-Behälter*

Collapsible food containers

Unternehmen — Company Tupperware France S.A. — Route de Monts — 37300 Joué-Les-Tours — France — www.tupperware.com

Design Tupperware Corporation — Tupperware Global Design — www.tupperware.com

Bei der Entwicklung der Produkte der „Chef-Familie" wurde besonders viel Wert auf Ergonomie gelegt. Obwohl die beiden „Chefs" ähnlich aussehen, haben sie doch ihre jeweiligen Besonderheiten: Der *Happy Chef* eignet sich zum Zerkleinern von Zwiebeln, Kräutern, Knoblauch, kleineren Mengen Kochschinken und Nüssen. Der *Quick Chef 2* ist die Alternative zur herkömmlichen elektrischen Küchenmaschine und zerkleinert Obst und Gemüse in Sekundenschnelle. The "Happy Chef" family has been designed with ergonomic in mind: both of them are easy to use and comfortable in the hands. Although they look quite similar they both have their own peculiarity: The *Happy Chopper* is a fantastic new ergonomic solution for chopping onions, herbs, garlic, small quantities of ham, nuts. The *Quick Chef 2* is a fantastic alternative to electronic food processors. Versatile, quick and easy to use, it chops fruit or vegetables quickly.

Quick-Chef 2 & Happy Chef
Quick-Chef 2 & Happy Chopper

Küchenmaschine und Zerkleinerer

Ergonomic food processor and chopper

Unternehmen — Company Tupperware Belgium N.V. — Pierre Corneliskaai 35 — 9300 Aalst — Belgium — www.tupperware.com

Design Tupperware General Services N.V. — Jan-Hendrik de Groote, Vincent Jalet — www.tupperware.com

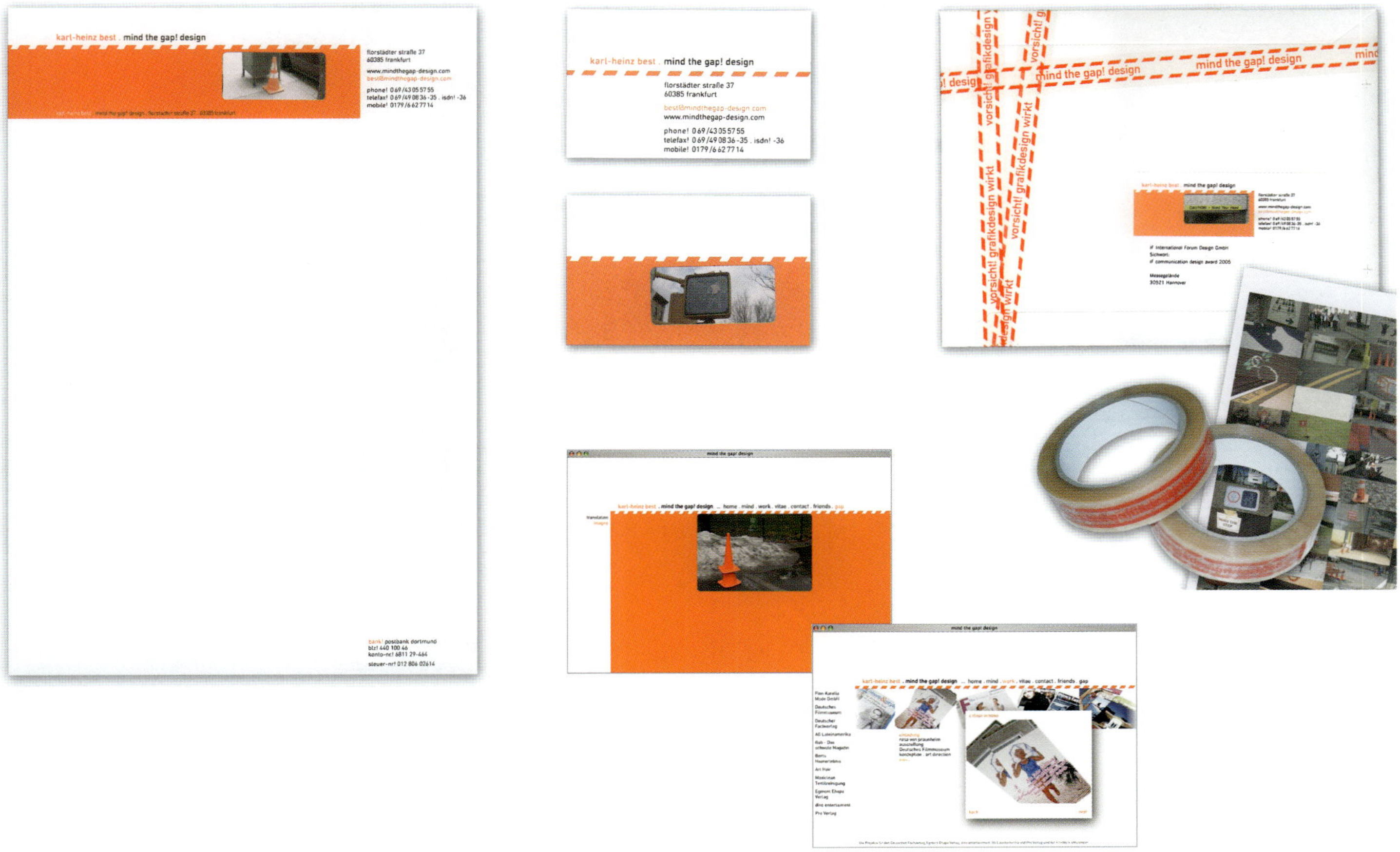

Grafikdesign sollte Aufmerksamkeit erregen, genau wie es der Claim bereits sagt: „mind the gap! design" („Pass auf!", „Achtung!" etc.). Daher erhält jeder seinen individuellen Briefbogen, Visitenkarte usw. durch Fotoaufkleber, die immer wieder mit neuen Motiven selbst bedruckt werden, welche Alltagssituationen abbilden, in denen Aufmerksamkeit erregt wird. Ebenso ist bedruckter Tesafilm, der wie ein Absperrband auf Briefumschläge, CD-Hüllen etc. geklebt wird, ein weiteres Mittel, um das Interesse des Betrachters zu wecken. Die dazu gehörende Internetseite www.mindthegap-design.com spielt mit den gleichen gestalterischen Mitteln, wie zum Beispiel dem Fotoaufkleber, der hier nach Zufallsprinzip gewechselt wird. Design has to raise attention as expressed in our claim "mind the gap! design" ("Attention!", "Stop!" etc.). Individual business cards, stationery and further items are created by using photo labels, which may be individually printed with motifs of everyday situations that raise attention. Another means to raise attention is to use printed Scotch tape, which can be adhered like barrier tape to envelops, CD covers etc. The Website www.mindthegap-design.com plays with the same graphic pattern, like the photo labels – here with random changes.

Corporate Design karl-heinz best . mind the gap! design

Corporate Design

Corporate Design

Unternehmen — Company karl-heinz best . mind the gap! design — Florstädter Straße 37 — 60385 Frankfurt — Germany — www.mindthegap-design.com

Design Werksdesign / Inhouse

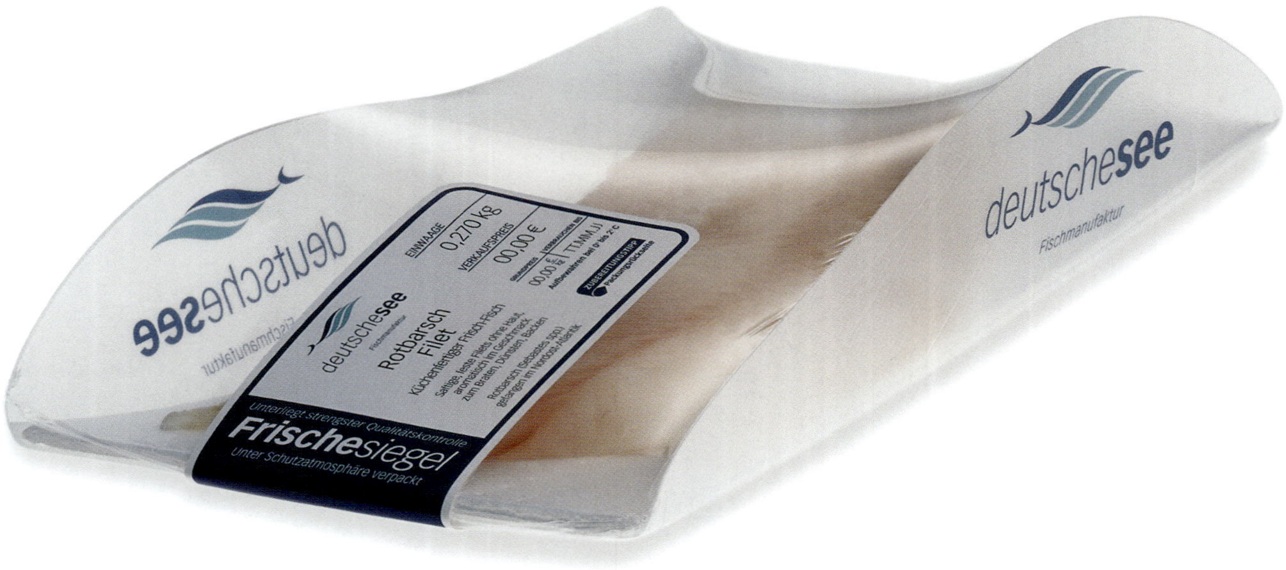

Die international patentierte Frischfischverpackung der "Deutschen See" Fischmanufaktur inszeniert die ultrafrischen Fischfilets als das Beste, was das Meer zu bieten hat. Die Transparenz der Materialien, die gegenläufige Wellenbewegung der Seitenflügel und die wie ein Wellental gespannte Folie erscheinen wie ein Stück bewegte See. Der Fisch liegt pur und klar wie im Medium Wasser. Die gesamte Qualität der Verarbeitung wird sichtbar. The internationally patented fresh fish packaging from fish processor "Deutsche See" Fischmanufaktur presents ultra-fresh fish fillets as the very best the oceans can offer. The transparency of the materials used, the opposing wave movements of the lateral wings and the foil stretched to form a trough in between the waves appear like a stretch of sea in motion. The fish lies in a pure, pristine state, as it does in its medium, water. The entire quality of processing becomes evident.

"Deutsche See" Wellenverpackung
"Deutsche See" wavepack

Frischfisch-Verpackung

Freshness packaging

Unternehmen — Company Deutsche See GmbH — Maifischstraße 3–9 — 27572 Bremerhaven — Germany — www.deutschesee.de

Design feldmann+schultchen design studios — www.fsdesign.de

Der Flaschenkühler ist mit einem speziellen Kühlmedium gefüllt. Vor Gebrauch wird der Kühler für drei Stunden ins Gefrierfach gelegt. Das Medium kristallisiert und speichert dadurch die Temperatur nachhaltig. Flaschen bleiben somit bis zu drei Stunden gekühlt und der Inhalt wird immer temperaturgerecht serviert. An der Außenseite entstehen während des Auftauens Eiskristalle. Die Bildung des sonst üblichen Kondenswassers wird so für ca. eineinhalb Stunden verhindert. This bottle cooler is filled with a special cooling medium. Prior to use, the cooler has to be placed for three hours in a freezer. The medium takes on crystal form and therefore stores the temperature durably. Bottles can thus be kept cool for up to three hours and the contents always served at the right temperature. Ice crystals form on the outside when the unit thaws, thus preventing the usual condensation from developing approximately for the first 1.5 hours.

KristallFrost*** *Flaschenkühler*

Bottle cooler

Unternehmen — Company ADOMA GmbH — Pettermandstraße 4 — 88239 Wangen / Allgäu — Germany — www.adoma.de

Design Werksdesign / Inhouse

Mit dem Kaffeebrüher *CafeSolo*™ wird Kaffee ohne Dampf, Druck und Zug sowie ohne Maschine und Filterpapier zubereitet, sondern einfach nur mit gemahlenem Kaffee und Wasser. Der Kaffee wird in der Glaskanne durch einen wärmeisolierenden Mantel warm gehalten. Der Filter verfügt über einen tropffreien Gießer und der im Lieferumfang enthaltene Kippdeckel öffnet sich beim Eingießen automatisch. The *CafeSolo*™ coffee-maker makes coffee without steam or pushing and pulling. Without machinery and filters. Simply coffee beans and water. The coffee in the glass jug is kept hot by the insulating cover. The filter has an integrated drip-free spout and the flip-top lid, which is included, opens automatically when you serve the coffee.

CafeSolo *Kaffeezubereiter*
Coffee maker

Unternehmen — Company Eva Denmark A/S — Maaloev Teknikerby 18–20 — 2760 Maaloev — Denmark — www.evadenmark.com
Design Tools Design — Henrik Holbaek — www.toolsdesign.com

Der Zuckerbehälter ist mit einem praktischen Einsatz ausgestattet, der bei jedem Umdrehen einen Teelöffel voll abmisst. Das Sahnekännchen mit tropffreiem Gießer wird nach Gebrauch mit einem dicht sitzenden Gummideckel geschlossen. Das Set besteht aus den Materialien Glas, Edelstahl, Silikongummi und Kunststoff. The sugar dispenser features an ingenious measurer, which dispenses one teaspoon of sugar every time the dispenser is inverted. The milk dispenser has a drip-free lip and can be closed after use with a snug-fitting rubber lid. The set is made of glass, stainless steel, silicone rubber and plastic.

Eva Solo *Zucker-Sahne-Set*

Milk and sugar set

Unternehmen — Company Eva Denmark A/S — Maaloev Teknikerby 18–20 — 2760 Maaloev — Denmark — www.evadenmark.com

Design Tools Design — Henrik Holbaek — www.tooldesign.com

Mit dem Teezubereiter lässt sich Tee auf zweierlei Weise zubereiten. Sie können die Teeblätter im Wasser (A) schwimmen lassen, wobei diese beim Eingießen vom Filter zurückgehalten werden, oder sie in den Filter (B) geben und – wenn der Tee lange genug gezogen hat – den Stempel ganz nach unten drücken. Darüber hinaus besteht die Möglichkeit zum Aufbrühen einer halben Kanne. Der Kippdeckel der Kanne öffnet sich beim Eingießen automatisch. Ein Neopren-Mantel hält den Tee warm, darüber hinaus besitzt die Kanne einen patentierten, tropffreien Gießer. The teamaker can brew tea in two different ways. You can allow the tealeaves to float freely in the water (A) and then the filter retains the leaves when you pour. Alternatively, you can put the tealeaves in the filter (B) and once the tea has brewed for long enough, fully depress the plunger. Besides that the tea maker includes the option of brewing half a jug. The flip-top of the jug lid opens automatically when you pour. A neoprene cover keeps the tea hot and beyond it the tea maker features a patented drip-free lip.

Eva Solo *Teezubereiter*
Tea maker

Unternehmen — Company Eva Denmark A/S — Maaloev Teknikerby 18–20 — 2760 Maaloev — Denmark — www.evadenmark.com
Design Tools Design — Henrik Holbaek — www.toolsdesign.com

Öl und Essig mit Kräutern eignen sich bestens zum Würzen. Zur Herstellung werden Kräuter und Gewürze in die Karaffe gegeben und mit Öl oder Essig aufgefüllt. Der tropffreie Gießer stellt eine präzise Dosierung sicher und der dichtschließende Stopfen bewahrt Aromen und Geschmack. Das Glas (0,5 / 0,25 l) ist hitzebeständig und fürs Aufbewahren, Zubereiten und Servieren geeignet. Aromatic oils and vinegars are the liquid spices of the kitchen. Place herbs and spices in the carafe and add either good oil or vinegar. The drip-free pouring spout pours well and precisely. The tightly fitting top preserves both aroma and flavour. The glass (0.5 / 0.25 litre) is heatproof and for storage, cooking and serving.

Eva Solo *Tropffreie Öl- und Essigkaraffen & Tropffreier Dressing-Shaker*
Drip-free oil and vinegar carafes & Drip-free salad dressing shaker

Unternehmen — Company Eva Denmark A/S — Maaloev Teknikerby 18–20 — 2760 Maaloev — Denmark — www.evadenmark.com
Design Tools Design — Henrik Holbaek — www.toolsdesign.com

Alles, was man von einem modernen Grill erwartet, und noch mehr. Das Äußere ist geprägt von sauberen Linien und glatten Oberflächen. Durch den flachen Deckel, der tagsüber den Grill schließt, wird er zu einem attraktiven Terrassentisch. Öffnet man aber die Haube, kommt ein robuster Rost aus Edelstahl – wie der ganze Grill – zum Vorschein. Der Holzkohlebehälter hängt unter dem Grill. Dazu ist ein Satz Tischbeine erhältlich, die in Kombination mit der flachen Haube den Grill in einen praktischen Serviertisch verwandeln. Everything you expect from a modern grill and then some. The outside reveals only clean lines and smooth surfaces. During the day, a flat lid covers the grill, converting it into an attractive patio table. But when the lid is removed, a heavy-duty grid made of stainless steel like the rest of the grill is revealed. The coal bucket hangs underneath the grid. A set of table legs which combined with the flat lid form a practical serving table is also available.

Eva Solo *Grillfass*
Barrel grill

Unternehmen — Company Eva Denmark A/S — Maaloev Teknikerby 18–20 — 2760 Maaloev — Denmark — www.evadenmark.com
Design Tools Design — Henrik Holbaek — www.toolsdesign.com

Der Blumentopf mit Wasserreservoir ist für „durstige“ Zimmerpflanzen und Gewürzkräuter geeignet. Durch die Löcher im Boden des Keramiktopfes wird ein Docht eingeführt. Über den mit Wasser gefüllten Glasbehälter nimmt die im Topf befindliche Pflanze die erforderliche Wassermenge auf. Der Blumentopf mit Wasserreservoir speichert genügend Wasser für etwa eine Woche – je nach Bedarf der Pflanze. Self-watering flowerpot is for “thirsty” indoor plants and herbs. Thread the wick through the holes in the base of the ceramic pot. Plant your indoor plant in the ceramic flowerpot and then place the pot on the glass container. Fill the glass container with water and the plant will now start drawing water up through the wick. The self-watering flowerpot holds enough water for about a week, depending on the requirements of the plant.

Eva Solo *Blumentopf mit Wasserreservoir*
Self-watering flowerpot

Unternehmen — Company Eva Denmark A/S — Maaloev Teknikerby 18-20 — 2760 Maaloev — Denmark — www.evadenmark.com
Design Tools Design — Henrik Holbaek — www.toolsdesign.com

Die *Okiu*-Schalen sind vor dem Hintergrund gestaltet, dass der ästhetische Anspruch nicht bei den Speisen Halt machen sollte. Sie sind ein Ausdruck der Vorstellung, dass das Tischambiente mit der Qualität des gereichten Glases wächst: „Die Glanzlichter brillanter Kanten, Gewicht und kühle Haptik von hochwertigem Manufakturglas bereichern das Empfinden noch über den Geschmack hinaus. Am Ende isst nicht nur das Auge mit, sondern auch der Kopf." The *Okiu* bowls have been designed with the idea in mind that sophistication should go beyond food. They represent the idea that table ambience increases with the quality of the glass presented: "The reflections from brilliant edges, the weight and the cool feeling of highquality hand-made glass enrich the experience beyond taste. In the end it is not only the eye that eats first, but also the mind."

Okiu *Schalen*
Bowls

Unternehmen — Company Zwiesel Kristallglas AG — Dr.-Schott-Straße 35 — 94227 Zwiesel — Germany — www.zwiesel-kristallglas.com
Design Prof. Michael Boehm

Um ästhetischen Inszenierungen von Speisen die geeignete Plattform zu geben, verabschieden sich bei der Kollektion *Signature* die Teller von der gewohnten runden Form der Außenkontur und setzen mit weichen, ovalen Linien neue Akzente – wobei sich die Essfläche weiterhin im bewährten Rund präsentiert. Die dazugehörigen Tassen überraschen durch ihren dynamisch geschwungenen Henkel, der der ovalen Tellerform nachempfunden ist.

In order to create a suitable stage for aesthetic food-presentations, the plates of the *Signature* collection no longer have the usual round shape, but set new trends with a soft, oval outline of the rim, whereas the well of the plate remains round. The corresponding tall cups surprise with dynamically curved handles that mimic the oval shape of the plates.

Signature Basic *Geschirrserie*

China series

Unternehmen — Company Porzellanfabrik Schönwald — Rehauer Straße 44-54 — 95173 Schönwald — Germany — www.schoenwald.com

Design Product Design - Interior Design — Carsten Gollnick — www.gollnick-design.de

Die Gestaltung von *The First* geschah vor dem Hintergrund, dass Wein Raum benötigt, um atmen zu können. Sie soll Kennern Interpretationsmöglichkeiten, die konservative Glaskollektionen oft vermissen lassen, vermitteln. Im Geiste einer „professionellen Könnerschaft, die sich Freiheiten leisten kann", erlauben die Gläser eine stärker assoziative Zuordnung zu den entsprechenden Aromen.
Die Formensprache von Glaskörper und Mundrand wurde in aufwändigen Manufakturverfahren zur Harmonie mit Spitzenweinen gebracht. The design of *The First* is based on the idea that wine needs space to breathe. It aims at offering connoisseurs the scope for interpretation often lacking in conservative glass collections. In the spirit of "professional connoisseurship that may allow itself certain liberties" the glasses permit a more pronounced associative attribution to a wine's particular flavours. The body and rim of the glasses have been designed and produced in an elaborate manufacturing process.

The First by Enrico Bernardo *Dekanter und Kelchgläser*

Decanters and stemware

Unternehmen — Company Zwiesel Kristallglas AG — Dr.-Schott-Straße 35 — 94227 Zwiesel — Germany — www.zwiesel-kristallglas.com
Design Werksdesign / Inhouse — Irmgard Braun-Ditzen

Die Serie aus hochwertigem Hartporzellan ist handgefertigt, handbemalt und gleichzeitig spülmaschinenfest. Die Dekore sind in Unterglasurmalerei in eigens entwickelter Maltechnik gefertigt und in Aufglasur mit hochprozentigen Poliergolden und Platin veredelt. Die echte Handarbeit zeigt sich in kleinen Variationen und Unterschieden, die jedes Einzelteil zum Unikat machen. The series manufactured out of high quality bone china is handmade, hand-painted and dishwasher safe. Decors applied under the glaze, using a painting technique developed by the artist herself, are refined over the glaze with high percentage polished gold and platinum. The genuine, handcrafted quality is proven by slight variations and distinctions, which make each piece unique.

Aladin und Drache
Aladin and Dragon

Porzellanserie
Porcelain series

Unternehmen — Company Barbara Flügel Porzellan — Unterweißenbacherweg 6 — 95100 Selb — Germany — www.fluegel-porzellan.de
Design Werksdesign/Inhouse — Barbara Flügel

Die Vasenserie *Blu Canela* von *Harry & Camila* erinnert an altbewährtes Spielzeug. Im Gussform-Verfahren werden diese Vasen hergestellt. Das Brennen der beiden kleinen Größen erfolgt durch das Aufstecken der Vasen auf Porzellanstäbe. Bei der großen Vase ist der Brand nur in einem Kammerofen möglich. Ein spezielles Plasma-Vakuum-Verfahren ermöglicht das dünne Beschichten mit einer speziellen Titanverbindung. Like a spinning top: The vase series *Blu Canela* by *Harry & Camila* is reminiscent of the traditional toy. These special vases are produced using a form casting process. Placing the vases on porcelain rods fires the two smaller sizes. The larger vase can only be fired in a kiln. The titanium coat incorporates glass, ceramic and stainless steel that has been metallized using a plasma vacuum process and thinly coated with a special titan compound.

Blu Canela *Vasenkollektion, Titanium bedampft*

Vase series with titanium décor

Unternehmen — Company Rosenthal AG — Philip-Rosenthal-Platz 1 — 95100 Selb — Germany — www.rosenthal.de
Design Harry & Camila // Werksdesign/Inhouse — Rosenthal AG — Rosenthal Creative Center

Die Schalen mit dem Titandekor auf weißem Porzellan wirken wie kleine weiche Wellen, die silbrig auf dem Meer glänzen. Durch einen langwierigen Prozess in einer mehrteiligen Gussform entsteht die besondere Form. Pro Tag kann jeweils nur eine Schale gefertigt werden. Die kleineren Größen erhalten ihre extravagante Form durch das Hohlgießverfahren. Ein Plasma-Vakuum-Verfahren ermöglicht das dünne Beschichten mit einer speziellen Titanverbindung. The dishes with the titanium décor on white porcelain seem like small gentle waves shining in silver on the sea. The distinctive shape of the dish is created by means of a lengthy casting process in a mould with several elements. Only one dish can be made per day. The smaller sizes have a hollow moulding procedure to thank for their extravagant shapes. The titanium coat incorporates glass, ceramic and stainless steel that has been metallized using a plasma vacuum process and thinly coated with a special titan compound.

Tectonics *Schalenserie aus Porzellan, Titanium bedampft*

Porcelain dishes with titanium décor

Unternehmen — Company Rosenthal AG — Philip-Rosenthal-Platz 1 — 95100 Selb — Germany — www.rosenthal.de

Design Werksdesign / Inhouse — Rosenthal Creative Center — Cairn Young

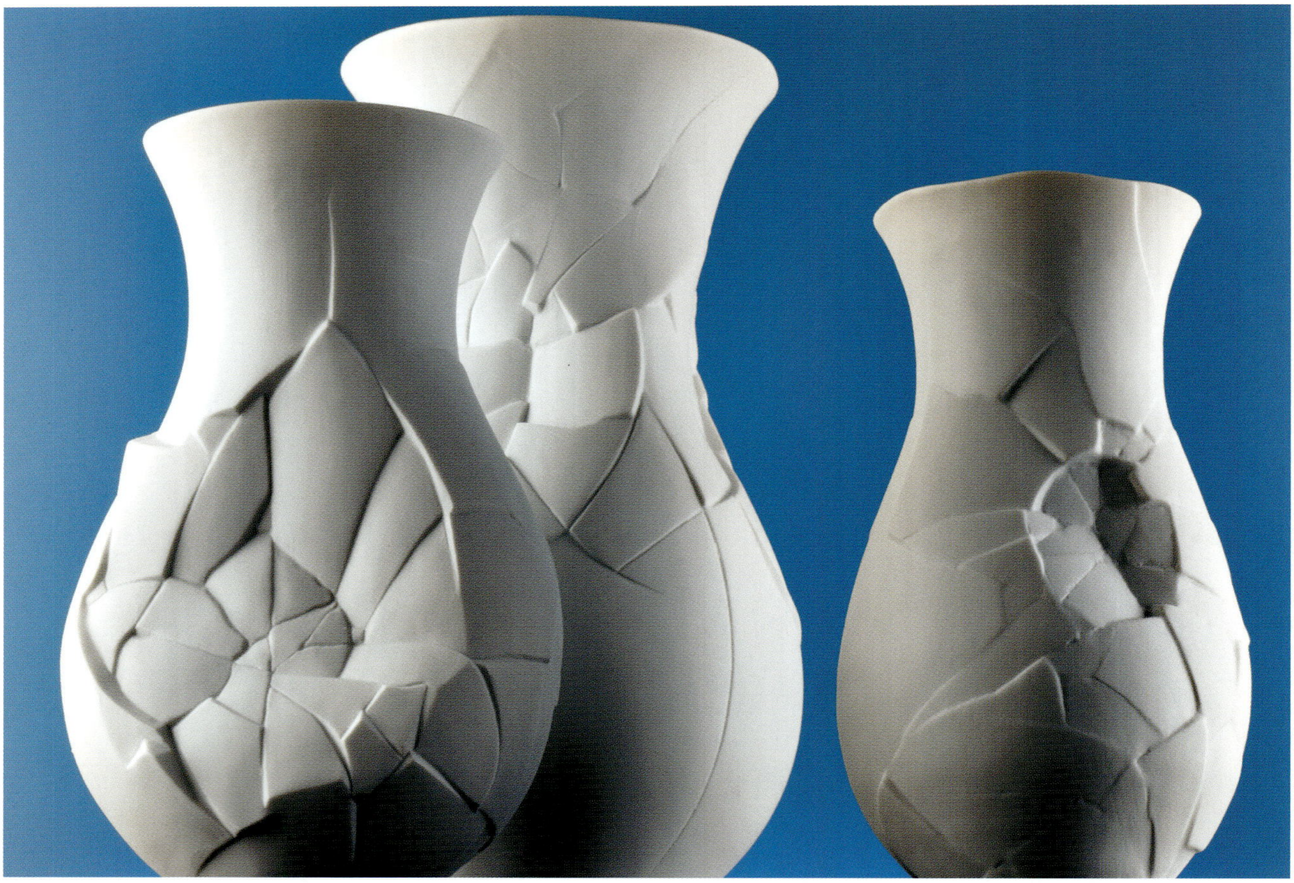

Für Dror Benshetrit symbolisiert die Vase die Erfahrung von Transformation im Leben. Die Basis bildet eine weiblich geformte Vase, die in eine experimentelle Phase tritt, wenn sie zerschmettert wird. Sie wird dadurch rauer, verführerisch, scharf und verlockend. Die Veränderungen werden durch verwischte Risse und geglättete Falten auf dem „zerbrochenen" Objekt sichtbar gemacht. Sie erhält dadurch eine neue Art von Schönheit, die Spuren der Phasen trägt, die sie durchlaufen hat. The vase symbolises the experience of changes in life for the artist. The basis is a feminine shaped vase, which enters into an experimental phase when it is shattered. It then becomes rougher, seductive, sharp and enticing. These changes are visualised on the "shattered" object by smudged cracks and smoothed out folds. The *Vase of Phases* is thus awarded a new type of beauty showing traces of the three phases it has passed through.

Vase of Phases *Vasenserie* *Vase series*

Unternehmen — Company Rosenthal AG — Philip-Rosenthal-Platz 1 — 95100 Selb — Germany — www.rosenthal.de
Design Studio Dror — Dror Benshetrit // Werksdesign / Inhouse — Rosenthal Creative Center

DR. FEEL GOOD könnte beinahe für einen kleinen Fassadenkletterer aus der Pop-Art-Kultur gehalten werden, würde nicht sein Name Medizinisches und Wohlbefinden assoziieren. In der Tat erfüllt die Figur eine äußerst praktische und wohltuende Aufgabe. Das formschöne Massagegerät eignet sich ideal für Menschen, deren Alltag sich mit Bewegungsmangel und in ungünstiger Körperhaltung am Schreibtisch abspielt. DR. FEEL GOOD could easily be mistaken for a small cat burglar from the Pop Art culture if its name wasn't associated with medicine and wellness. The nicely designed massaging device is great for people who don't get enough exercise and for those who sit at a desk all day. A true wellness experience for those who have let DR. FEEL GOOD go to work.

DR. FEEL GOOD *Massagegerät*

Massaging device

Unternehmen — Company Berendsohn AG — Alte Sülldorfer Landstraße 400 — 22559 Hamburg — Germany — www.berendsohn.com
Design Nicole Syben — www.artworklab.de

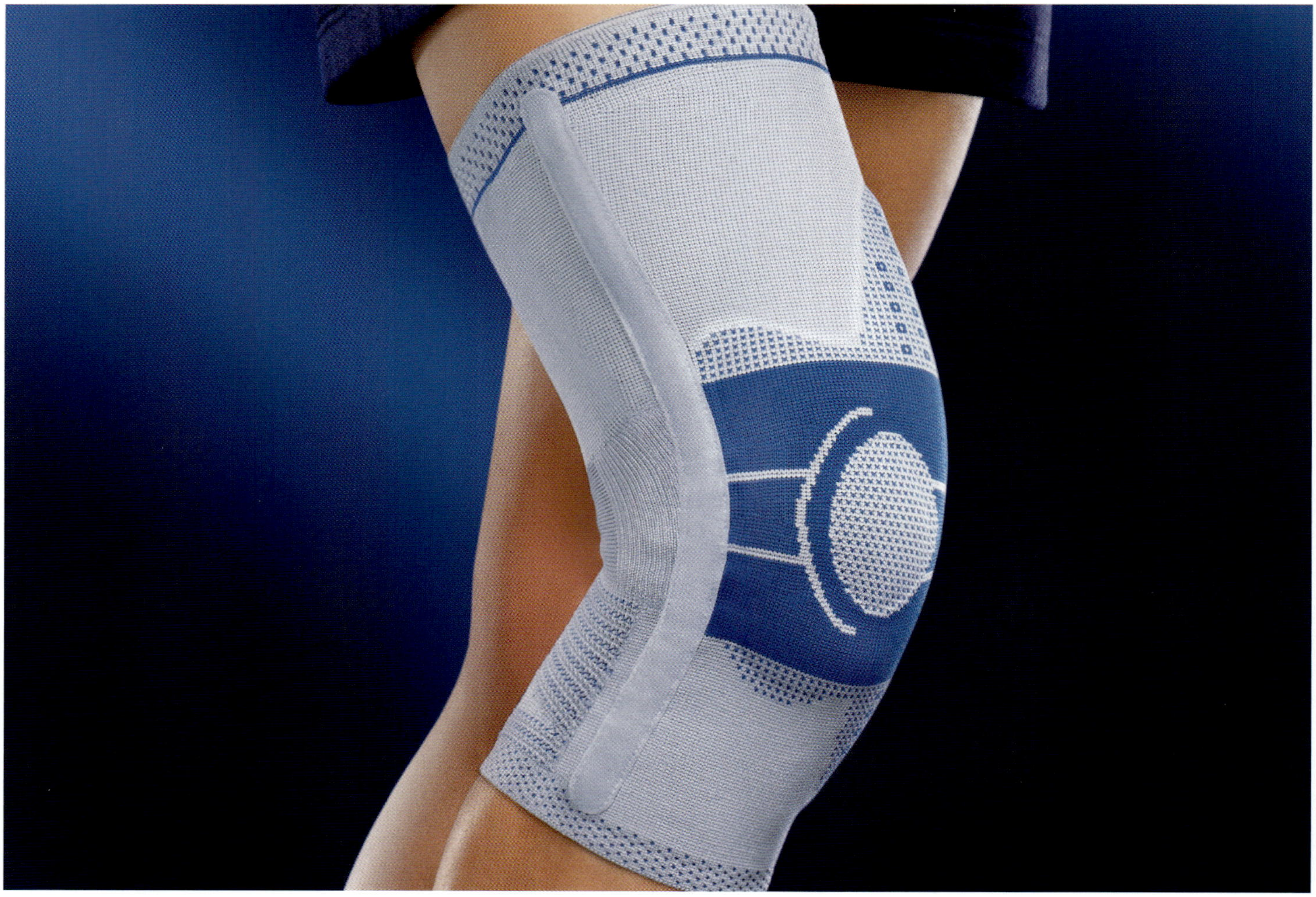

Bauerfeind Genu Train A3, die Aktivbandage zur komplexen Behandlung bei Kniebeschwerden, insbesondere bei Gonarthrose. Atmungsaktives Train-Gestrick, mit Komfort- und Dehnungszonen, einer Pelotte mit Friktionsnoppen und integriertem Korrekturzügel, die die medizinische Wirksamkeit, den Tragekomfort und die exakte Passform dieses Produktes prägen und die den Heilungsverlauf sowie die Schmerzlinderung beschleunigen. *Bauerfeind Genu Train A3*, the active support for complex treatment of knee pain, especially gonarthrosis. Breathable materials with comfort and stretch zones, a pad with friction dots and an integral corrective guide characterise the medical effectiveness, the wearer comfort and the correct fit of this product accelerate the recovery and alleviation of pain.

GenuTrain A3 *Train Aktivbandage*

Train active supports

Unternehmen — Company Bauerfeind AG — Triebeser Straße 16 — 07937 Zeulenroda-Triebes — Germany — www.bauerfeind.com

Design Rokitta Produkt & Markenästhetik — Hermann Rokitta — www.rokitta.de

Frieren und Schwitzen sind natürliche, aber Kräfte zehrende Methoden des Körpers, seine Temperatur konstant auf 37 °C zu halten. Der *Energy Accumulator* kühlt beim Schwitzen und wärmt bei Frieren über das patentierte *3D-BionicShpere*®-System. Es nutzt die Verdunstungskälte eines feinen Schweißfilms. So wird erst über das Ansteigen der Körperwärme und den Beginn des Schwitzens die Kühlung aktiviert. Bleibt der Schweiß aus, setzt die Isolierung ein. Zusätzlich wirken Kompressionsstufen, Polster, Wärmeabgabezonen und Frischluft-Belüftungskanäle. Freezing or sweating are natural methods the body uses to regulate its temperature at a constant 37 °C: but they cost energy. The *Energy Accumulator* cools you when you sweat and warms you up when you freeze – thanks to the patented *3D-BionicSphere*®-System. It uses the evaporation cool of a fine layer of sweat. Thus, the cooling is first activated when the body heat starts to rise and the body starts to sweat. If there is no sweat, then insulation commences. In addition, there are compression stages, padding, heat-release zones, and fresh-air ventilation channels.

X-Bionic®

Bekleidung / Unterwäsche
Clothing / Underwear

Unternehmen — Company Trerè s.r.l. — Via Modena, 18 — 46041 Asola-Mantova — Italy — www.x-bionic.com //
X-Technology Swiss — Samstagernstrasse 45 — 8832 Wollerau — Switzerland — www.x-technology.com
Design X-Technology Swiss — Prof. Bodo W. Lambertz — www.x-technology.com

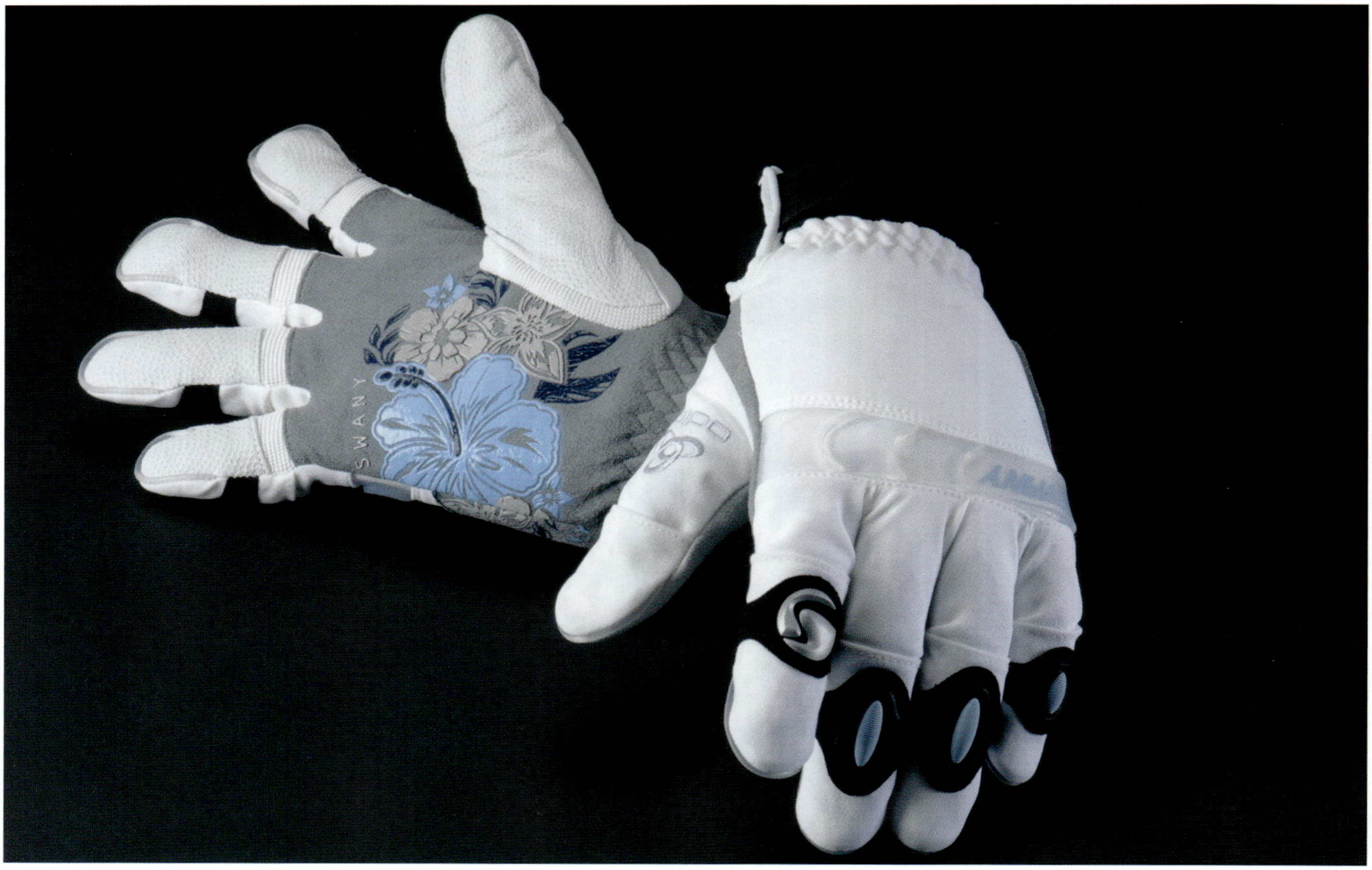

Lange mussten sportliche und modebewusste Frauen darauf warten, einen Handschuh zu finden, welcher sowohl visuell überzeugend als auch hochfunktionell und äußerst formschön ist. Dank des biomorphen Konzepts wird die Fingerbeweglichkeit um 30 – 50 % gegenüber herkömmlichen Handschuhen verbessert. In Kombination mit der innovativen und dehnbaren *Schoeller Nanosphere*®-Außenhülle und dem wärmenden ODLO „effect“®-Lining ist ein angenehm eng anliegender, aber dennoch sehr warmer Handschuh mit wunderschöner Silhouette entstanden. It was in vain that sporty and fashion savvy women were looking for a glove that would combine function, slim looks, comfort and performance in one product. Due to the biomorphic concept the dexterity is improved by 30 – 50 % compared to regular gloves. In combination with the stretchable, innovative *Schoeller Nanosphere*® shell and the warm odour free ODLO “effect”® lining the NFX-8 is a smooth and comfortably tight fitting glove with nevertheless perfect warmth properties and a beautiful slim looking silhouette.

NFX-8 Lunar

Technische Schneesporthandschuhe
Technical glove for snowsports

Unternehmen — Company SWANY Europe GmbH — Tösstalstraße 212 — 8405 Winterthur — Switzerland — www.swanyeurope.com
Design Werksdesign / Inhouse — Markus Kellermüller, Masahiko Kato, Ichiro Kuwahara

Ein modischer Skihandschuh der neuesten Generation mit einer speziellen Ergonomie und schlanker Silhouette. Die 2-in-1-Konstruktion beinhaltet einen leichten und atmungsaktiven Trikot-Innenhandschuh und eine funktionale wie attraktive 4-Way Stretch-Außenhaut. Dünne, aber leistungsfähige Mesh-Materialien sorgen für eine besonders schlanke Silhouette. Die aufwändigen Silikonprints bieten in der Innenhand optimalen Halt. Der Neoprene-Schaft schließt den Handschuh satt am Handgelenk und ist mit einer Zugschlaufe versehen, die das Anziehen zusätzlich erleichtert. A futuristic looking glove that combines ergonomic function with a contemporary slim and stylish silhouette. The 2 in 1 construction features a removable moisture wicking and breathable tricot liner. The thin but functional 4-way stretch shell provides both, a comfortably tight fit and technical look. The chamude palm with 4 colour print offers a perfect grip sensation. The neoprene cuff closes tight to the wrist and with its pull loop is a comfort helper to get smoothly into the gloves.

Renegade AX-16L *Skihandschuh*

Glove for snowsports

Unternehmen — Company Swany Europe GmbH — Tösstalstrasse 212 — CH-8405 Winterthur — Switzerland — www.swanyeurope.com

Design VISTAPARK GmbH — Joachim Rao, Matthias Ocklenburg — www.vistapark.de

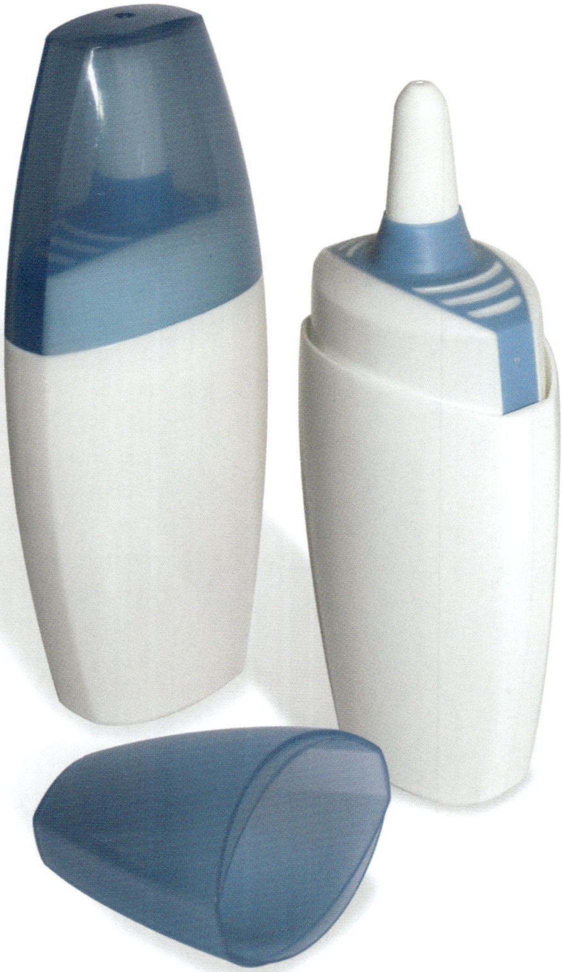

Das Dosiersystem sichert eine stets gleich bleibende und exakte Dosiermenge, unabhängig vom Benutzer. Das System erweitert existierende Nasensprays ohne Modifizierung der Primärverpackung. Die überspannten Flächen ermöglichen ein einfaches Handling, erleichtern die Mitnahme in der Tasche und bieten Platz für Branding und Produktbeschreibung. Das komfortable Handling wird ebenfalls durch die TPE-Griffzone sowohl auf visuelle als auch taktile Weise unterstützt. The innovative dispensing system ensures a defined spray performance every time, independent of the user. The system is an outer package that upgrades existing multi-dose nasal spray systems with this highly practical feature. The surfaces enhance the user comfort, the transportation into a pocket and allow the placement of branding and product description labels. The TPE grip zone offers visual and tactile support for comfortable and selfexplanatory handling.

User Independent Package *Mikrodosiersystem*

Micro spray system

Unternehmen — Company Ing. Erich Pfeiffer.de — Oeschlestraße 54-56 — 78315 Radolfzell — Germany — www.pfeiffer-group.com
Design yellow design | yellow circle — www.yellowdesign.com

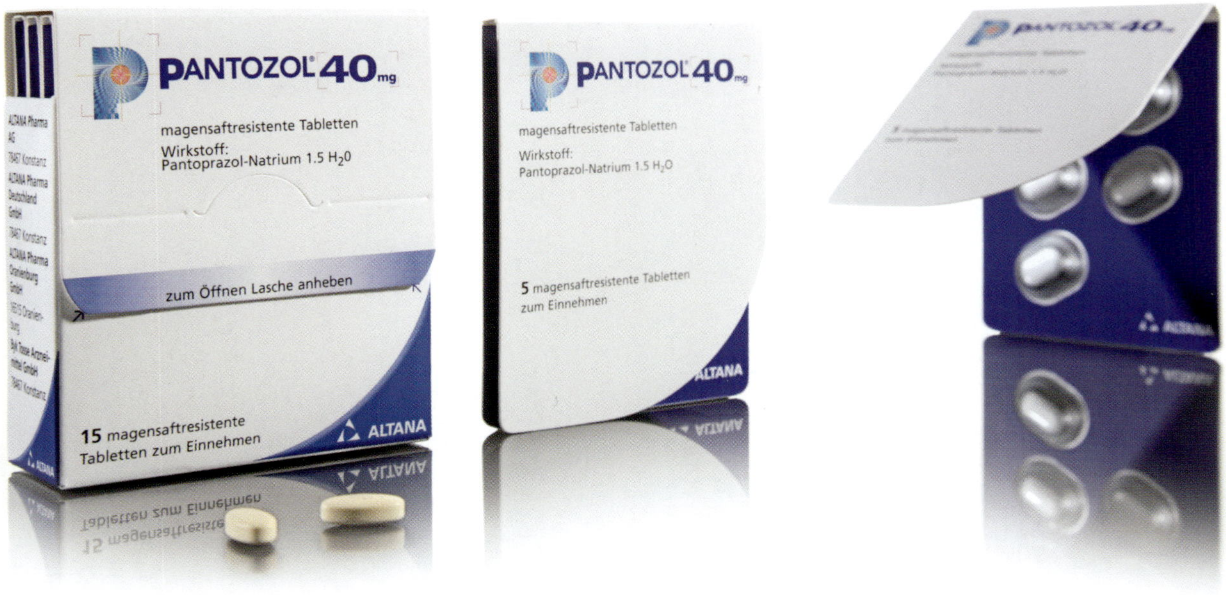

Pantozol® wird zur Akut- und Dauermedikation eingesetzt. Die Aufgabenstellung an die Entwickler der neuen Faltschachtel beinhaltete zwei wesentliche Forderungen: Die Mobilität des Patienten sollte durch praktische Blistergrößen unterstützt werden. Die zweite Forderung lautete, die Attraktivität der Verpackung bei Arzt und Patient solle mit dazu beitragen, dass das Original-Produkt *Pantozol®* gegen die starke Konkurrenz seine Marktposition weiter ausbauen kann.

Pantozol® is used for acute and continous medication. The brief for the designers of the new folded box contained two main requirements: To support the mobility of patients via convenient blister sizes, and to further increase the market position of the original *Pantozol®* against its strong competitors via attractive packaging for doctors and patients.

Pantozol® *Medikamentenverpackung*

Pharmaceutical packaging

Unternehmen — Company ALTANA Pharma AG — Byk-Gulden-Straße 2 — 78467 Konstanz — Germany

Design Designconnection — Andreas Klatt, Klaus Hinderegger — www.designconnection.eu

Der Audio-Verstärker *soulution 710* erfüllt höchste klangliche Anforderungen. Er vereint die Stärken von klassischen Röhren- wie auch modernen Transistor-Verstärkern. Dafür ist Spitzentechnologie Voraussetzung und nicht Selbstzweck. Die formale Gestaltung genügt ebenso wie die Elektronik dem Credo der High Fidelity „nichts hinzufügen, nichts weglassen". Kurzum: *soulution 710* vereint Schnelligkeit, Präzision, Stabilität und Kraft im Dienste der Musik. Audio power amplifier *soulution 710* fulfils highest sound requirements. It combines the advantages of classic tube amplifiers with those of modern transistor amplifiers. For the natural reproduction of music state-of-the-art technology is a prerequisite and not an end in itself. Both the design and the electronics satisfy High Fidelity's credo of "nothing superfluous and nothing omitted". In a nutshell: *soulution 710* for the first time combines speed, precision, stability and power in the service of music.

soulution 710 *Audio-Verstärker*

Audio amplifier

Unternehmen — Company Spemot AG — Industriestraße 70 — 4657 Dulliken — Switzerland — www.soulution-audio.com

Design greutmann bolzern designstudio — Urs Greutmann, Carmen Greutmann — www.gbdesign.ch

Ein perfekter Vorverstärker, welcher höchste klangliche Anforderungen erfüllt. Er verfügt über zwei symmetrische, drei unsymmetrische Eingänge wie auch über einen zusätzlichen Phono-Vorverstärker. Um die Reinheit des Klanges reproduzieren zu können, ist innovative Spitzentechnologie Voraussetzung, welche mit hochwertigsten Komponenten realisiert wird. The perfect pre-amplifier, fulfilling highest sound requirements. It boasts two symmetrical and three asymmetrical inputs plus an additional phono pre-amplifier. All digital (CD / SACD) and analogue music sources can be connected. For reproducing the purity of sound, innovative state-of-the-art technology is a prerequisite, and this has been achieved using top-class components.

soulution 720 *Audio-Vorverstärker*
Audio preamplifier

Unternehmen — Company Spemot AG — Industriestraße 70 — 4657 Dulliken — Switzerland — www.soulution-audio.com
Design greutmann bolzern designstudio — Urs Greutmann, Carmen Greutmann — www.gbdesign.ch

Universelle *iPod*-Dockingstation mit Anschlüssen für externe Tonquellen. Inklusive Funkfernbedienung für die Steuerung des *iPod*. Der *iPod* kann geladen und mit dem Computer synchronisiert werden. Der Sound wird über vier Odyssey-Treiber mit Neodym-Antrieb wiedergegeben. Univeral *iPod* docking station with line in for external sources. The dock comes with a RF remote control for the docked *iPod*. The *iPod* can be charged and synchronised with a computer when docked. Four advanced Odyssey Neodymium drivers provide full-spectrum sound.

JBL OnStage II *iPod-Dock mit Lautsprechern*

iPod docking station with speakers

Unternehmen — Company Harman Deutschland GmbH — Hünderstraße 1 — 74080 Heilbronn — Germany — www.harmanmultimedia.com // JBL, Inc. — 8500 Balboa Blvd. — CA 91329 Northridge — USA — www.jbl.com

Design Werksdesign/Inhouse — Kurt Solland

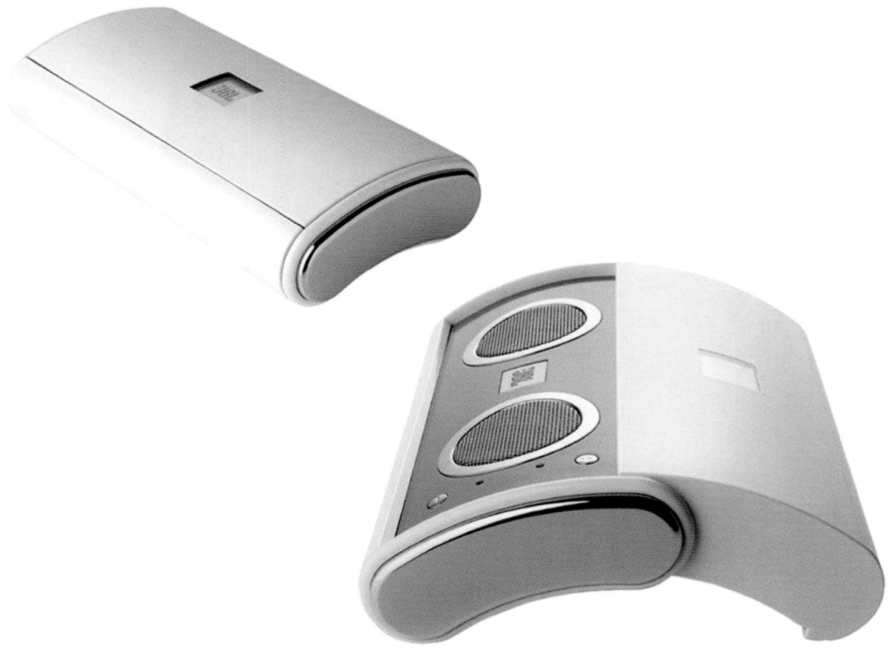

Kompaktes, taschengroßes Lautsprechersystem – verleiht jedem CD-Player, Laptop, *iPod* oder einer Spielekonsole den legendären JBL-Sound. Breites Frequenzspektrum durch zwei Odyssey Neodym-Treiber. Lässt sich mich AAA-Batterien oder mitgeliefertem Netzteil betreiben.

Compact, pocket-sized speaker system for legendary JBL sound from any *iPod*, CD-player, laptop or game console. Delivers a broad frequency range due to Odyssey Neodymium drivers. Runs on AAA-batteries or AC power for worldwide portability.

JBL OnTour *Portables Stereo-Lautsprechersystem*

Portable stereo speaker system

Unternehmen — Company Harman Deutschland GmbH — Hünderstraße 1 — 74080 Heilbronn — Germany — www.harmanmultimedia.com
JBL, Inc. — 8500 Balboa Blvd. — CA 91329 Northridge — USA — www.jbl.com
Design Werksdesign / Inhouse — Kurt Solland

Das Bluetooth-Headset *Discovery*™ *640* verbindet ein kleines und ultraleichtes Design mit einer einzigartigen Auswahl an Ladeoptionen. Das innovative Ladegerät mit einer AAA-Batterie erhöht die Sprechzeit auf bis zu 15 und die Standby-Zeit auf bis zu 180 Stunden. Das Set aus vier Ladeadaptern ermöglicht das direkte Aufladen des Headsets über die Ladegeräte der gängigsten Mobiltelefone (Nokia, Motorola, Sony Ericsson, Siemens). Dank *MultiPoint*-Technologie kann das Headset mit zwei Bluetooth-Geräten gleichzeitig verbunden werden. The Bluetooth-Headset *Discovery*™ *640* combines small and ultra light-weight design with a unique choice of charging options. The innovative AAA battery charger boosts talk time up to 15 and standby time up to 180 hours. The set of four charging adapters allows direct charging from mobile phone chargers (Nokia, Motorola, Sony Ericsson, Siemens). *MultiPoint* functionality allows connection of the headset with two Bluetooth-enabled devices at once.

Discovery™ 640 *Bluetooth-Headset*
Bluetooth headset

Unternehmen — Company Plantronics GmbH — Gildenweg 7 — 50354 Hürth — Germany — www.plantronics.com
Design Werksdesign / Inhouse — Soohyun Ham, Darrin Caddes

Ein Navigationssystem mit kostenlosem Stauwarner (TMC) und umfangreichem Kartenmaterial für 28 Länder Europas. Die integrierte TMC-Funktion ermöglicht die automatische Alternativroutenberechnung bei Verkehrshindernissen. Dank optionalem Mobility Pack lässt sich *Naviflash* auch außerhalb des Fahrzeuges bis zu neun Stunden nutzen. A navigation system with free traffic info (TMC) and detailed maps of 28 European countries. The integrated TMC function provides automatic and alternative re-routing for missed turns/directions and deviation possibility. Thanks to the optional mobility pack the system can be used up to nine hours out of the car.

Naviflash 1020 *Navigationssystem*

Navigation system

Unternehmen — Company Naviflash Sp.z.o.o. — ul. Wojska Polskiego 4 — 39-300 Mielec — Poland — www.naviflash.de

Design Werksdesign/Inhouse — Andre Bury

Mit dem *Merten Bed Side Panel* als zentraler Steuereinheit können Hotelgäste vom Bett aus sämtliche Raumfunktionen steuern. Per Tastendruck kann der Gast seine Wünsche signalisieren, die über Lichtsymbole außen am Hotelzimmer angezeigt werden. Die Steuerung aller elektrischen Verbraucher und die Regelung der Heizungs- und Lüftungsanlage durch die *Merten Power Box* ermöglichen vielfältige Funktionen. Das *Bed Side Panel* gibt es als Einbau- und Tischvariante in den Farben Aluminium und Polarweiß sowie in drei Ausstattungsvarianten. With the *Merten Bed Side Panel* as central control unit, guests can remotely control all room functions from their bed. The hotel guest can signal his or her wishes by simply pressing buttons. This wish is then displayed on light symbols outside the room door. The centralised control of all electrical loads and of heating and ventilation systems offered by the *Merten Power Box* means a wide range of functions is available. *Merten's Bed Side Panel* comes in three different models. The control device is available in both a built-in model and a table model, and in the colours aluminium or polar white.

Merten Bed Side Panel

Zentrale Steuereinheit für Hotelzimmer
Central control unit for hotel rooms

Unternehmen — Company Merten GmbH & Co. KG — Fritz-Kotz-Straße 8 — 51674 Wiehl — Germany — www.merten.de — www.bedsidepanel.de
Design Werksdesign/Inhouse

Dieses neue wetterfeste und robuste Aufputz-Terminal mit berührungslosem Proximity-Leser fügt sich harmonisch und zeitlos in jede Umgebung im Innen- und Außenbereich ein und ist das jüngste Mitglied einer Gerätefamilie. Das Terminal ermöglicht eine berührungslose Identifikation per Karte oder Schlüssel und eine benutzergerechte PIN-Eingabe und Information. Das IF-701 vereint eine stabile Konstruktion für leistungsstarke Hightechelektronik und maximalen Anwendungskomfort mit einem Design, das klar und sachlich auf das Wesentliche reduziert ist.

The new access terminal with contactless proximity reader for indoor and outdoor use. The weatherproof and robust terminal with its timeless design blends in naturally with any environment. It allows for contactless identification via a card or key tag and features a user-friendly keypad for entering PIN codes and an LED display for user information. The IF-701 terminal combines high-tech electronics with maximum user-friendliness in a robust housing with a simple and practical design.

IF-701 *Terminal für Zutrittskontrolle*
Terminal for access controll

Unternehmen — Company Interflex Datensysteme GmbH & Co. KG — Zettachring 16 — 70567 Stuttgart — Germany — www.interflex.de
Design Neumeister & Partner — www.neumeister-partner.de

*smart*FIS ist ein biometrischer, ethernetfähiger Fingerprintsensor zur Speicherung und Authentifizierung von bis zu 500 Fingerabdrücken ohne Zusatzgeräte. Durch integrierte Schnittstellen, Sabotageschutz, internes Relaismodul und Softwareapplikationen bestehen vielseitige Einsatzmöglichkeiten. Die Materialien Edelstahl und Glas mit komplementärer Innenbeleuchtung sorgen für ein ergonomisches, zeitloses Design. *smart*FIS is a biometric, Ethernet-ready fingerprint sensor to store and authenticate up to 500 fingerprints without any additional device required. The integrated interfaces, sabotage protection, internal relay modules, and software applications make for highly flexible utilization. The stainless steel and glass with the complementary inside lighting provide an ergonomic, timeless design.

smartFIS *Biometrische Zutrittskontrolle*
Biometry access control

Unternehmen — Company Doepke Schaltgeräte GmbH & Co. KG — Stellmacherstraße 11 — 26506 Norden — Germany — www.doepke.de
Design Riemann Design — Herbert Riemann, Dörte Waldöstl — www.riemanndesign.de // Werksdesign / Inhouse — Johann Meints, Gerold Roolfs

Mit dem *visiPad*®SL 18 können alle Geräte und Medien in einem Gebäude gesteuert, kontrolliert und visualisiert werden. Alle Dienste, Informationen und Funktionen aus den Bereichen Heimautomation, Entertainment, Infotainment und Kommunikation liegen gebündelt und zentralisiert in einer Hand. Über das ultraflache *visiPad*®SL 18 wird die Beherrschung von Technik und Medien spielend einfach und zum visuellen Ereignis. Exklusive Schaltzentrale und multimediales Fenster zur eigenen Welt nach Maß. Exclusive switchboard and multimedia window to your own personalized world, adding new dimensions in terms of luxury, security and effectiveness. The *visiPad*®SL 18 makes it possible to control, monitor and visualize all of the devices and media in a building. All of the services, information and functions from the sectors home automation, entertainment, infotainment and communication are bundled, centralised and put in the palm of your hand.

visiomatic Home System/visiPad®SL 18

Display

Display

Unternehmen — Company visiomatic GmbH — Weissacher Straße 1 — 70499 Stuttgart — Germany — www.visiomatic.de

Design wiege Entwicklungsgesellschaft mbH — Michael Englisch, Torsten Wittenberg, Britta Clausen

Das *e-info-Terminal* bietet dem Nutzer im öffentlichen Raum Kommunikationsmedien wie Internet, E-Mail, Videotelefon bzw. Webcam und WLAN-Station. Gestaltet wurde ein rollstuhlgerechtes Bedienpult, das die Funktionen der jeweiligen Schnittstellen logisch und eindeutig hervorhebt. Durch die klare Gestaltung und Verwendung hochwertiger Materialien wird Vandalismus und witterungsbedingten Einflüssen im Outdoorbereich entgegengewirkt. The *bluespot-terminal* provides users in areas accessible to the public access to communication media, such as the Internet, e-mail, videophone, or a web cam as well as WLAN. The design features a wheelchair-adjusted operating panel with logical and unambiguous references to the functions of the relevant interfaces. Its uncluttered design and the use of high-quality materials protect it against the weather and vandalism in public spaces.

e-info Terminal
bluespot terminal

Multimedia Terminal für den öffentlichen Raum

Street furniture

Unternehmen — Company Wall AG — Friedrichstraße 118 — 10117 Berlin — Germany — www.wall.de

Design ION industrial design berlin — Christoph Fleckenstein, Anja Götz — www.iondesign.de

Der *tt7000* ist ein modular aufgebauter industrietauglicher PDA, der trotz seiner Schutzklasse und Stoßfestigkeit extrem variabel einsetzbar ist. Alle Schnittstellen des *tt7000* sind wasser- und staubdicht nach IP 64 ausgeführt. Die radikale Reduktion auf eine zentrale und mehrere optische Schnittstellen macht das Gerät außergewöhnlich robust und stabil. Die seitliche Rippung garantiert eine hohe Schlag- und Kratzfestigkeit des Gerätes und erlaubt gleichzeitig die Umsetzung eines kompakten Gehäusekonzeptes. *tt7000* is a modular industry-ready PDA, that is highly versatile despite its protection features and shock-proofing. All the *tt7000* interfaces are waterproof and dustproof in line with IP 64. The radical reduction to a central interface and several optical interfaces makes the device unusually robust and stable. The ribbing on the sides guarantees high impact- and scratchproofing, also keeping the shell compact in size.

tt7000 *Industrie-PDA* *Industrial PDA*

Unternehmen — Company t + t netcom — Elektrastraße 6 — 81925 München — Germany — www.tt-netcom.com

Design SquareOne GmbH — Gert Trauernicht, Eckhard Ruoff, Stefan Mostert — www.squareonedesign.de

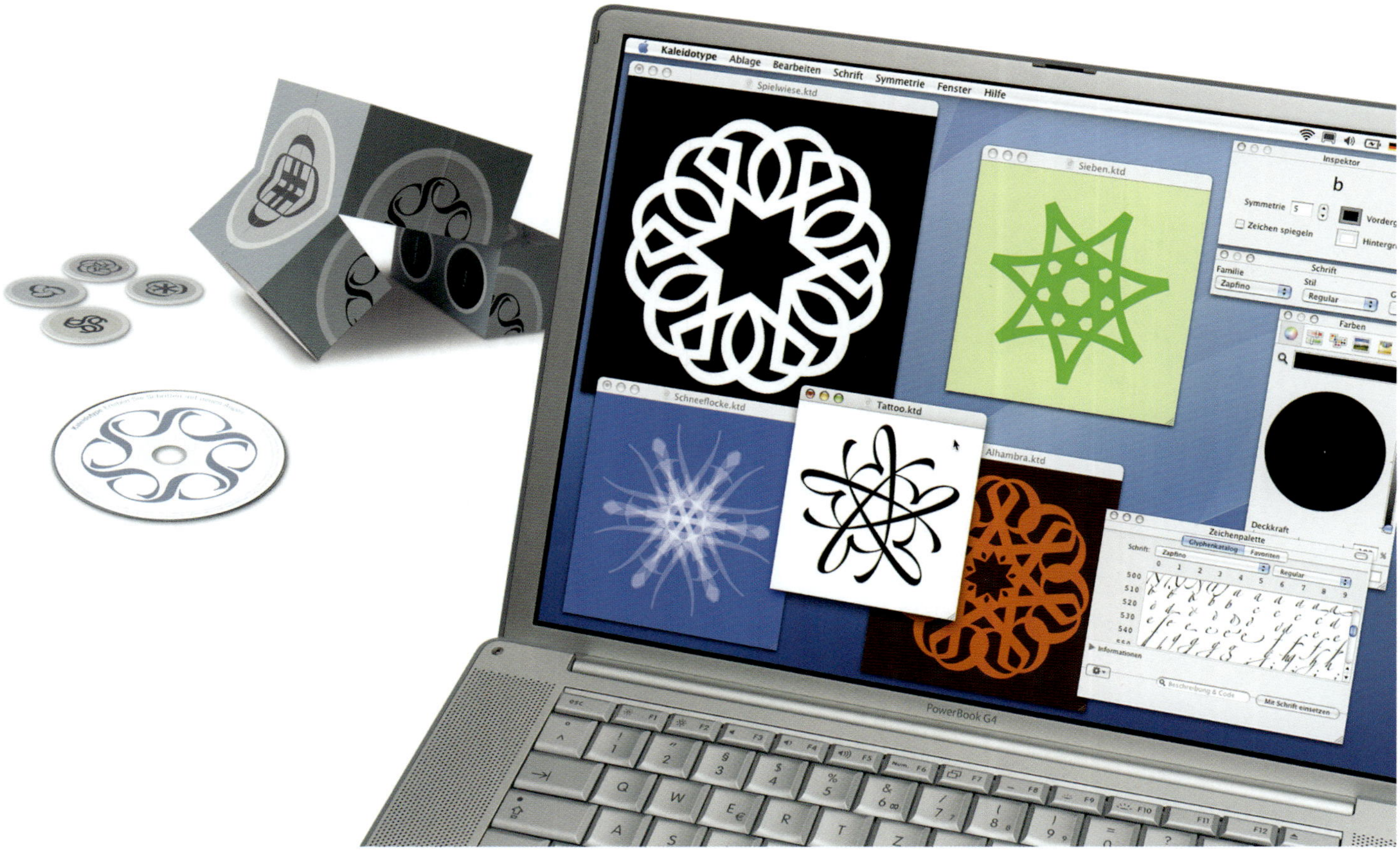

Kaleidotype ist Computersoftware und Spiel in einem. Die aufwändige Produktverpackung – ein Klappwürfel mit Typo-Memory – vermittelt den besonderen Anspruch der Produktidee. Im Kern ist *Kaleidotype* eine Software (Mac OS X), mit der symmetrische Muster aus Schriftzeichen in einem interaktiven Prozess dynamisch erzeugt werden können. Die erzeugten Muster können in unterschiedlichen Dateiformaten ausgegeben und in anderen Programmen weiterverarbeitet werden. *Kaleidotype* is computer-software and game in one. The elaborate packaging – a retractable cube with a typographic card game – communicates the special claim of the products idea. Basically *Kaleidotype* is a computer-programme (Mac OS X), which allows creating symmetric patterns consisting of letters in an interactive process. The patterns can be exported into different file formats and processed in different programmes.

Kaleidotype *Computerprogramm / Software*

Computer programme / software

Unternehmen — Company Verlag Hermann Schmidt Mainz — Robert-Koch-Straße 8 — 55129 Mainz — Germany — www.typografie.de

Design Buttgereit und Heidenreich GmbH — Michael Buttgereit, Wolfram Heidenreich, Martin Winter, Jörg Fassmann — www.b-und-h.de, www.kaleidotype.de

Der é-PUG 2 ist ein mobiler Videorekorder für die Speicherung von Videodaten im Krankenhaus. Ein klares Design der Front soll die Darstellung des Farbdisplays unterstützen, wohingegen die Rückseite des Gerätes der Ergonomie Rechnung trägt und mit weichen Formen der Hand schmeichelt. Bewusst wurde auf einen medizinischen „Look" verzichtet, um die Umgebung des Patienten freundlich zu gestalten und durch ein modernes Design eine angstfreie Kommunikation zu ermöglichen. The é-PUG 2 is a mobile video recorder for storing video data in hospitals. A clear design of the front supports the presentation of the colour display, whereby the back of the device is ergonomic and caresses the hand with its soft forms. A medical "look" was consciously avoided in order to humanise the patient's environment and allow communication without fear through a modern design.

é-PUG 2 (personal users gadget) *Digitales Aufzeichnungsgerät*

Digital recording device

Unternehmen — Company WISAP Gesellschaft für wissenschaftlichen Apparatebau mbH — Rudolf-Diesel-Ring 20 — 82054 Sauerlach — Germany — www.wisap.de

Design IMAGO DESIGN — Bernhard Schwanitz, Stefan Thalhammer — www.imago-design.de

Ein Ladegerät für Hörgerätebatterie-Akkus NiMH der Größe 13, betrieben mit zwei AAA Alkaline-Batterien. Somit ist das Gerät unabhängig von ortsfesten Energiequellen. Es handelt sich um ein Schnellladesystem, welches innerhalb von ca. fünf Stunden den Akku auflädt. Der Lader wird durch einen Controller mit integrierten Sicherheitsfunktionen gesteuert, welcher ein Aufladen ungeeigneter Batterien sowie ein Verpolen unterbindet. Das Gehäuse besteht aus Edelstahl. A charger for rechargeable hearing aid batteries size 13 NiMH powered by two AAA round cells. Therefore the device is independent from any local power source. It is a fast charging system, the NiMH hearing aid battery is fully charged within ca. five hours. It is microprocessor controlled with safety functions; the device will work only with suitable electrochemical systems and if batteries are inserted correctly. The charger is made of stainless steel and format is a convenient pen style design.

power one pen charger *Mobiles Ladegerät für Hörgerätebatterien*

Mobile charger for hearing-aid batteries

Unternehmen — Company VARTA Microbattery GmbH — Daimlerstraße 1 — 73479 Ellwangen — Germany — www.powerone-batteries.de
Design Werksdesign/Inhouse — Herbert Schein, Michael Kron, Eduard Pytlik, Eugen Rechtenbacher

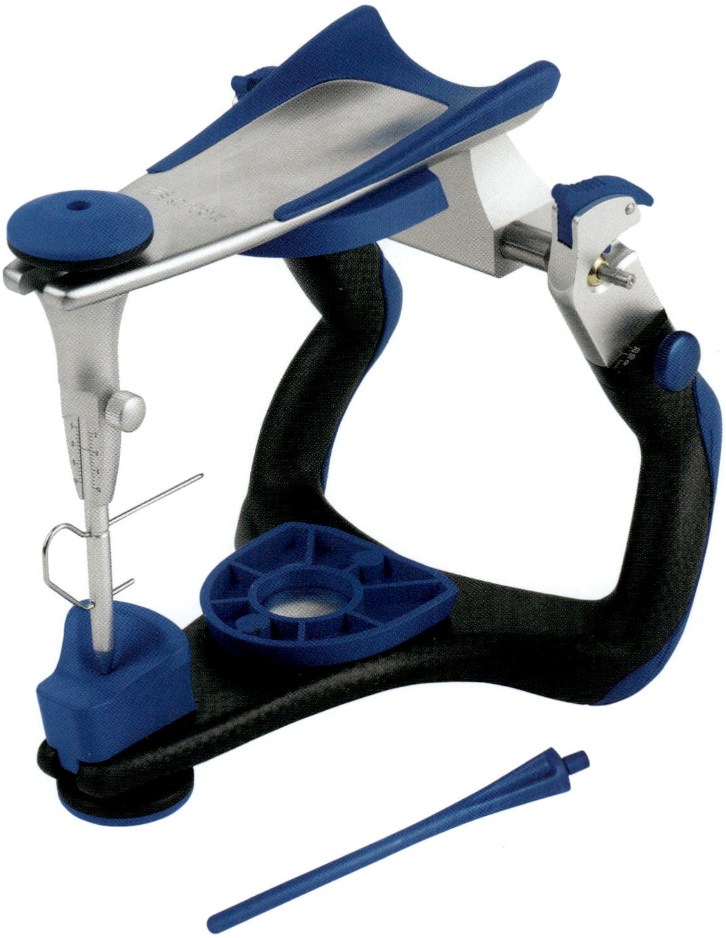

Artex® *Carbon*-Artikulatoren sind Bewegungssimulatoren. Für Zahnarzt und Zahntechniker sind sie unentbehrliche Hilfsmittel zur Herstellung von funktionierendem Zahnersatz. *Artex®* gibt es seit 25 Jahren am Markt. Bei dem vorliegenden neuen Design wurde vor allem Wert auf ergonomische Form und Vereinfachung im Handling gelegt. Durch den Einsatz von Karbon konnte bei gleich bleibender Funktionalität das Gewicht entscheidend reduziert werden. *Artex®* *Carbon* articulators are sensitive movement simulators. Dentists and dental technicians need this instrument to make functional dentures. Since 25 years *Artex®* is used on the market. The new design attached importance to an ergonomic shape and to simplify the handling. Using Carbon has reduced the weight by unchanging the functionality.

Artex® Carbon Kieferbewegungssimulator für die Zahnmedizintechnik
Artex® Carbon Sensitive jaw movement simulator for dental technology

Artikulator

Articulator

Unternehmen — Company Amann Girrbach GmbH — Herrschaftswiesen 1 — 6842 Koblach — Austria — www.amanngirrbach.com

Design Ingenieurbüro Wäger GmbH & Co. KEG — Karl-Heinz Wäger, Michael Schrom — www.ib-waeger.at

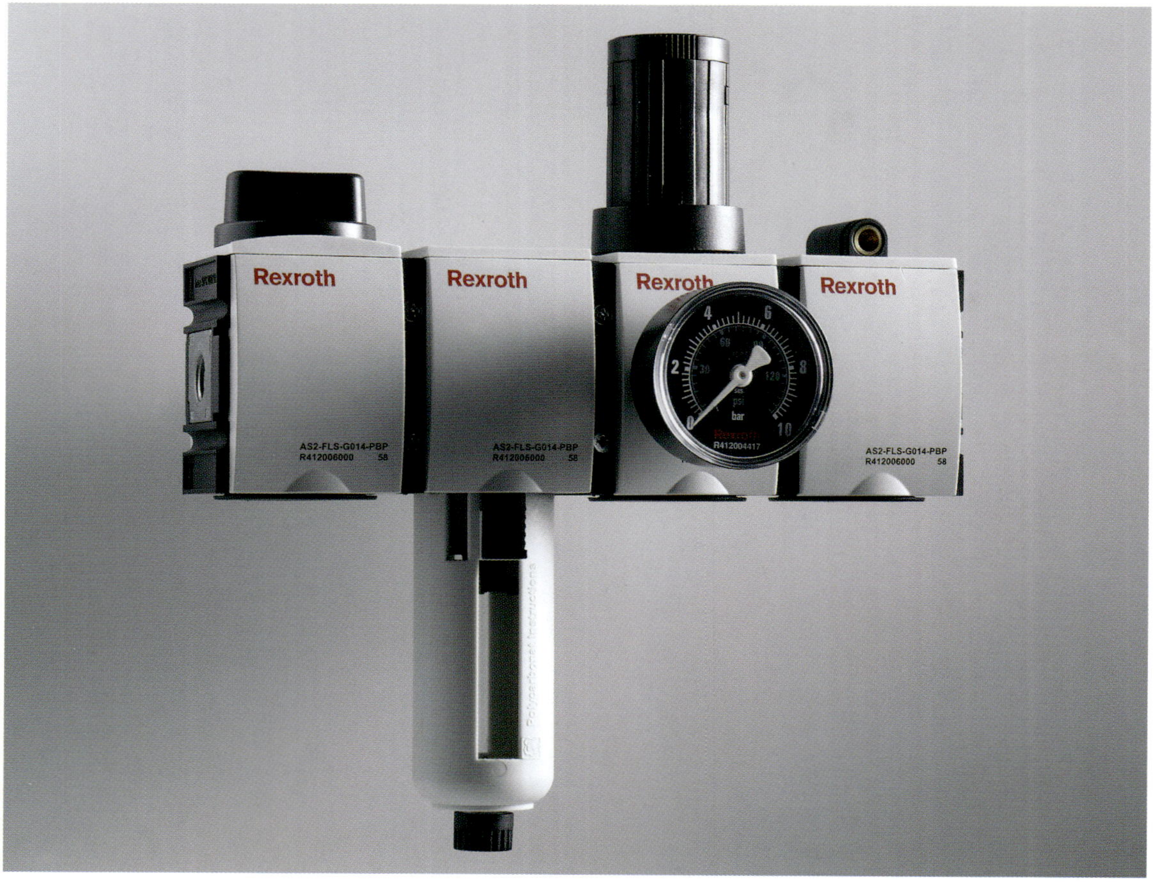

Die neuen Wartungseinheiten der Serie AS2 erfüllen höchste Ansprüche an Druckluftaufbereitung seitens des Anwenders. Aufbereitete Druckluft als Werkstoff schützt die Betriebsmittel vor Korrosion und Schäden durch Schmutzpartikel und senkt so Ausfälle und Verschleiß. In der Serie AS2 sind die Einzelprozesse der Luftaufbereitung Filtern, Regeln, Ölen und Entwässern in einer Wartungseinheit zusammengefasst. Die Ausstattung orientiert sich an den Anforderungen der Anwender wie einfacher Demontage im eingebauten Zustand oder der Zulassung nach ATEX. The new AS2 series maintenance units can fulfil the highest user demands for compressed air preparation. Used as a preventative measure, prepared compressed air protects equipment from corrosion and damages caused by contaminants, thus decreasing downtimes and wear. In the AS2 series, the individual processes for air preparation (filtering, regulating, lubricating, and draining) are combined into a single maintenance unit, which is oriented towards user requirements, such as easy component removal in an assembled state or ATEX approval.

AS2 *System zur Druckluftaufbereitung*
Air preparation unit

Unternehmen — Company Bosch Rexroth AG — Maria-Theresien-Straße 23 — 97816 Lohr am Main — Germany — www.boschrexroth.com
Design Henssler & Schultheiss Fullservice Productdesign GmbH — Martin Schultheiss — www.henssler-schultheiss.de

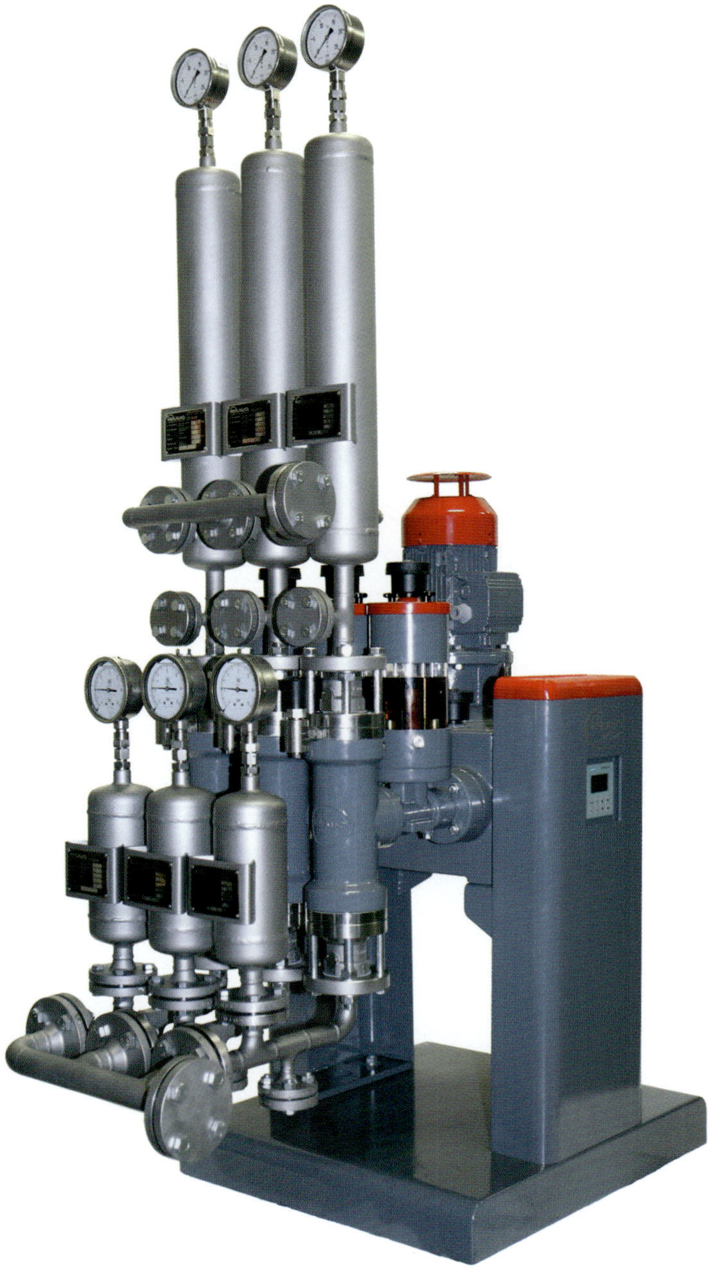

Das Herz der MULTISAFE-Pumpe bilden zwei ineinander angeordnete Schlauchmembranen. Ziel der Designmaßnahme war eine charakteristische Form-/Farbsystematik, die Gestalt- und Funktionsqualität in Einklang bringt. Ein Trägersystem lässt die Pumpe frei schweben und verschafft ihr eine materielle/optische Leichtigkeit. Die auf zwei Farben reduzierte Gestaltung kombiniert den technischen Charakter mit lebendigen Details und bewirkt einen hohen Wiedererkennungswert. At the heart of the MULTISAFE pump are two hose-diaphragms, which are arranged one inside the other. Aim of the design measure was to provide MULTISAFE pumps with typical design and colour characteristics that bring design and functional quality in line. Two beams allow for suspended arrangement and give an impression of physical/visual ease. Design is reduced to two colours, which combine the technical unit character with vivid details and allow for a high level of distinctive recognition features.

MULTISAFE Doppel-Schlauchmembranpumpe
MULTISAFE Double Hose-Diaphragm Pump

Prozesspumpe

Process pump

Unternehmen — Company FELUWA Pumpen GmbH — Beulertweg — 54570 Mürlenbach — Germany — www.feluwa.com

Design Fachhochschule Trier — Jonas Kartenbeck — www.fh-trier.de // Werksdesign/Inhouse — Heinz M. Nägel

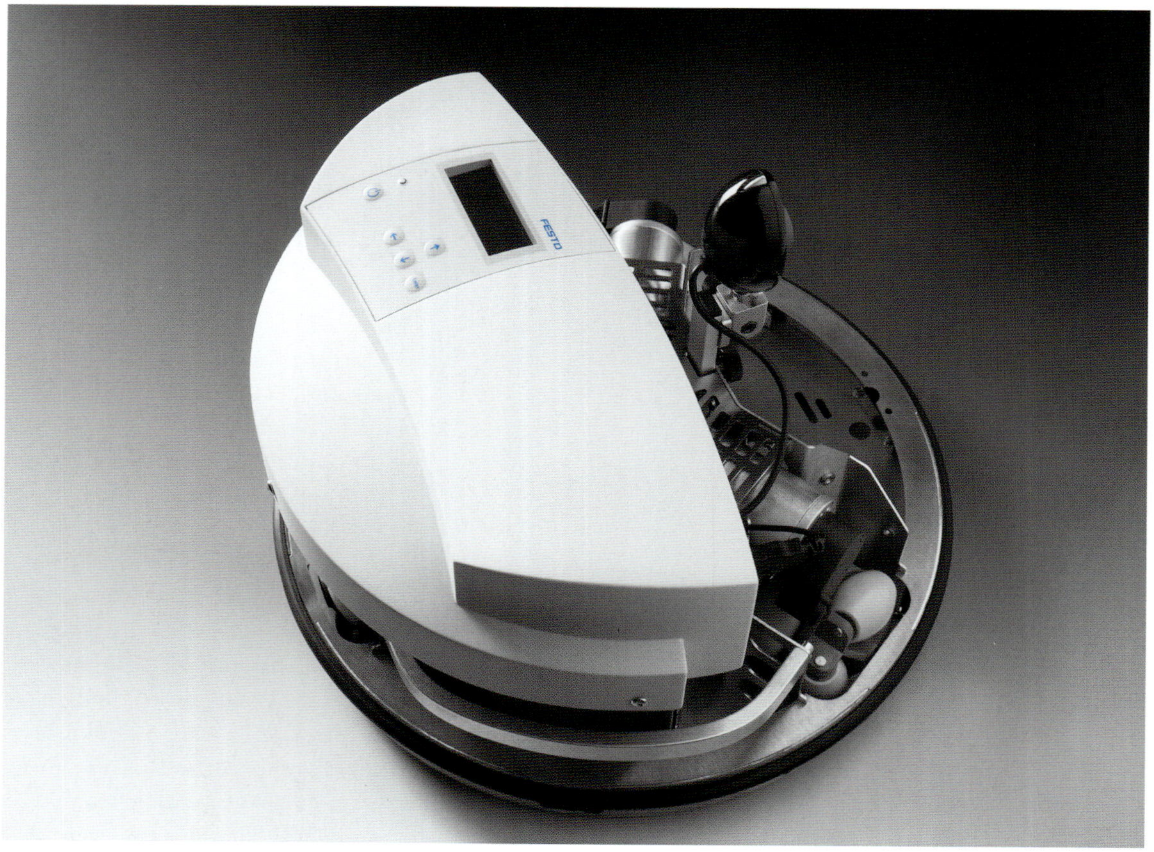

Mit *Robotino*® spielend erlernt: Retten, Bergen, Reinigen, Prüfen, Reparieren, Programmieren oder Beobachten. Dieser autonome, mobile Lernroboter ist ein komplexes mechatronisches System mit innovativem Antrieb und robustem Chassis aus lasergeschweißtem Edelstahl. Motoren, Räder und Elektronik liegen geschützt im Inneren der kreisrunden Grundform, die in Arbeitsbereich und Kommandobrücke geteilt ist. Für Schülerinnen und Schüler entsteht eine faszinierende Aufgabenwelt mit futuristischer Anmutung. Its very easy to learn with *Robotino*®: Rescuing, recovering, cleaning, inspecting, repairing, and monitoring. This autonomous, mobile learning robot is a complex mechatronic system with an innovative drive unit and a robust chassis made of laser-welded stainless steel. Motors, wheels and electronics are safely located within the circular base, which is divided into an operating area and a command bridge. It provides schoolchildren with a fascinating, futuristic world of things to do.

Robotino® *Mobiler Roboter*

Training robot

Unternehmen — Company Festo Didactic GmbH & Co. KG — Rechbergstraße 82 — 73770 Denkendorf — Germany — www.festo.com
Design Werksdesign/Inhouse — Karoline Schmidt, Jan Kleffmann

Der ASM 9-2 ist ein Mittelhandgriff-Schrauber, der für den industriellen Dauereinsatz konzipiert wurde. Bei der Gestaltung wurde großer Wert auf eine Ergonomie gelegt, die ermüdungsfreies Arbeiten erlaubt. Die Gewichtverteilung des kompakten Gerätes ist ausgeglichen, wobei der schonende Anlauf und die sanfte Abschaltung des wartungsfreien Motors Mensch und Maschine entlasten. Durch einen Linkslauftaster auf jeder Seite kann der ASM 9-2 von Rechts- und Linkshändern mit nur einer Hand bedient werden. The ASM 9-2 is a centre-grip screwdriver conceived for continuous use. Particular emphasis was placed upon an ergonomic design facilitating fatigue-free work. The weight distribution of the compact unit is balanced, and both start-up and switch off the maintenance-free motor occur gently, thus relieving man and machine from excessive stress. The ASM 9-2 includes two identical buttons for counter-clockwise rotation on either side of the housing so that the screwdriver can be operated with just one hand by both right- and left-handers.

ASM 9-2 *Akku-Schrauber*

Battery screwdiver

Unternehmen — Company C. & E. Fein GmbH — Hans-Fein-Straße 81 — 73529 Schwäbisch Gmünd — Germany — www.fein.de

Design Henssler und Schultheiss Fullservice Productdesign GmbH — www.henssler-schultheiss.de

Es handelt sich um eine Windkraftturbine mit aerodynamisch geformtem Gehäuse. Hierdurch wird eine 2,5-fach höhere Energieausbeute gegenüber frei fahrenden Anlagen erreicht. Die Vorteile sind, neben einer kleineren Bauweise und einem vereinfachten Aufbau, der geringe Wartungsaufwand, da Getriebe- und Pitsch-Systeme nicht notwendig sind, Reduzierung des Stroboskopeffekts, Schutz vor Eisabwurf und Vogelschlag. Anlagen dürfen in bebauten Gebieten aufgestellt und betrieben werden, da nur eine geringe Geräuschentwicklung entsteht. This is a wind generator with an aerodynamically shaped cowling. The result is a 2.5-fold increase in power output compared with free-running rotor generators. The advantages include: Smaller size and simplified structure; lower maintenance thanks to the ability to dispense with pitch control and gear mechanism, reduction of stroboscopic effect, as well as zero ice release and no threat to birds. Ideal turbine for urban and nears-to-building sites because of the low noise emission.

enflo 0060/0.5 *Windkraftturbine*
Wind generator

Unternehmen — Company KBE Engineering GmbH — Robert-Bosch-Straße 5 — 72172 Sulz a. N. — Germany — www.kbe.de
Design Werksdesign/Inhouse — Michael Ehrenberger

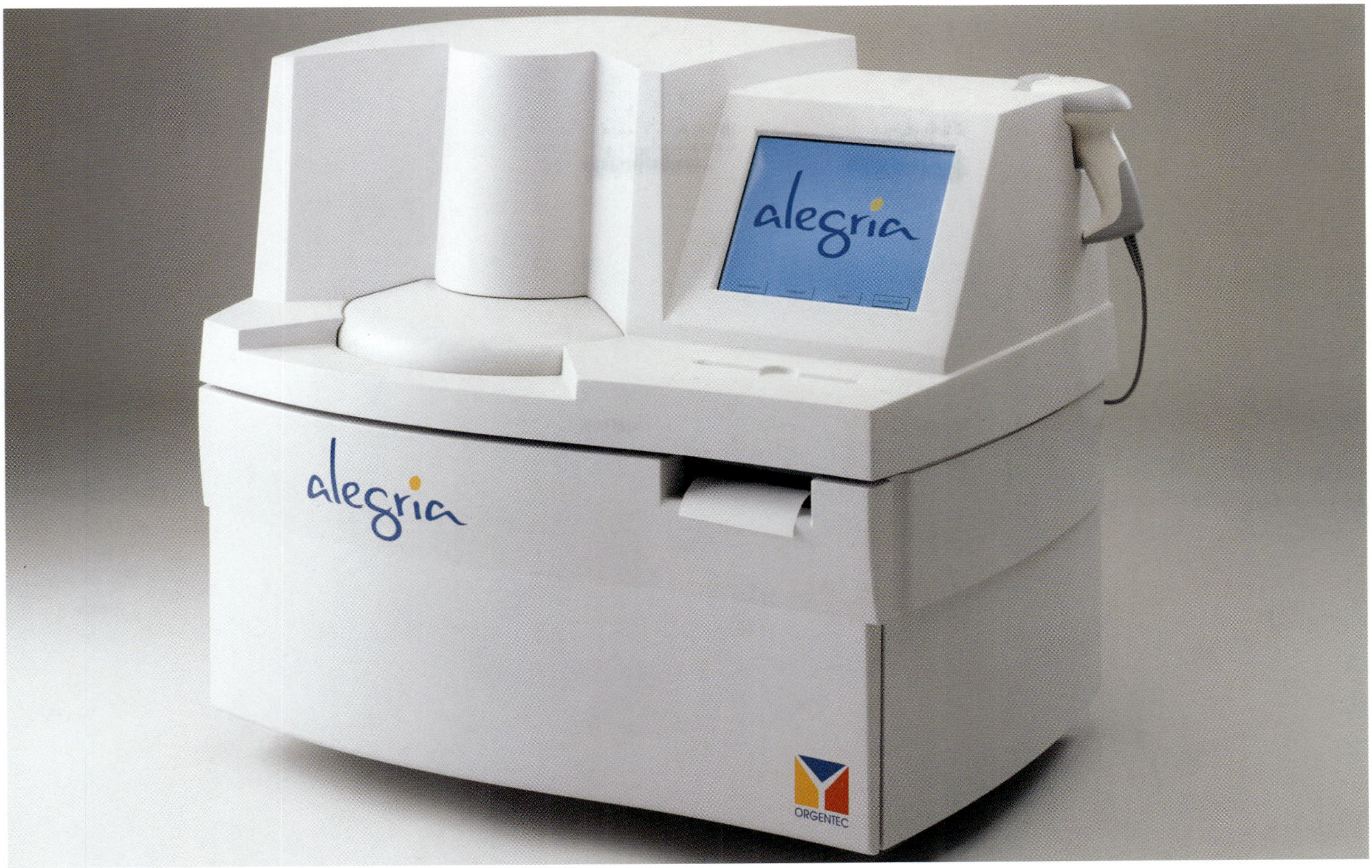

Alegria, basierend auf der SMC®-Technologie stellt eine neue Dimension in der Autoimmundiagnostik dar; ein vollständig automatisiertes Detektionsverfahren in einer bislang nicht zur Verfügung stehenden Flexibilität. *Alegria* ermöglicht eine spezielle, individuell angepasste Diagnostik. Jede Probe wird spezifisch prozessiert und benötigt dennoch nur minimalen Aufwand. Der somit erzielte rasche Befund trägt maßgeblich zum Wohle des Patienten und zur Kostensenkung im Gesundheitswesen bei. *Alegria* the random access analyser, based on SMC® Technology, represents a new dimension in autoimmune diagnostics. It provides a fully automated process for detecting auto antibodies and offers a new level of flexibility. *Alegria* allows users to process samples in a way that is especially adapted to the individual needs of their own particular environment. Each patient will be handled individually, which saves costs, thus freeing up resources to the benefit of the patient and health systems.

ALEGRIA *Diagnosegerät*
Diagnostic equipment

Unternehmen — Company ORGENTEC Diagnostika GmbH — Carl-Zeiss-Straße 49 — 55129 Mainz — Germany — www.orgentec.com // BIT Analytical Instruments GmbH — Am Kronberger Hang 3 — 65824 Schwalbach — Germany — www.bit-instruments.com

Design Teleca System Design AB

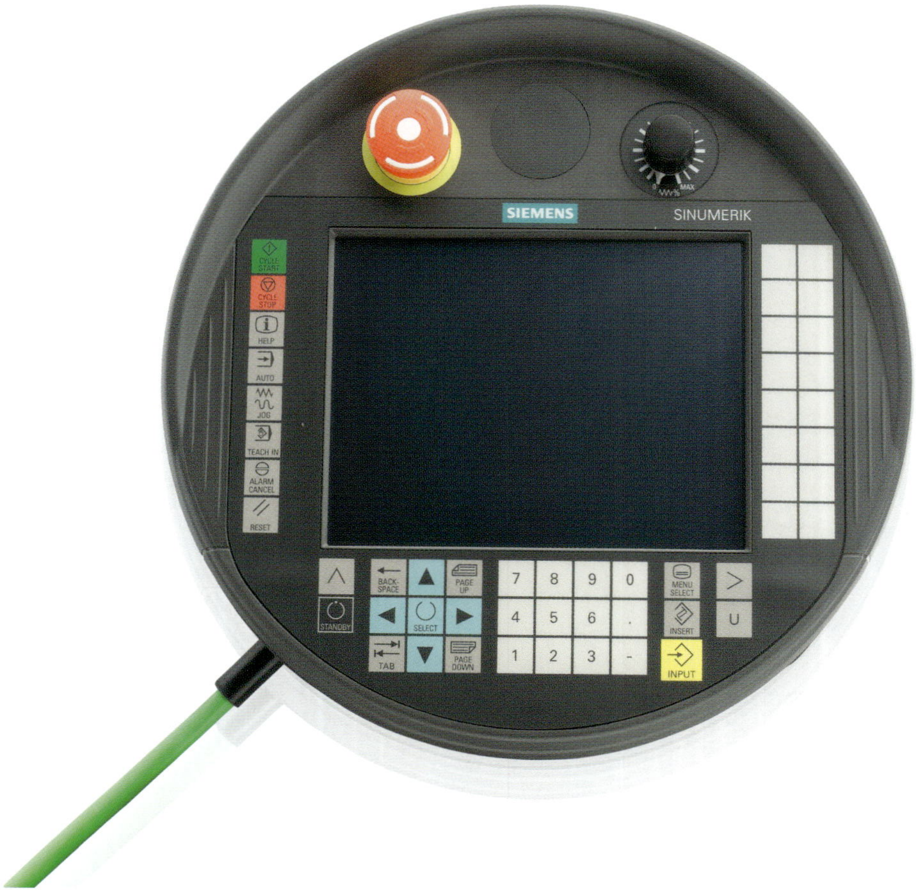

Ein mobiles Handbediengerät, mit dem die komplette Bedienung, Programmierung und Visualisierung von Handhabungsgeräten und Robotern innerhalb des Herstellungsprozesses erfolgen kann. Das Design des Handheld Terminals HT8 des Produktsystems SINUMERIK zielt auf die verbesserte Unterstützung der Bedienung, die optimale Handhabbarkeit des Geräts, das homogene Erscheinungsbild mit anderen Produktlinien sowie die optimierte Integration in das Corporate Design von *Siemens Automation and Drives.* A mobile hand control unit for operating and programming of handling-machines or robotsystems that enables the user to perform the complete control, programming and visualization of manufacturing processes. The design of the HT8 handheld terminal of the product system SINUMERIK aims at the improved support of the operation, the optimized handling of the device, the homogenous corporate appearance in accordance with other product systems and the optimized integration in the Corporate Design of *Siemens Automation and Drives.*

SINUMERIK HT8 *Mobiles Handbediengerät*

Mobile hand control unit

Unternehmen — Company Siemens AG, Bereich Automation and Drives — Gleiwitzerstraße 555 — 90475 Nürnberg — Germany — www.siemens.com/automation
Design at-design — Jan Andersson, Christoph Tomczack — www.atdesign.de // Werksdesign/Inhouse — Siemens AG A & D Design Management — Gunter Ott, Nina Wendt

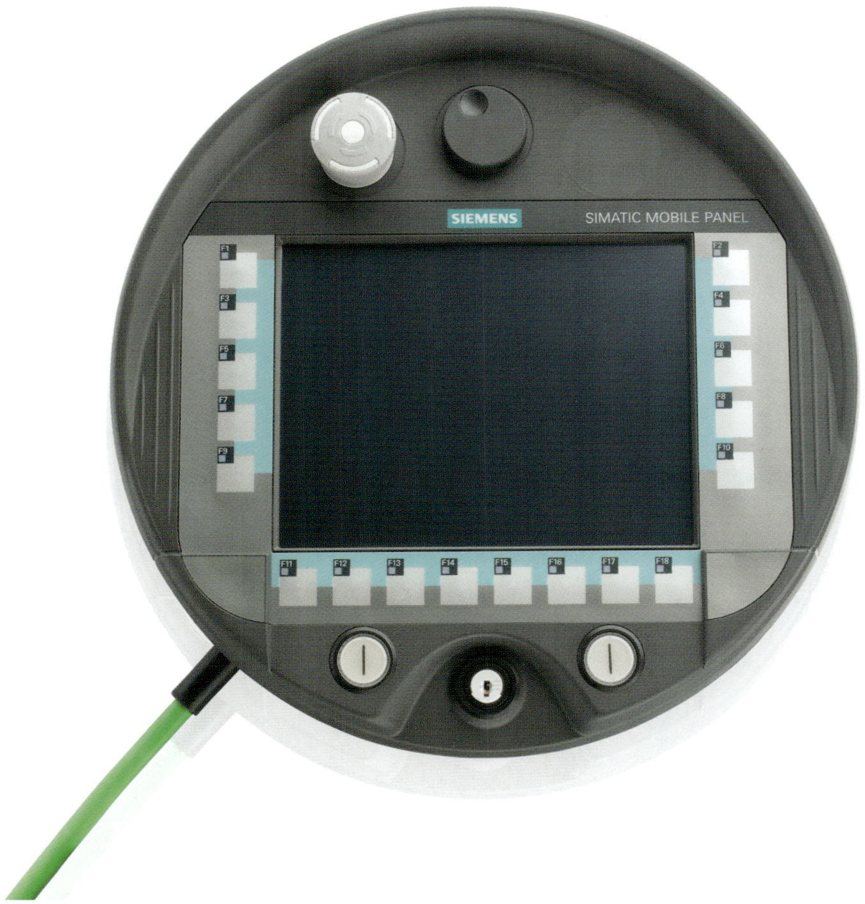

Mobiles Handbediengerät für Produktionsprozesse zum Verfolgen und Steuern beweglicher Maschinen- und Anlagenteile. Das Design des MOBILE PANEL 277 des Produktsystems SIMATIC zielt auf die verbesserte Unterstützung der Bedienung, die optimale Handhabbarkeit des Geräts, das homogene Erscheinungsbild mit anderen Produktlinien und die optimierte Integration in das Corporate Design von *Siemens Automation and Drives.* Mobile hand control unit for factory automation processes for the observation and the controlling of moveable machine parts and parts of production plants. The design of the MOBILE PANEL 277 of the product system SIMATIC aims at the improved support of the operation, the optimized handling of the device, the homogenous corporate appearance in accordance, with other product systems and the optimized integration in the Corporate Design of *Siemens Automation and Drives.*

SIMATIC MOBILE PANEL 277 *Mobiles Handbediengerät*

Mobile hand control unit

Unternehmen — Company Siemens AG, Bereich Automation and Drives — Gleiwitzerstraße 555 — 90475 Nürnberg — Germany — www.siemens.com/automation

Design at-design — Jan Andersson, Christoph Tomczack — www.atdesign.de // Werksdesign / Inhouse — Siemens AG A & D Design Management — Gunter Ott, Nina Wendt

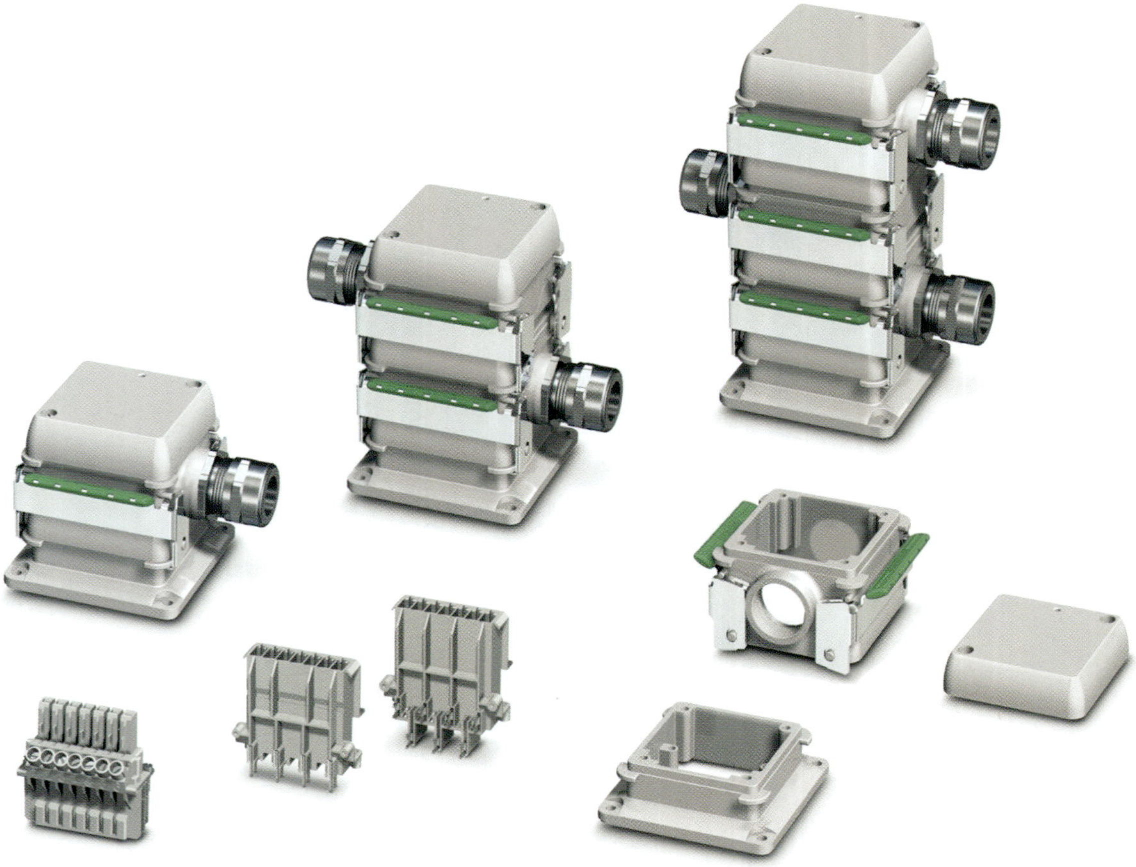

DUPLICON ist die innovative Steckverbinder-Lösung für Ihren IP67-Energie-Bus für 7 x 4 mm^2-Leiter / 25 A / 400 / 690 V. Einzigartig ist die Durchgängigkeit von der Stromschiene über steckbare Reihenklemmen ST-COMBI im Schaltschrank bis zur Leiterplatte bzw. Steckverbindung am Feldgerät. Die robusten, EMV-geschützten Aluminiumgehäuse mit ergonomischen Verriegelungsbügeln qualifizieren DUPLICON für härteste Einsatzbedingungen. DUPLICON is the innovative plug connector solution for your IP67 power bus for 7 x 4 mm^2 conductors / 25 A / 400 / 690 V. Seamless connection from the power rail via pluggable ST-COMBI modular terminal blocks in the control cabinet and on to the PCB or plug connection on the field device. The sturdy, EMC-protected aluminium housing with ergonomically designed locking latches makes DUPLICON qualify for the harshest of conditions.

DUPLICON *Steckverbinder für die Energieverteilung*

Plug connector for power distribution

Unternehmen — Company PHOENIX CONTACT GmbH & Co. KG — Flachsmarktstraße 8 — 32825 Blomberg — Germany — www.phoenixcontact.com
Design Tecform — Gerwin Müller

Die neue Hochdruckkompressorreihe wird zur Luft- und Gasverdichtung eingesetzt. Haupteinsatzbereiche sind die Kunststoffindustrie, Erdgastankstellen und die Atemlufterzeugung zum Beispiel für Feuerwehren. Das Produkt überzeugt durch einfachste Wartung und lange Wartungsintervalle von 2.500 Betriebsstunden. Hierdurch lassen sich für den Betreiber die anfallenden Betriebskosten signifikant reduzieren. Das innovative Produktkonzept setzt sich in der ergonomischen Anordnung der einzelnen Baugruppen fort: Alle Komponenten wie Filter, Entwässerung und Antrieb sind über seitliche Wartungsklappen und die vordere Tür schnell zugänglich. Informationen über den Betriebszustand sind von außen klar erkennbar. This new range of high-pressure compressors is deployed for air and gas compression. The main fields of application are the plastics industry, natural gas fuelling stations and for the production of oxygen e.g. for fire brigades. The product with its easy maintenance in long intervals of up to 2,500 operating hours is convincing. In this way running costs can be significantly reduced for the operating company. The innovative concept is extended through ergonomic subassembly arrangement: All components e.g. filters, demisters and drive train are easily and quickly accessible using forward and lateral maintenance doors. Operating status information is clearly externally visible.

Tornado *Hochdruckkompressor*

High pressure compressor

Unternehmen — Company J.P. Sauer & Sohn — Maschinenbau GmbH — Brauner Berg 15 — 24159 Kiel — Germany — www.sauersohn.de

Design mehnert corporate design gmbh & co. kg — Prof. Kurt Mehnert, Jan Eickhoff, Alasdair Wright — www.mehnertdesign.de

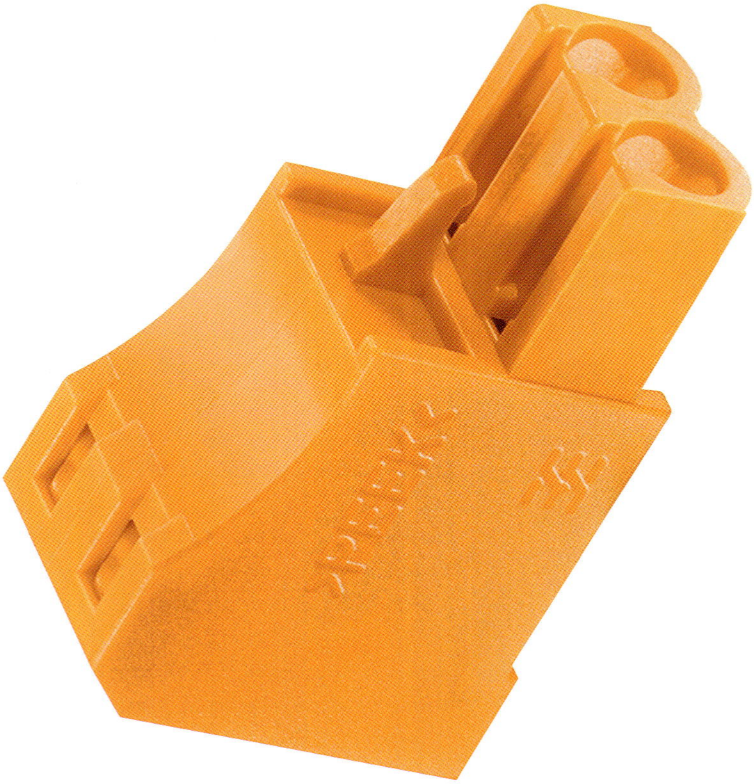

Mit der BLZF ist es erstmalig gelungen, Standards in einem neuen, eigenständigen Design zu realisieren. Der Clou: die ergonomisch geformte Silhouette ist an die Fingerkuppe angepasst. So wird das Handling wesentlich vereinfacht. Der Hochleistungskunststoff PEEK garantiert zudem eine lange Lebensdauer. Das Ergebnis: Ein Produkt, das Maßstäbe setzt – vom innovativen Design bis zum hohen Nutzwert. The BLZF is the first time existing standards have been given a new, distinct look. The key to the design lies in its ergonomic shape, which perfectly fits the fingertip, making the connector much simpler to handle. The PEEK high-performance plastic also guarantees a long life-span. The result is a product that sets standards from its innovative design to its exceptional functional benefits.

Buchsenleisten mit Zugfederanschlusstechnik
Socket block with tension clamp connection

Steckverbinder
Omnimate range

Unternehmen — Company Weidmüller Interface GmbH & Co. KG — Klingenbergstraße 16 — 32758 Detmold — Germany — www.weidmueller.com
Design Werksdesign / Inhouse

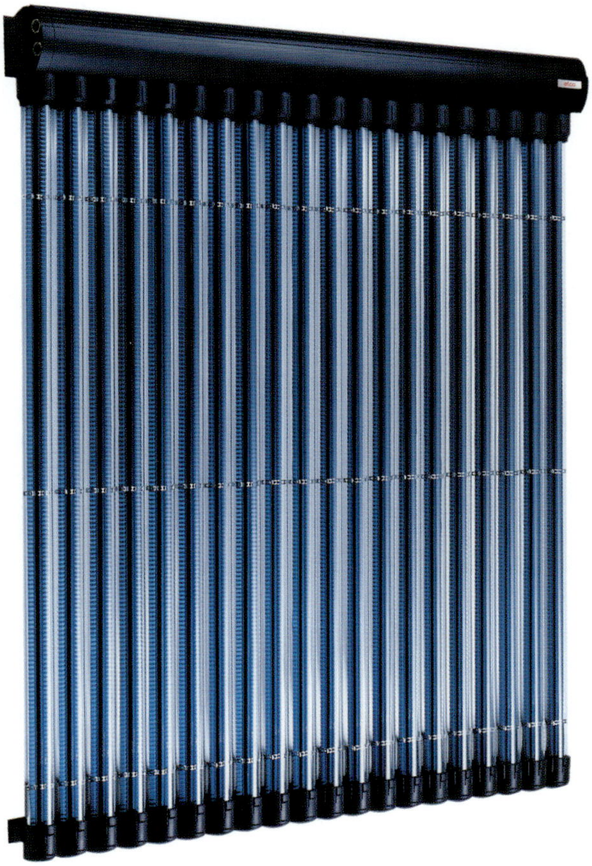

AURON ist ein hoch effizienter Vakuumröhren-Solarkollektor, der zur Trinkwassererwärmung und Heizungsunterstützung eingesetzt wird. Konstruktion und Design des Befestigungssystems folgen konsequent den Anforderungen der Montage. Die einzelnen Kollektormodule lassen sich nahtlos koppeln, so dass eine homogene Fläche ohne Zwischenräume entsteht. AURON kann nahezu verschattungsfrei sowohl im Quer- als auch Hochformat auf Flach- und Schrägdächern und selbst an Fassaden montiert werden. AURON is a high efficient vacuum-tube solar collector for sanitary water heating and space heating support. The concept and design of the mounting system meet consequently the installation needs. The individual modules can be joined seamlessly without any gaps to create an optically harmonious solution. AURON allows virtually shadow-free installation in horizontal or vertical orientation, on flat or sloped roofs or even on facades.

AURON *Vakuumröhrenkollektor*

Vacuum tube solar collector

Unternehmen — Company ELCO Shared Services GmbH — Hohenzollernstraße 31 — 72379 Hechingen — Germany — www.elco.net
Design Weinberg & Ruf — Andreas Weinberg — www.weinberg-ruf.de // Werksdesign/Inhouse — Markus Hödl, Thomas Weidemann

Auf dem Museumsuferfest Frankfurt 2004 erhielten die Museen eine zentrale Bühne: Die Untermainbrücke wurde zur KLANG|PASSAGE. Das mediale Portal spannte eine Spur aus Licht, Video und Sound über die Brücke hinweg und rückte damit das aktuelle Thema der Museen ins Blickfeld: die Musik in der Kunst und die Kunst in der Musik.
At the 2004 Frankfurt Museum Riverbank Festival, the Untermain bridge became a central stage for the museums – a SOUND|PASSAGE. The media portal spun a bright trail of light, video and sound across the bridge, turning the main theme of the museums into the main theme of the festival: Music in art and art in music.

KLANG|PASSAGEN – Museumsuferfest Frankfurt 2004
SOUND|PASSAGES – 2004 Frankfurt Museum Riverbank Festival

Licht-und Medieninszenierung   Light and media installation

Unternehmen — Company Tourismus+Congress GmbH Frankfurt am Main — Kaiserstraße 56 — 60329 Frankfurt am Main — Germany — www.tcf.frankfurt.de
Design Atelier Markgraph GmbH — Roland Lambrette, Andreas Lorenschat, Jürgen Zeller — www.markgraph.de // Meso — www.meso.net // Parviz Mir-Ali — www.sl15.com // Procon MuliMedia AG — www.procon-online.de // XL Video GmbH // Procédés Chénel Werbegesellschaft — www.procedes.de

Der Symbolbau des BMW-Hochhauses München symbolisiert die Kraft des Motors, das Herzstück der Freude am Fahren. Derzeit kernsaniert, verliert er vorübergehend seine innere Funktion. Die Außenhaut wird durch die *Aktion 4Z* zur architektonischen Metapher – der „Vierzylinder" zu einem gigantischen Modell. Die Lichtinstallation simuliert die Verbrennungsphasen, Stockwerke werden im Takt illuminiert, Lichtscheiben bewegen sich. Das Motorherz ist lebendig. The emblematic BMW high-rise in Munich is symbolizing the power of the engine, the very joy of driving. The office block is being overhauled and has temporarily lost its core function. The façade is being turned into an architectural metaphor – and the four cylinders into an oversized model. Precisely controlled light simulates the four-stroke combustion; floors are illuminated in sync with the motor and light discs move. The heart of the engine is alive.

BMW Aktion 4Z – Zylindersimulation
BMW Aktion 4Z – cylinder light simulation

Lichtdesign
Light design

Unternehmen — Company DESEO GmbH, Agentur für Gestaltung — Koppstraße 2/RGB — 81379 München — Germany — www.deseo.de

Design Werksdesign / Inhouse — Manfred Beck, Philipp Stegmüller

Zum Frankfurter Licht-Kultur-Festival „Luminale 2004“ rückte das Forschungsinstitut und Naturmuseum Senckenberg seine Naturschätze in ein neues Licht. Mit Videoinstallationen, Licht und Sounddesign führte die Inszenierung „ILLUMINATION: EVOLUTION“ durch vier Milliarden Jahre Evolution. Das Museum wurde zur Netzhaut der Forschung, auf der sich Fundobjekte längst vergangener Zeiten abbildete. For the Frankfurt light culture festival “Luminale 2004” the Senckenberg Natural History Museum was showcasing its natural history treasures in a new light. Featuring video installations, light and sound design “ILLUMINATION: EVOLUTION” guided visitors through four billion years of evolution. The idea was to transform the Museum into a retina reflecting the research on objects of times long past.

ILLUMINATION: EVOLUTION – Eine Inszenierung zu Licht und Leben
ILLUMINATION: EVOLUTION – A light-and-life-production

Licht- und Medieninszenierung
Light and media spectacle

Unternehmen — Company Forschungsinstitut und Naturmuseum Senckenberg — Senckenberganlage 25 — 60325 Frankfurt am Main — Germany — www.senckenberg.de
Design Atelier Markgraph GmbH — www.markgraph.de // FOUR TO ONE: scale design — Institut für Licht-und Bühnensimulation GmbH — www.4-to-1.com // SHOWTEC Beleuchtung-und Beschallungs GmbH — www.showtec.de // Praviz Mir-Ali — www.sl15.com // 235 Media GmbH — www.235media.com

Die Schlüsselbotschaft „Enhance Peoples's Lives with Lighting“ steht bei diesem Messeauftritt einmal mehr für die Bedeutung von Licht im alltäglichen Leben: „Mehr Leben mit Licht“ gilt es gerade auch im Shopbereich zu beweisen. Die Antwort auf die Frage, wie das perfekte Einkaufserlebnis am besten zu beschreiben ist, ist eine inspirierende Shopping Mall und Einkaufsszenerie, die Shopgestaltung in einem ganz neuen Licht zeigt. Anwendungsorientierte Einkaufswelten mit Licht stehen im Mittelpunkt. So entsteht ein farbenfrohes Patchwork und ein bühnenreifes Kontrastprogramm aus elf Shopping-Spaces. The key message “Enhance People's Lives with Lighting” again stands for the significance of light in everyday life: “the more light, the more life!” The idea is made especially evident in the area of shop lighting. What makes the perfect shopping experience? The answer is an inspiring shopping mall and retail environment, showing shop design in an entirely new light. Application-oriented shopping worlds of light are the centre of attention. The result: a patchwork of eleven shopping spaces rich enough in colour and contrast to grace the stage.

Philips Lighting “Lighthouse Experiences“

Messeauftritt auf der Euroshop 2005, Düsseldorf

Trade fair appearance at the Euroshop 2005, Düsseldorf

Unternehmen — Company Philips Lighting – UB der Philips GmbH — Postfach 104929 — 20034 Hamburg — Germany — www.philips.com

Design D 'ART DESIGN GRUPPE — www.d-art-design.de

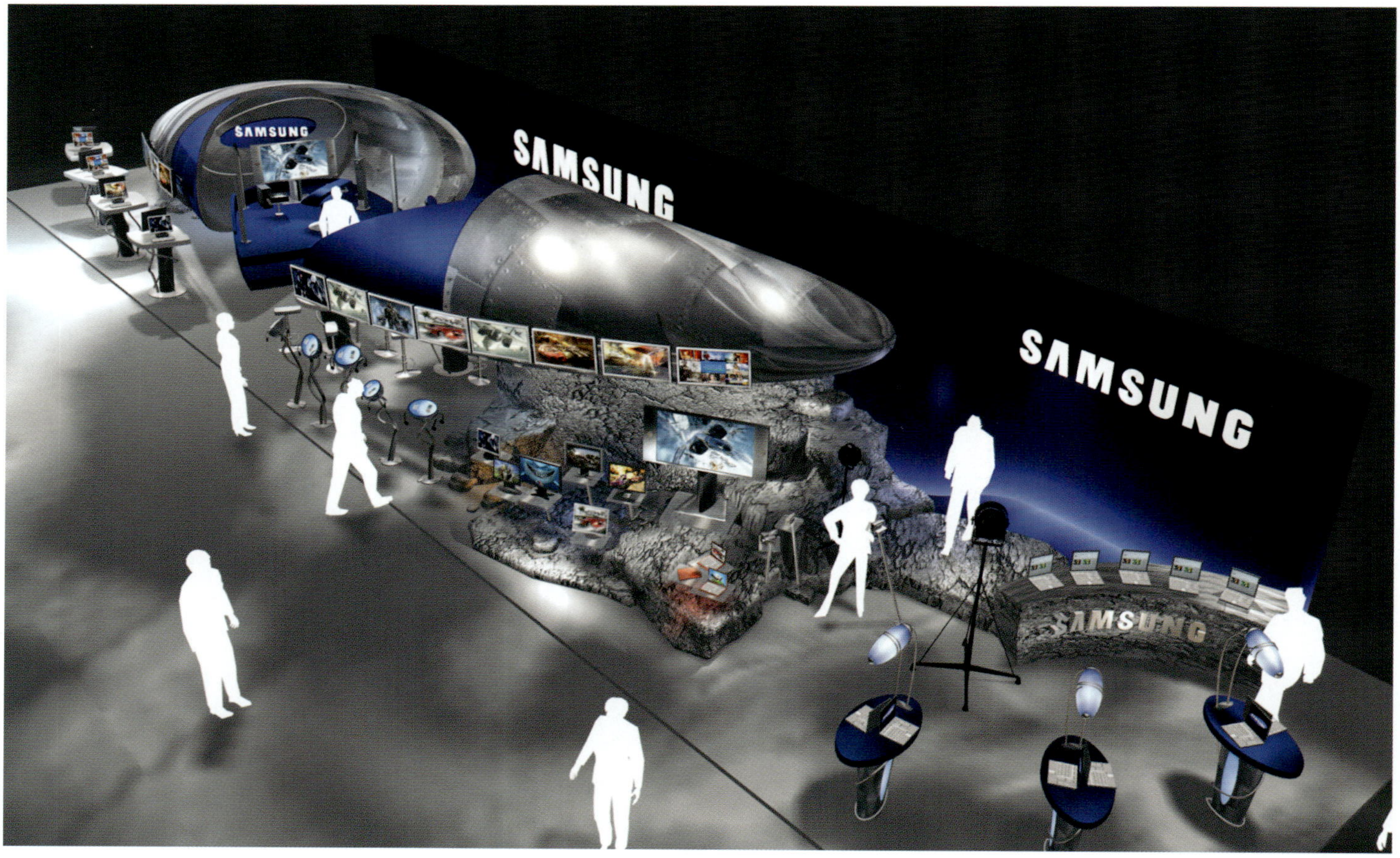

Der *Samsung Pod* wurde als begehbares, dreidimensionales Objekt für einen nachhaltigen Image-Auftritt der Marke *Samsung* im Rahmen der World Cyber Games 2005 auf der CeBIT Hannover entwickelt. Wie ein Flugobjekt aus einem Science-Fiction-Film war der *Pod* der Markenbotschafter Samsungs. Zielgruppengerecht bot er die Plattform, um den Spielern die vielfältigen Anwendungsmöglichkeiten von *Samsung*-Produkten emotional nahe zu bringen. Für viele junge Besucher wurde er zum „ultimate place to be". The *Samsung Pod* was developed as three-dimensional walkable object for a long-lasting image campaign of the *Samsung* brand in the context of the World Cyber Games 2005 during CeBIT in Hannover. Like a science fiction object the *Pod* became the brand ambassador of *Samsung*. In keeping with the target group it provided an opportunity for players to get emotionally acquainted with the wide-ranging applications of *Samsung* products. For many young people the *Pod* became the ultimate place to be.

„Beyond The Games"

Messeauftritt von Samsung Electronics bei den World Cyber Games – Samsung European Championship 2005 während der CeBIT 2006

Trade fair stand of Samsung Electronics at the World Cyber Games – Samsung European Championship 2005 at CeBIT 2006

Unternehmen — Company Samsung Electronics GmbH — Am Kronberger Hang 6 — 65824 Schwalbach/Ts. — Germany — www.samsung.com

Design ideea, Messe-und Dekorationsbau GmbH — www.ideea.de // Werksdesign/Inhouse — Jürgen Schmidt-André

Triad Berlin konzipierte, plante und realisierte den Messeauftritt der *mobilcom GmbH* auf der CeBIT 2005. Der Auftrag beinhaltete sowohl das Design des Messestandes als auch die Entwicklung und Umsetzung der Medienbespielung sowie der Showacts. Der Messestand hatte eine Größe von insgesamt 1.500 m^2. Das Konzept ordnete die drei Ebenen Marke, Produkt und Service in kommunikativer Reihenfolge und vermittelte sie mit unterschiedlichen Medien. *Triad Berlin* has been responsible for concept, planning and realization of the *mobilcom* booth at the CeBit 2005 in Hanover. The job included the booth design but also the development and realization of media use and show acts. The booth had a size of all together 1,500 square metres.

Messestand von mobilcom auf der CeBIT 2005
Fair stand mobilcom CeBIT 2005

Messestand
Fair stand

Unternehmen — Company mobilcom Communicationstechnik GmbH — Hollerstraße 126 — 24782 Budelsdorf — Germany — www.mobilcom.de
Design Triad Berlin Projektgesellschaft mbH — Daniel Strauß, Ulf Eberspächer — www.triad.de

Triad Berlin ist für die Konzeption, Gestaltung und Gesamtplanung der rund 1.500 m² großen *Dr. Oetker Welt* am Stammsitz des Unternehmens in Bielefeld verantwortlich. Es handelt sich um eine Markenausstellung, in der die Entstehung und Entwicklung von Marke und Produkten multimedial erlebbar werden. Der Besucher taucht mit allen Sinnen in eine traumhafte, faszinierende Erlebniswelt ein. *Triad Berlin* has been responsible for concept, design and the whole planning of the *Dr. Oetker World* at the company's headquarters in Bielefeld. The brand world is a brand exhibition. Visitors experience with the help of films and pictures the founding and development of *Dr. Oetker* products in a multimedia-based style. They plunge with all senses into a fascinating and dreamlike brand world.

Dr. August Oetker KG | Markenausstellung 2005: „Dr. Oetker Welt"
Dr. August Oetker KG | Brandworld 2005: "Dr. Oetker World"

Ausstellung
Exhibition

Unternehmen — Company Dr. August Oetker KG — Lutterstraße 14 — 33617 Bielefeld — Germany — www.oetker.de
Design Triad Berlin Projektgesellschaft mbH — www.triad.de

Stückwerk oder *Sample* das ist die Kernaussage. Wer D'ART finden will, muss auf die Suche gehen und eine Entdeckungsreise durch das Labyrinth der Samples starten. Dabei kann er einem Instinkt vertrauen, der den Menschen seit jeher treibt: Jagen und Sammeln. Im Fall der D'ART DESIGN GRUPPE kann der Besucher das Unternehmen en miniature selbst sammeln und mitnehmen. Am Ende des Parcours hält der Besucher dann endlich das „ganze" Unternehmen in den Händen. *Stückwerk* ("patchwork") or *Sample* are the key message. To find D'ART, some searching is required. Visitors are invited on a voyage of discovery through a labyrinth of samples. The experience awakens the instincts of hunting and gathering. Visitors collect treasures along the way, elements of D'ART DESIGN GRUPPE in miniature – at the end of the course the guest at last holds the complete company in his hands.

D'ART DESIGN „Sample" *Messeauftrit auf der Euroshop 2005, Düsseldorf*

Trade fair appearance at the Euroshop 2005, Düsseldorf

Unternehmen — Company D'ART DESIGN GRUPPE — Am Zollhafen 5 — 41460 Neuss — Germany — www.d-art-design.de

Design Werksdesign/Inhouse

Die neue Counter-Architektur für *Europcar* durchbricht die trennende, streng-gerade Linie klassischer Counter. Die Kommunikation erfolgt auf Augenhöhe und jeder Kunde wird exklusiv beraten. Direkter Service und nachvollziehbare Abläufe sind die Folge. Das ebenfalls von f+s entwickelte neue Corporate Design ist der emotionale Hintergrund der hochfunktionalen Servicearchitektur. Der knallgrüne, gelb und schwarz gestreifte Fond, der an eine vorbeirauschende Landschaft erinnert, wird als Keyvisual von *Europcar* in der Airport-Kommunikation genutzt. The design of the new service counters for *Europcar* is a radical departure from the alienating, rigid straight-line design of traditional service counters. Communication takes place at eye level, and each customer receives the benefits of an exclusive advisory situation. This results in increasingly direct service and procedures that are readily understandable. The new brand appearance also developed by f+s represents the emotional background to the highly functional service architecture. The *Europcar* logo banded in bright green, yellow and black, which evokes the image of a landscape rushing past the observer, is used as key visual in *Europcar* communication at the airport.

Europcar Point of Service

Corporate Interior und Branding
Corporate interior and branding

Unternehmen — Company Europcar Autovermietung GmbH — Tangstedter Landstraße 81 — 22415 Hamburg — Germany — www.europcar.de
Design feldmann+schultchen design studios — www.fsdesign.de

In die eher nüchternen Räume der „permanent collection" wurde mit monumentalen Vektorgrafiken, bewegten Linienprints auf Plexiglas und Neonlicht ein „zweiter Raum" eingefügt. In diesem neuen Rahmen zeigte *Die Neue Sammlung* Arbeiten des New Yorker Designers Karim Rashid. Die Raumornamente betonen den stilistischen Wechsel und den Balanceakt zwischen populärem Image und Design. Das Verschwimmen der Grenzen zwischen analog und digital, high und low ist das Thema der temporären „change"-Präsentation. Monumental vector graphics, moving line prints on acrylic glass and neon light were used to insert a "second room" into the essentially sober rooms of the permanent collection. In this new frame, the *Neue Sammlung* presented works by the New York-based designer Karim Rashid. The spatial ornaments emphasise the change in style and the balancing act between popular image and design. The underlying theme of the temporary "change" presentation is the blurring of the borderlines between analog and digital, high and low.

Karim Rashid – Change *Ausstellungsinstallation*

Exhibition installation

Unternehmen — Company Die Neue Sammlung – Staatliches Museum für angewandte Kunst | Design in der Pinakothek der Moderne — Türkenstraße 15 — 80333 München — Germany — www.die-neue-sammlung.de

Design bangertprojects — Dr. Albrecht Bangert — www.bangertverlag.com // Karim Rashid

Mit einem Messestand in eigener Sache sollte der Unternehmens-Schwerpunkt „Entwurf" unter dem Motto „Wir bauen Atmosphäre" präsentiert werden. Und das auf besonders spektakuläre Weise. Ein 250 Jahre alter Olivenbaum im Mittelpunkt des 42 Quadratmeter großen Standes schaffte beim Betrachter vielfältige positive Assoziationen. Kombiniert mit interaktivem Lichtdesign dominierte er einen innovativen Kommunikationsraum. An exhibition stand for own purposes was meant to present the company's focus on sketching and projecting under the catch phrase "we build atmosphere". In a very spectacular way, well understood. An original 250 years old olive-tree located in the centre of the 42 square meters measuring stand creates numerous positive associations in the viewer's eye. Combined with an interactive light-design it dominates an innovative surrounding for profound communication.

Eigendarstellung auf der EuroShop: Ueberholz. GmbH, Büro für temporäre Architektur Self-Presentation at Euroshop 2005: Ueberholz. GmbH, Office for temporary architecture

Messestand
Fair stand

Unternehmen — Company Ueberholz GmbH — Warndtstraße 7 — 42285 Wuppertal — Germany — www.ueberholz.de
Design Werksdesign/Inhouse — Nico Ueberholz, André Füsser

Was macht die Faszination des Motorsports aus? Antworten auf diese Frage finden die Besucher der Wanderausstellung *Faszination Motorsport* von *Mercedes-Benz*. Entstanden ist eine atmosphärisch dichte Inszenierung voller Energie und Emotionen. Herzstück des Auftritts bildet eine bumerangförmige, rasant nach oben strebende Steilkurve. What accounts for the fascination of motorsports? Visitors find answers to this question at the *Mercedes-Benz* exhibition *Fascination Motorsports*. An atmospheric staging was developed, that is full of energy and emotion.

„Faszination Motorsport" *Ausstellung* *Exhibition*

Unternehmen — Company DaimlerChrysler AG — Epplestraße 225 — 70546 Stuttgart — Germany — www.daimlerchrysler.com
Design design hoch drei mit Ludwig architekten — Wolfram Schäffer, Almut Weinecke-Ludwig — www.design-hoch-drei.de

Die Gesamtkonzeption des zur Gedenkstätte transformierten Konzentrationslagers Sachsenhausen hat zum Ziel, den verstellten Blick auf die Wirklichkeit des Ortes wiederzuerlangen. Durch die angestrebte artifizielle Leere des Geländes und den als abstraktes Objekt konzipierten Schutz- und Kontemplationsbau, der sich über den Relikten des ehemaligen Krematoriums der *„Station Z"* erhebt, wird bewusst jede Art von Symbolik und Pathos vermieden. The conception for the transformation of the former concentration camp into Sachsenhausen Memorial pursues the aim of regaining the original view of the former location, which was distorted by various interventions. The building serves as a protective cover and as a contemplative space. It is designed as an abstract object that floats above the relics of the former crematorium of *"Station Z"*. The artificial emptiness of the grounds avoids any kind of symbolism and pathos.

Gedenkstätte Sachsenhausen „Station Z"
Sachsenhausen Memorial "Station Z"

Planung des Gedenkortes der Vernichtungsstation
Planning of the extermination place memorial

Unternehmen — Company Stiftung Brandenburgische Gedenkstätten — Heinrich-Grüber-Platz 3 — 16515 Oranienburg — Germany
Design hg merz architekten museumsgestalter — www.hgmerz.com

constructiv TELVIS I und *constructiv* TELVIS II sind hochvariable Teleskopsysteme, die sich leicht, schnell und werkzeuglos montieren lassen. Mit *constructiv* TELVIS I lassen sich beliebig große Formate von 1,6 bis 4,6 m² in variablen Höhen beidseitig anbringen. *constructiv* TELVIS II kann bis zu vier Präsentationsflächen in den Größen von 375 x 1.900 mm bis 1.554 x 3.700 mm tragen. Zubehör wie Ablagen, Monitorhalterung, Vitrinen sowie Beleuchtung machen das System zu einem vollwertigen Display- und Präsentationssystem. *constructiv* TELVIS I and *constructiv* TELVIS II are highly versatile telescopic systems that can be assembled quickly and easily without tools. With *constructiv* TELVIS I, screens of any format from 1.6 to 4.6 m² can be fastened in varying heights. *constructiv* TELVIS II can carry up to four presentation screens from 375 x 1,900 mm to 1,554 x 3,700 mm. Accessories such as shelves, monitor fastenings, showcases and lighting make the system a complete display and presentation system.

constructiv TELVIS I und II *Präsentations- und Displaysystem*

Presentation and display system

Unternehmen — Company Burkhardt Leitner constructiv GmbH & Co. KG — Breitwiesenstraße 17 — 70565 Stuttgart — Germany — www.burkhardtleitner.de

Design Werksdesign / Inhouse

constructiv PON ist ein Knoten-Stab-System mit magnetischer Verbindung. Extrem leicht und filigran in der Erscheinung und gleichzeitig äußerst stabil, ist *constructiv* PON ein Architektursystem für Messe-, Display- und Shop-in-Shop-Anwendungen. Durch den kleinen Querschnitt des Stabs und seinen intelligenten Verbindungsknoten aus Edelstahl-Feinguss erlaubt *constructiv* PON die werkzeuglose Montage und das Einsetzen aller Füllmaterialien von 5 bis 19 mm Stärke und garantiert damit ebenso kostengünstige wie schnelle Auf- und Abbauten. *constructiv* PON is a system employing connectors and rods that are joined using magnetic push-fit connections. Light and graceful in appearance, *constructiv* PON is nevertheless extremely stable. It is a complete architectural system for use in trade fairs, displays or shop-in-shop systems. Thanks to its slender rods and the clever connectors made of precision-cast stainless steel, *constructiv* PON can be assembled without tools and all kinds of panels from 5 to 19 mm can be installed allowing the system to be set up and taken down quickly and economically.

constructiv PON

Messe-, Ausstellungs-, Display-, Shop- und Office-System
Trade fair-, exhibition-, display-, shop- and office system

Unternehmen — Company Burkhardt Leitner constructiv GmbH & Co. KG — Breitwiesenstraße 17 — 70565 Stuttgart — Germany — www.burkhardtleitner.de
Design Werksdesign / Inhouse

Auf der *Mondial de l'Automobile* 2004 in Paris zeigte *DaimlerChrysler Corporate* sein Engagement auf dem Weg zu einer emissionsfreien Mobilität. Unter dem Leitmotiv *Energie für die Zukunft* konnten die Besucher in einem stilisierten Labor erforschen, wie aus Biomasse Kraftstoff gewonnen wird. Der Stand selbst bestand aus Biomasse. Diese wurde nach Messeschluss in 500 Liter SunDiesel umgewandelt. Aus dem immobilen Messestand entstand Mobilität. At the 2004 *Paris Motor Show, DaimlerChrysler Corporate* demonstrated its commitment to the development of emission-free mobility. Under the motto *Energy for the Future*, a stylised laboratory showed how biomass is converted into fuel. The booth itself was built from biomass and later converted to 500 litres of SunDiesel. From show booth to mobility.

Energy for the Future – DaimlerChrysler-Konzernpräsentation auf der Mondial de l'Automobile Paris, 2004
Energy for the Future – DaimlerChrysler Corporate Presentation at the 2004 Mondial de l'Automobile, Paris 2004

Messeauftritt
Fair stand

Unternehmen — Company DaimlerChrysler AG — Epplestraße 225 — 70546 Stuttgart — Germany — www.daimlerchrysler.com

Design Atelier Markgraph — www.markgraph.de // Kauffmann Theilig & Partner Freie Architekten BDA — www.ktp-architekten.de // Design Hoch Drei GmbH & Co.KG — www.design-hoch-drei.de // TLD Planungsgruppe — www.tldplan.com // Freisteel — www.freisteel.de // Ernst F. Ambrosius & Sohn — www.ambrosius.de

,simple schuf mit einfachsten Mitteln und geringem Budget spannende Räume und Zwischenräume. In Augenhöhe reduzieren sich die Schallplatten allmählich zu schmalen Linien und werden vom räumlichen zum zweidimensionalen, fragmentarischen Erlebnis, das von weitem durch den unrhythmischen Rapport an einen Strichcode oder eine DNA-Kette erinnert. Auf weniger als 50 m² Popkultur in seiner reinsten Form: raffiniert, plakativ, ironisch und gefühlvoll. *,simple* created with the simplest means and only a small budget, eye-catching spaces and interspaces. At eye level, the disks gradually diminish to narrow lines, turning from a spatial into a two-dimensional, fragmentary experience. From a distance, the irregular pattern is reminiscent of a bar code or a DNA-chain. On less than 50 m² pop culture at its best: sophisticated, striking, ironic and soulful.

Plattenküche
Vinyl kitchen

Messestand

Fair stand

Unternehmen — Company Ministry of Sound Recordings — Pfuelstraße 5 — 10997 Berlin — Germany — www.ministryofsound.de

Design ,simple GmbH — Felix Hansen, Tobias Miethke — www.simple.de

Das klare wie einfache CI des Unternehmens sollte im Erscheinungsbild in eine für die nächsten Jahre tragende Gestaltungsvision einfließen. Schwarz und Weiß sind die einzigen Materialfarben, die eingesetzt werden. Realisiert wurde der Effekt eines blendfreien Kubus durch semitransparente Plexiglasflächen an Wänden und Decken. Der Messestand ist nicht nur geschäftlicher Treffpunkt, sondern auch Markenzeichen und Standpunkt. The distinct as well as simple CI of the company should be incorporated in the overall design and depict a vision that will stand for the next few years. Black and white are the only material colours to be used. The effect of a dazzle-free cube was created with semitransparent Plexiglas sections on walls and ceilings. The exhibition booth is not only the social meeting point but also a trademark and viewpoint.

Lonza *Messestand*
Fair stand

Unternehmen — Company Lonza AG — Muenchensteinerstraße 138 — 4002 Basel — Switzerland — www.lonza.com

Design Tribolet arch — Bernardo Tribolet — www.tribolet-arch.ch // Ueberholz GmbH — Nico Ueberholz — www.ueberholz.de

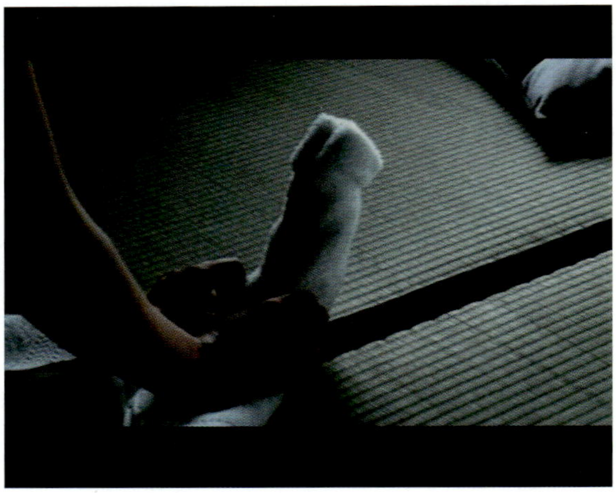

Im Ambiente eines japanischen Teehauses wird ein weißes Handtuch auf der Bastunterlage ausgebreitet. Zu mystischen Klängen falten es Frauenhände origamiartig zunächst in der Mitte, legen sorgfältig die Ecken nach innen und schlagen die Spitze des entstandenen Dreiecks um. Nach weiteren geschickten und gewissenhaften Schritten bäumt sich schließlich langsam ein Frottee-Penis in die Luft. „Die hohe Kunst der Lust" meldet der Abspann mit dem Verweis auf das Erotikunternehmen ORION.

In the ambience of a Japanese teahouse, a white towel is spread out on bast mat. Accompanied by mystical sounds, women's hands fold it like origami, starting in the middle, turning the corners carefully inwards, and fold the tips of the triangle that thus arises over. After a few more skilful and careful steps, a towel-penis gradually rises upwards. "The high art of lust" the trailer concludes, referencing sex-toy company ORION.

Handtuch
Towel

Werbespot
Commercial

Unternehmen — Company ORION Versand GmbH & Co. KG — Schäferweg 14 — 24941 Flensburg — Germany — www.orion.de

Design Heye & Partner — Jan-Florian Ege, Nicole Oberberger, Christiane Seifert — www.heye.de

Die Sendung *Made in Germany*, das Wirtschaftsmagazin auf DW-TV, richtet sich an Entscheider, Geschäftsreisende und an alle Interessierten, die das Wirtschaftsgeschehen hautnah verfolgen wollen – in Deutschland, in Europa und weltweit. Das Magazin *Made in Germany* bringt wöchentlich aktuelle Reportagen aus der Wirtschaft und wird in deutscher und englischer Sprache produziert. *Made in Germany* is DW-TV's weekly business magazine for decision-makers, business travellers and anybody interested in first-hand coverage of economic developments in Germany, Europe and across the globe. *Made in Germany* features in-depth business reports, and is broadcasted in German and English.

Made in Germany *Identität für TV Magazin*

Identity for TV magazine

Unternehmen — Company DW-TV — DEUTSCHE WELLE — Voltastraße 6 — 13355 Berlin — Germany — www.dw-world.de

Design Velvet Mediendesign GmbH — Holger Geisler, Andi List, Torsten Lippstock, Jochen Kraus — www.velvet.de // Werksdesign / Inhouse — Holger Zeh

Merten ist einer der führenden Hersteller von intelligenten elektrotechnischen Steuerungsoberflächen und Systemlösungen. Mit der CD stellt das Unternehmen Elektroinstallateuren ein anschauliches Hilfsmittel zur Beratung von Investoren und Endverbrauchern zur Verfügung. Anhand typischer Situationen in Eigenheim, Büro, Hotel, Krankenhaus und in einer Schule wird vermittelt, wie intelligente *Merten*-Lösungen das Leben und Arbeiten leichter, flexibler, wirtschaftlicher und sicherer machen. *Merten* is one of the leading manufacturers of intelligent electrotechnical control interfaces and system solutions. The CD provides electrical installers with a useful demonstration tool when advising investors and end users. Typical situations in homes, offices, hotels, hospitals and in a school show how intelligent *Merten* technology makes life and work easier, more flexible, more productive and more secure.

Lösungen für intelligente Gebäude
Solutions for intelligent buildings

CD-ROM / Digital media animation

CD-ROM / Digital media animation

Unternehmen — Company Merten GmbH & Co. KG — Fritz-Kotz-Straße 8 — 51674 Wiehl — Germany — www.merten.de

Design Zink & Krämer — Markus Junker, Hendrik Junker — www.zuk.de

Vibracoustic präsentiert sich mit seiner neuen Corporate Website emotional, designorientiert und technikbegeistert. Die Kernkompetenz von *Vibracoustic* „We convert noise and vibration into sound and comfort" wird in den unterschiedlichsten Facetten online erlebbar. Die komplexen, multilingualen Inhalte sowie interaktive, animierte Grafiken, Videos und Sounds werden in Flash verwaltet und über das Content-Management-System auf Basis von PHP / XML komfortabel editiert.

Vibracoustic presents itself on its new corporate website as emotional, design-oriented and technologically inspired. The core competency of *Vibracoustic* "We convert noise and vibration into sound and comfort" can be experienced in a wide variety of forms online. The complex, multilingual contents plus interactive and animated graphics, videos and sound are managed in Flash and conveniently edited using a content management system based on PHP / XML.

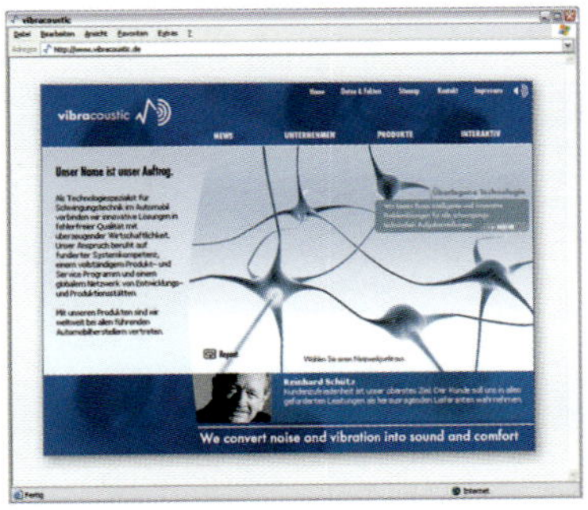

Vibracoustic Corporate Website *Internetseite* *Website*

Unternehmen — Company vibracoustic GmbH & Co. KG — Höhnerweg 2-4 — 69465 Weinheim — Germany — www.vibracoustic.com

Design designverign GmbH — Christoph Scheufeld, Christian Hohn, Stefan Tomski — www.designverign.de

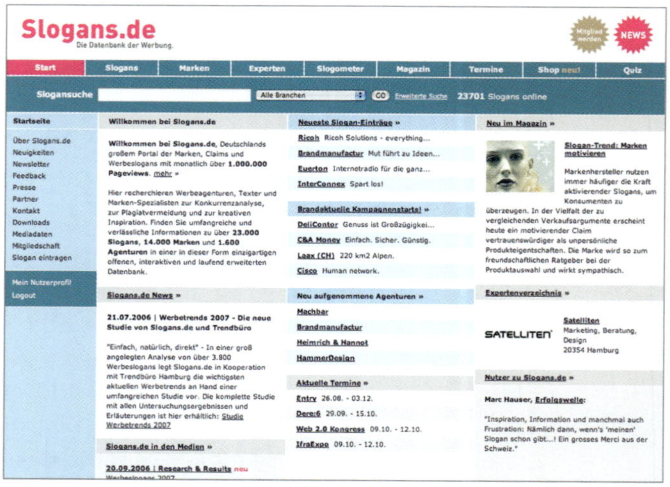

Slogans.de – Die Datenbank der Werbung ist ein offenes, laufend erweitertes Infoportal zur Recherche von Marken und Werbeslogans als Spiegel von Wirtschaft und Gesellschaft. Es richtet sich an Spezialisten aus Werbung, Marketing, Medien und Kommunikation zur Konkurrenzanalyse, Plagiatvermeidung und zur kreativen Inspiration. Die Datenbank bildet eine Brücke zwischen Unternehmen, Agenturen und Verbrauchern. *Slogans.de – the advertising database* is an open information portal. It is used to search for brands and advertising slogans, and serves as a mirror on business and society. Its target audience consists of specialists from advertising, marketing and communication who use it not only to analyse the competition and avoid plagiarism, but also as a source of creative inspiration.

www.slogans.de – Datenbank der Werbeslogans
www.slogans.de – Database of advertising slogans

Internetseite
Website

Unternehmen — Company Satelliten Media Design — Hohe Bleichen 5 — 20354 Hamburg — Germany — www.satelliten.de
Design Werksdesign / Inhouse — Alexander Hahn, Inga Wermuth

DER BLAUE HUMMER steht für innovatives Catering auf gehobenem Niveau. Dementsprechend soll sein Webauftritt sowohl kompetent und professionell, als auch frisch und appetitlich wirken – gemäß dem Claim: Fusion der Sinne. Dieses wird über hochwertige Fotoanimationen erfahrbar gemacht. Klare Gliederungen und eine reduzierte Aufbereitung der Inhalte verschaffen eine schnelle Orientierung. Als zusätzlicher Service werden Anfrage-, LogIn- und Download-Optionen geboten.
DER BLAUE HUMMER offers innovative catering at a high level. The challenge was to give their internet presence a fresh and appetising appeal according to their claim: fusion of the senses. The website showcases the wide range of expertise with premium photo animations. It features a clear structure and a reduced content for quick orientation. Various quoting-, login- and download-options provide additional benefit to the user.

Caterer DER BLAUE HUMMER *Internetseite* *Website*

Unternehmen — Company DER BLAUE HUMMER — Große Elbstraße 212 — 22767 Hamburg — Germany — www.der-blaue-hummer.de

Design weissraum.de(sign) — Lucas Buchholz, Bernd Brink, Irena Celecki, Patric Trautwein — www.weissraum.de

Im April 2006 eröffnete das neue *Mercedes-Benz Museum* in Stuttgart. Die dazugehörige Website steht im Zentrum der Kommunikationsmaßnahmen und macht die Geschichte der Marke *Mercedes-Benz* im Netz erlebbar. Der User bewegt sich virtuell durch die Mythos- und Kollektionsräume und erkundet so die Höhepunkte des Museums. In Anlehnung an das futuristische Gebäude visualisieren 3D-Bänder die Doppelhelix-Architektur. Nur im Netz: „Der virtuelle Raum" mit von Usern generierten Inhalten. The new *Mercedes-Benz Museum* opened in Stuttart in April 2006. The Museum's website is the focus of its communication activies, making it possible to experience the *Mercedes-Benz* brand's history in the Internet. Inspired by the futuristic building, 3D films make the double helix architecture visual. Users discover the Museum's highlights in a virtual tour through the Myths and Collection Halls. Moreover – only available in the Internet – there is a "virtual room" with stories contributed by users.

Mercedes-Benz Museum

Internetseite Mercedes-Benz Museum

Website Mercedes-Benz Museum

Unternehmen — Company DaimlerChrysler AG — Abteilung Brand Communications. HPK 000/G328 — 70546 Stuttgart — Germany — www.mercedes-benz.com/museum

Design Scholz & Volkmer — Heike Brockmann, Katja Rickert, Thorsten Kraus, Rainer Eidemüller, Philippe Just, Carla Meinhard, Peter Reichard, Ron Kellermann — www.s-v.de

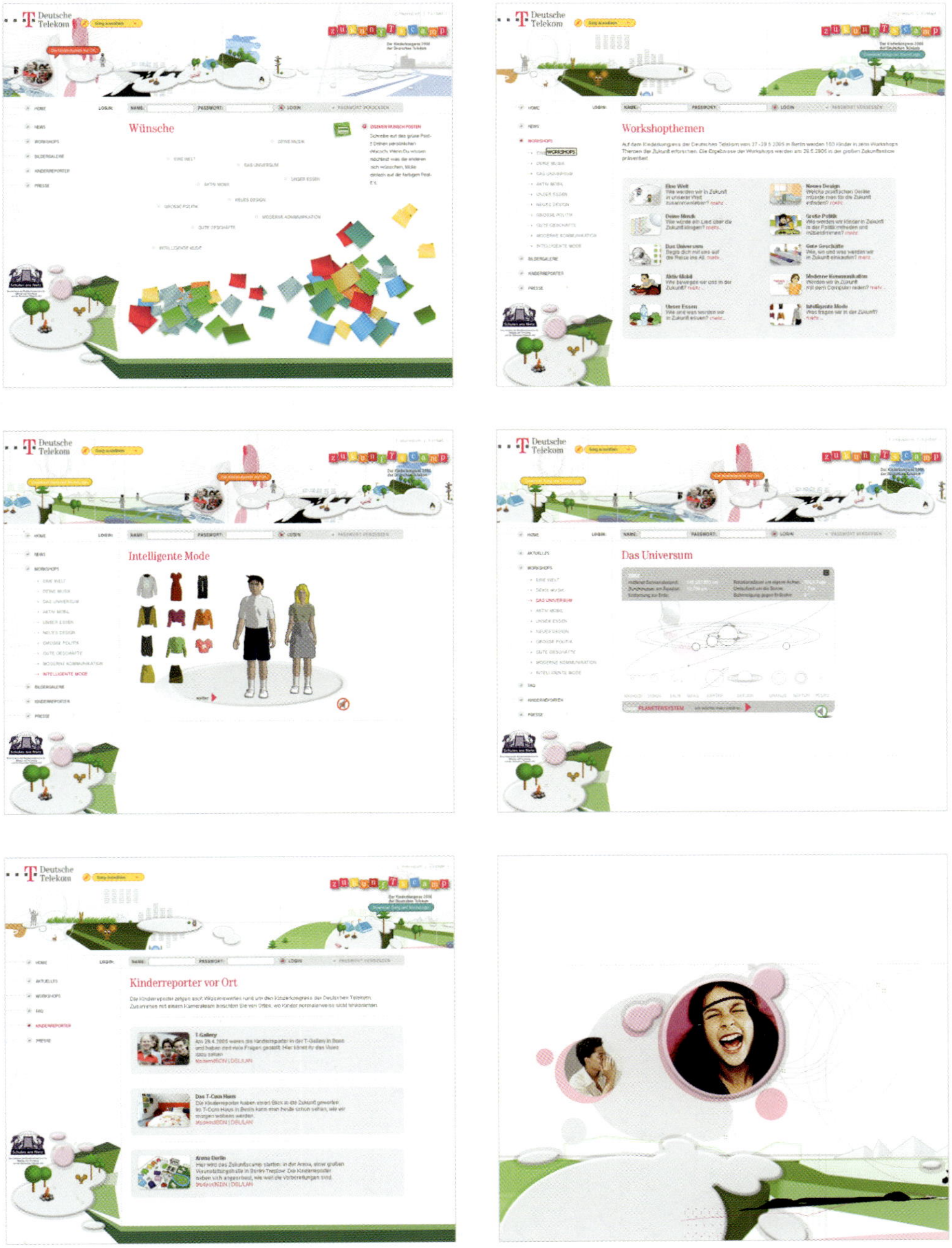

Die Website begleitet als zentrales und zielgruppenadäquates Medium das 2005 initiierte *Zukunftscamp*, ein Kongress der *Deutschen Telekom* für Kinder und Jugendliche. Eine Header-Animation führt die jungen Nutzer zu den Workshop-Themen, die spielerisch aufbereitet sind. Viele innovative Features und Formate wie die Videoreportagen von Kindern für Kinder, Online-Spiele, Live-Streams und virtuelle Post-its animierten zahlreiche Kinder zur Teilnahme. In 2005 the *Deutsche Telekom* initiates the *Future Camp*, a convention for children and teenagers. As the central medium to accompany the event the website is adapted to the young target group. The header-animations lead the young users to the workshop topics that are put into a playful, hands-on context. Many innovative features such as video reportages from kids for kids, online games, live streams and virtual post-its incited many children to participate.

Zukunftscamp http://zukunftscamp.telekom.de/
Future camp *Internetseite*

Website

Unternehmen — Company Deutsche Telekom AG — Friedrich-Ebert-Allee 140 — 53113 Bonn — Germany — www.telekom.de

Design people interactive GmbH — www.people-interactive.de

SINNSALON beschenkte seine Kunden und Dienstleistungspartner zum Jahresende 2004 mit einer Eintrittskarte zu einer ganz besonderen Veranstaltung: dem Jahr 2005 voller Glück. Wert: 365 Glückstage. Mit dem Abtrennen an der Perforation zum angegebenen Einlasstermin aktivierte der „Nutzer" sein persönliches Glück 2005. SINNSALON gave its clients and business partners a ticket for a very special event: a happy new year 2005. Value: 365 lucky days. Ticket holders cashed in on their personal luck for 2005 by tearing off the stub along the perforated line at midnight of New Years eve.

Ticket zum Glück 2005 *Neujahrskarte*

New Years greeting card

Unternehmen — Company SINNSALON — Jarrestraße 20 — 22303 Hamburg — Germany — www.sinnsalon.de
Design Werksdesign/Inhouse

Die Zeitschrift *A&W Architektur & Wohnen* beauftragte *Factor Design* mit der Entwicklungeines Merchandising-Artikels zumThema Design. Das *A&W-Design-Merkspiel* greift das Thema spielerisch auf und vermittelt Wissen über Design-Klassiker auf plakative, attraktive und unterhaltsame Weise. The magazine *A&W Architektur & Wohnen* commissioned *Factor Design* to develop a merchandising article on the subject of design. The design memory game playfully deals with the subject and imparts knowledge about design classics in a striking, attractive, and enjoyable way.

A&W-DESIGN-MERKSPIEL
A&W-Design-Memo-Game

Design-Merkspiel

Design-memo

Unternehmen — Company A&W Architektur & Wohnen — Poßmoorweg 2 — 22301 Hamburg — Germany — www.awmagazin.de

Design Factor Design — www.factordesign.com

Ein variables, einfaches und kostengünstiges Bindesystem zur Präsentation und Archivierung von Bindegut unterschiedlicher Formate. Es besteht aus einem V2A-Stäbchen und einem textilummanteltem Kautschukring. Komplettiert wird das System durch Deckblätter aus handelsüblichen Materialien wie Karton oder Folie, die individuell bedruckt werden können. Eine individuelle und vielfältig einsetzbare Lösung für Produkt- oder Projektpräsentationen, Notizblöcke, Fotoalben oder Speisekarten. A various, low-tech and low-cost binder system for archiving and presentation for all kind of papers. It comprises thin stainless steel rods and fabric covered rubber bands. The system is completed by commercially available covers like transparencies, cardboards or films. The binder system allows many kinds of using: project or product presentations, notebooks, photo albums, menu etc.

keep well 163 *Bindesystem*

Binder system

Unternehmen — Company Ute Soelch — architektur.innenarchitektur.design — Hopfenstraße 2 — 95652 Waldsassen — Germany
Design Werksdesign / Inhouse — Ute Soelch

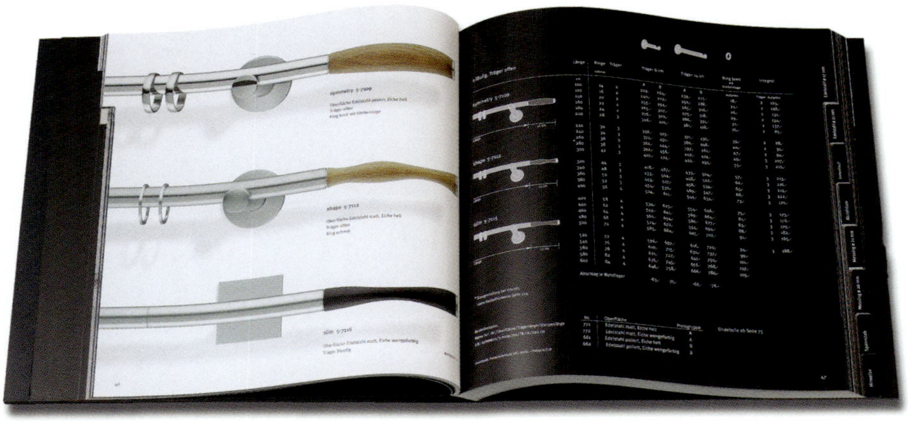

Komplexität und Variabilität des *Metropolitan* werden ebenso unkompliziert wie leicht überschaubar dargestellt. Die Einteilung der Produkte in Programme, Sets und Einzelteile gibt einen logischen Überblick über die zahlreichen Komponenten und ermöglicht das mühelose Zusammenstellen individueller Kombinationen. Anhand von Originalmaterialien, mit großformatigen Abbildungen, exakten technischen Zeichnungen, klaren Produktbeschreibungen und verbindlichen Preisangaben liefert das Buch alle nötigen Informationen rund um das Vorhangstangensystem. This book portrays the complexity and variability of the *Metropolitan* system in a clear, straightforward manner. The book breaks the products down into programmes, sets and individual parts, offering a logical overview of the numerous components and making it easy to put together individual combinations. The book uses original materials, precise technical drawings, concise product descriptions and binding prices to supply all the necessary information on the *Metropolitan* curtain pole system.

Metropolitan *Produkt- und Bestellkatalog*
Product and order catalogue

Unternehmen — Company JAB Josef Anstoetz KG — Potsdamer Straße 160 — 33719 Bielefeld — Germany — www.jab.de
Design Peter Maly — www.peter-maly.de // Carsten Gollnick — www.gollnick-design.de // Werksdesign/Inhouse — Maria Loer-Reuschel

Vorstellung und Dokumentation der bis 2005 im *Büro für Gestaltung Wangler & Abele* konzeptionell und gestalterisch entwickelten Projekte. *gestalten drei* und *gestalten vier* umfasst die Themen Erscheinungsbilder, Gestaltungsrichtlinien, Publikationen, Informations- und Orientierungssysteme, Gestaltungskonzepte Architektur sowie Produktgestaltung. Presentation and documentation of the projects developed in concept and design by *Büro für Gestaltung Wangler & Abele* until 2005. *gestalten drei* and *gestalten vier* are covering the themes corporate identities, design guidelines, publications, information and orientation systems, architecture design concepts and product design.

„gestalten drei" und „gestalten vier" *Büroberichte* *Office reports*

Unternehmen — Company Büro für Gestaltung Wangler & Abele — Hohenzollernstraße 89 — 80796 München — Germany — www.bfgest.de
Design Werksdesign/Inhouse — Ursula Wangler, Frank Abele, Amelie von Büdingen, Kerstin Ullrich

Auf den rechten Sportschuh der Anzeige VITA VM 13 wurde ein Schnürsenkel so aufgeklebt, dass er aus der Zeitschrift ragte und der Anzeige zu maximaler Aufmerksamkeit verhalf. Über diesen Eyecatcher gelangte der Betrachter in eine Perspektive, als stünde er selbst an der Startlinie. Die innovativen Produkteigenschaften des zahntechnischen Materials spiegeln sich in diesen kreativen Elementen wider. In the VITA VM 13 ad, a shoe lace was attached to the right shoe in such a way that it poked out of the magazine helping the advertisement to draw maximum attention. By means of this eye-catcher the observer landed in a perspective as if he himself was standing on the start line. The innovative product properties of materials used in dental technology reflect in these creative elements.

VITA VM 13 – Auf die Gerüste! Fertig! Los!
VITA VM 13 – On your frameworks! Set! Go!

Anzeige
Advertisement

Unternehmen — Company VITA Zahnfabrik H. Rauter GmbH & Co. KG — Spitalgasse 3 — 79713 Bad Säckingen — Germany — www.vita-zahnfabrik.com
Design Studio Laeis, Design & Werbung — www.laeis.com

Die im Bereich Neue Produkte der Süddeutschen Zeitung erscheinende Edition *Klavier Kaiser* beinhaltet vierzehn Audio-Booklets mit 20 CDs. Es handelt sich um die von dem Musikkritiker Joachim Kaiser ausgewählten Lieblingspianisten mit ihren schönsten Einspielungen. Die repräsentative Sammelbox transportiert die Wertigkeit des Produktes. Für die Einzel-CDs wurde ein neues Verpackungsformat entwickelt. Durch ein innovatives Gestaltungskonzept wird ein Höchstmaß an Informationen präsentiert, ohne auf Lesbarkeit und Ästhetik zu verzichten. The newly published edition consists of 14 Audio booklets with some 20 CDs in total. The music critic Joachim Kaiser has carefully assembled a selection of the most beautiful recordings from these popular pianists. A representative Collector's box displays the high quality of the product. For individual CDs a new packaging format has been developed. Through the innovative design concept a maximum of information can be displayed without infringing on legibility and aesthetics.

Klavier Kaiser *Piano Editon*
Audio booklets

Unternehmen — Company Süddeutsche Zeitung GmbH — Sendlinger Straße 8 — 80331 München — Germany — www.sueddeutsche.de
Design Werksdesign/Inhouse — Eberhard Wolf

Die Künstlerin Leni Hoffmann arbeitet ortsbezogen. Ihre skulptural architektonischen Werke verzahnen Raum und Wand, Bild und Boden, greifen auf den Betrachter über und sprengen die Gattungsgrenzen. Die Gestaltung der Publikation thematisiert das Buch als Medium an der Schnittstelle zwischen handwerklicher Tradition und industrieller Machbarkeit. Buchbinderische Handarbeit wird in industrielle Serienfertigung umgesetzt. Gestaltung, die Ästhetik, Funktionalität und künstlerisches Credo eint. The artist Leni Hoffmann works in situ. Her sculptural architecture related works link space and wall, image and ground, seize the viewer and burst any genre. The publication reflects on the book as a medium, on the intersection between artisan tradition and industrial feasibility. Handmade book-binding was translated into industrial serial production. Choreography, graphic design and production were to satisfy documentary purposes as well as highly aesthetic and functional-haptic claims.

leni hoffmann
beautiful one day – perfect the next

Künstlerbuch
Artist book

Unternehmen — Company Leni Hoffmann — Uhlandstraße 23a — 40237 Düsseldorf — Germany
Design Werksdesign / Inhouse — Leni Hoffmann

thema communications – verantwortlich für erfolgreiche Geschäftsberichte – fördert die fachübergreifende Diskussion über die aus Analystensicht wichtigste Publikation des Unternehmens. Die *Stimmen zum Geschäftsbericht* geben den Verantwortlichen in Investor Relations und Unternehmenskommunikation aktuelles Know-how für ihre Arbeit. thema communications responsible for successful Annual Reports promotes overarching discussion about what analysts consider to be a company's most important publication. The *Opinions on the Annual Report* provide those responsible in Investor Relations and Corporate Communications with up-to-date know-how for their work.

Stimmen zum Geschäftsbericht – Der Geschäftsbericht zwischen Image und Information
Opinions on the Annual Report – The Annual Report between Image and Information

Buch
Book

Unternehmen — Company thema communications ag — Hanauer Landstraße 190 — 60314 Frankfurt — Germany — www.thema-communications.de
Design Werksdesign/Inhouse

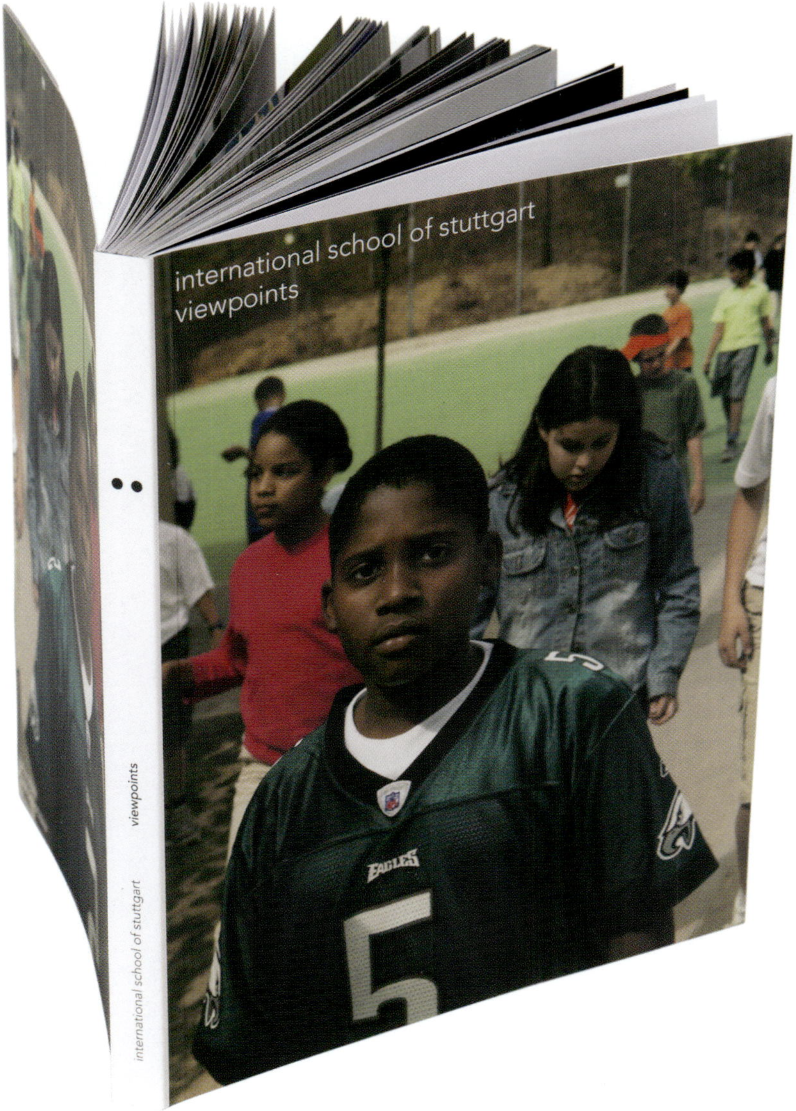

Bilder sind nonverbale Kommunikatoren. Sie erzählen Geschichten, dokumentieren, informieren, entführen und verführen. Die internationale Identität verlangt eine internationale Funktionalität: ein barrierefreies Erleben der Philosophie in wenigen Worten und vielen Bildern. Das Buch dient als Informationsmedium für Eltern und Interessenten der *International School of Stuttgart*. Es kommuniziert die Philosophie der Schule und gestattet einen Einblick in den Alltag. Pictures are non-verbal communicators. They tell stories, document things, serve as information vehicles, allow room for interpretation, transport you to other places and seduce you. An international identity demands an international functionality: a barrier-free access to the philosophy with a modicum of words and pictures. This book serves as an information medium for parents and anyone interested in the ISS. It communicates the school's philosophy and offers a glimpse into its everyday practice.

International School of Stuttgart, „viewpoints" *Buch* *Book*

Unternehmen — Company International School of Stuttgart e.V. — Sigmaringerstraße 257 — 70597 Stuttgart — Germany — www.international-school-stuttgart.de
Design Engenhart Visuelle Kommunikation — Marc Engenhart — www.engenhart.com // Lothar Bertrams Fotograf DGPh — www.lotharbertrams.com

Das Buch dient der Präsentation und Vermarktung des *Park Hyatt*-Projekts in Zürich. In diesem neu erbauten Gebäudekomplex befindet sich das Fünf-Sterne-Hotel *(Park Hyatt Zürich Hotel)*, ein komplett vermieteter Büroflügel sowie eine öffentliche Tiefgarage. Durch Design, Fotografie, Typografie und Verarbeitung werden die Gestaltungsmerkmale und Materialien des Gebäudes aufgegriffen und in ihrer Exklusivität visualisiert. This book serves solely to present and market the *Park Hyatt* complex in Zurich. This new building contains the five-star *Park Hyatt Zurich Hotel*, a fully rented office wing and a public underground parking lot. The individual design, photography, typography and production are inspired by the architecture and materials of the building and underline its exclusive nature.

Park Hyatt Hotel, Zürich

Buch

Book

Unternehmen — Company Jones Lang Lasalle Holdings GmbH & Co. KG — Mies-van-der-Rohe-Straße 6 — 80807 München — Germany — www.joneslanglasalle.de

Design UNID – One Voice Branding — Ralph Missy, Martina Donat, Stanislav Erma — www.unid.de // Werksdesign / Inhouse

Dieses Buch wurde als Dokumentation und zur Unterstützung des Verkaufsprozesses der *Residenzpost*, München, entwickelt. Die Residenzpost bietet die historische Chance, an einem der exklusivsten Immobilienprojekte Deutschlands teilzuhaben. Das Gebäude befindet sich direkt gegenüber der Residenz und der Oper. Es zeichnet sich durch eine markante historische Fassade aus, die den Opernplatz und die Maximilianstraße dominiert. This book is designed to assist the sales process, while serving as a documentation of the *Residenzpost* project, Munich. Occupying the most exclusive site in the historic centre, the *Residenzpost* represents a unique opportunity to participate in the development of one of the country's top locations. It is located opposite the Residenz and the Opera House, and has a prominent historic façade that dominates the Maximilianstraße.

Residenzpost München *Buch* *Book*

Unternehmen — Company Sireo – Real Estate Asset Management — Seizstraße 16 — 80538 München — Germany — www.sireo.de
Design UNID -One Voice Branding — Ralph Missy, Martina Donat, Stanislav Erman — www.unid.de

William Blakes (englischer Lyriker, 1757–1827) Hauptwerk *Die Hochzeit von Himmel und Hölle* in einer luxuriösen Sammleredition mit Buch, Hörbuch-CD und einem Glasobjekt. Das Buch enthält die englischen Originaltexte mit einer zeitgemäßen Übersetzung ins Deutsche. William Blake's (English poet, 1757–1827) magnum opus *The Marriage of Heaven and Hell* in a luxurious collector's edition with book, audio book on CD and a glass object. The book includes both the original English version and a contemporary translation into German.

William Blake, Die Hochzeit von Himmel und Hölle
William Blake, The Marriage of Heaven and Hell

Buchobjekt
Book object

Unternehmen — Company edition GALERIE VEVAIS — Verlagsgesellschaft mbH — An der Dornbuschmühle 7 — 16269 Vevais — Germany — www.minotaurusverlag.de
Design die Typonauten® — Krepinsky Krömer GbR — Stefan Krömer, Ingo Krepinsky — www.typonauten.de

Midnight Ride … oder die Macht herausragender Kommunikation ist eine Broschüre zur Eigenkommunikation. Wie kann Kommunikation beurteilt werden – und durch welche spezifischen Qualitätsmerkmale zeichnet sich Kommunikationsdesign, das Maßstäbe setzt, aus? Auf diese Fragen gibt die Broschüre *Midnight Ride* Antworten, am Beispiel einer historischen Begebenheit, die auf den amerikanischen Freiheitskämpfer Paul Revere Bezug nimmt. *Midnight Ride* … or the power of outstanding communication is a house advertising brochure. How can communication be assessed and what are the intrinsic quality criteria of standard setting communication design? The brochure *Midnight Ride* answers these questions by the example of a historic event, which refers to the American freedom fighter Paul Revere.

Midnight Ride ... oder die Macht herausragender Kommunikation
Midnight Ride ... or the power of outstanding communication

Broschüre zur Eigenkommunikation

A house advertising brochure

Unternehmen — Company BRANDIT — Marketing Kommunikation — Mittelstraße 12-14 — 50672 Köln — Germany — www.brandit.de

Design Werksdesign / Inhouse — Peter Specht, Angela Strecker

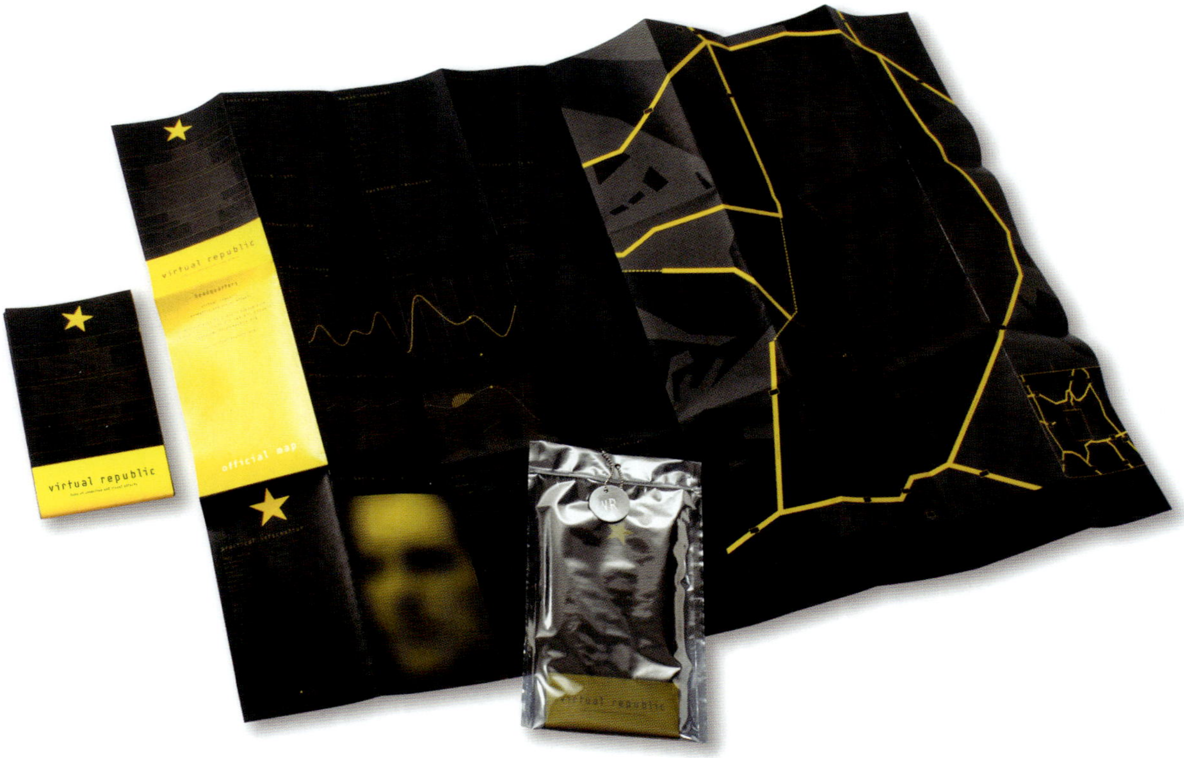

Ein Unternehmen für Computeranimation und visuelle Effekte gründet die *virtual republic* mit merkfähigem CI und CD. Das Geschäftspapier ist staatstragend, die Visitenkarten ähneln Ausweisen, die eigens komponierte Hymne schafft akustische Erkennbarkeit, die Homepage ist präsidial. Allen Einreisewilligen präsentiert sich die Republik detailliert in Landkartenform. Mit einem Tonband-Mailing werden Partner und potenzielle Kunden zu den Eröffnungsfeierlichkeiten geladen. Als Dank für ausgezeichnete bilaterale Geschäftsbeziehungen versendet die „vr" Hissflaggen. With a catchy CI and CD a company for computer animation and visual effects sets up *virtual republic.* Its business paper highlights the republic's status, its visiting cards look like identity cards, the self-composed anthem serves as an acoustic logo, and its homepage is presidential. For those who want to enter the republic, detailed maps are available. By an acoustic invitation mailing, partners and potential customers will be invited to the opening ceremony. In recognition of excellent bilateral business relations, the "vr" sends out flags.

Virtual Republic – Staatsgründung
Virtual Republic – foundation

Unternehmensbroschüre Mailings, Internetseite
Corporate folder, mailings, website

Unternehmen — Company Gute Gesellschaft mbH — Grafenberger Allee 126 — 40237 Düsseldorf — Germany — www.gutegesellschaft.com
Design Werksdesign / Inhouse — Raimund Schmelzer, Ruben Cuellas Is

Die *Imagebroschüre* vermittelt das bestehende Angebot des Buchbinders und spricht bestehende sowie zukünftige Kunden an. Ein Bestandteil der Broschüre ist die Philosophie, in welcher der Buchbinder seine Arbeitseinstellung wiedergibt und einen Einblick in seine Arbeitsweise gewährt. Die äußere Darstellung durch Rillung, Falzung und Bindung zeigt die Vielfältigkeit des Buchbinderhandwerks und dient somit gleichzeitig als Musterbuch. Das Augenmerk wird auf die Bilder gelenkt. The *image brochure* is supposed to show the existent range of products and to address already existing and future customers. One element of the brochure is the philosophy in which the bookbinder gives an account of his attitude towards his work and provides an insight into his operation. The outer portrayal by creasing, folding and binding shows the manifoldness of the craft of bookbinding. At the same time it serves as a sample book. The attention is directed towards the images.

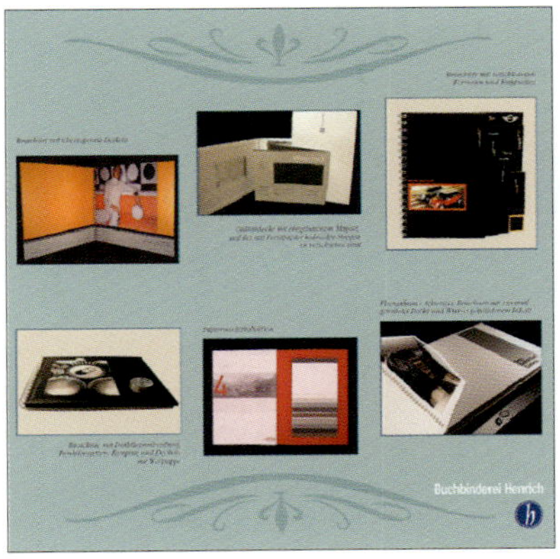

Imagebroschüre *Imagebroschüre*

Image brochure

Unternehmen — Company Raymund Henrich — Ammerthal 3a — 85622 Weißenfeld — Germany — www.buchbinderei-henrich.de

Design Fa-Ro Marketing GmbH — Sabrina Ruchti — www.fa-ro.de // Werksdesign / Inhouse — Raymund Henrich

„Das hat Format. Erstens eine Neupräsentation von Erwerbungen und Sammlungsbestand mit dem Ausstellungstitel *What's New, Pussycat?* zu überschreiben, nach dem Gassenhauer des Waliser Damenunterwäschesammlers Tom Jones. Zweitens dazu passend ein Poster zu benutzen, auf dem eine Art Bond-Gespielin im schwarzen Catsuit gelangweilt herumsitzt und eine mausetote weiße Stoffkatze hochhält, gleich so, als habe sie ihr – der Kunst natürlich – gerade den ohnehin schon strohgefüllten Hals umgedreht." SZ *(Süddeutsche Zeitung)* "Now, there's class. First: headline a presentation of acquisitions and well known items of the collection by the title *What's New, Pussycat?* and thus evoke the tune of Tom Jones, a notorious Welsh collector of ladies' underwear. Second: use an appropriate subject for the poster which depicts a kind of Bond-Girl looking bored while hanging out in her black catsuit and holding up a stone dead toy animal, as if she had just broken its – and art's – neck that was stuffed merely with straw anyway." SZ *(Süddeutsche Zeitung)*

„What's New, Pussycat?" *Erscheinungsbild für eine Ausstellung zeitgenössischer Kunst*
Design for an exhibition of contemporary art

Unternehmen — Company MMK Museum für Moderne Kunst — Domstraße 10 — 60311 Frankfurt am Main — Germany — www.mmk-frankfurt.de
Design Wolff Kommunikation — Carsten Wolff, Thomas Rott, Christiane Wolff, Wim Schafspelz — www.wolff-kommunikation.de

Dr. Ernest de Ruiter, Senior Scientist

4

Hasso, Paul and Philip talk to Ernest

Hasso: Ernest, where were you born?
Ernest: In the sleeping compartment of a train just before it arrived in Vienna. I was known in the family as the "kid on wheels" for many years!
Hasso: An unusual start and afterwards an even more unusual school career.
Ernest: Yes, I lived with my parents in Vienna for my first 13 years and then we moved to Brussels. I was able to use the confusion of the different school systems to my advantage and graduated from school at the tender age of 15. I had my Doctorate in Chemistry at 22.
Philip: Were you a child prodigy?
Ernest: No, I just had an incredibly good memory – an eidetic memory. If I read a page very slowly, I was able to reproduce the page in my mind. The combination of this along with the ability to structure information logically was ideal for a successful chemistry degree.
Paul: So, you studied during the war in Belgium?
Ernest: Well, that's a bit complicated. One of the conditions of enrolling at Löwen University was declaring loyalty to the occupying authorities. I refused and consequently, I was not allowed to enroll. So, from that time on, up until Brussels was liberated by the Allies in 1944, I lived a semi-secluded life in the city without valid papers. However, during that time I read a great deal, had many discussions, tinkered around and from time to time went to the Löwen University by bicycle and managed to complete some courses in the laboratories.
During September to December 1944 I took all the examinations I had missed from the last 3 years and I was soon able to start working on my Doctorate.
Hasso: You stayed in Belgium for 33 years altogether and you have been in Germany for 33 years since then. Do you feel a stronger affiliation with one nationality than the other?
Ernest: Do you know Grillparzer? He was a remarkable Austrian poet in the 19th century. He once wrote: "from humanity, over nationality, to Bestiality" Modern age has proved him to be correct. I have a Dutch passport because my father was in the Dutch diplomatic service; my mother was Viennese, therefore an Austrian. I am a European and therefore in the fortunate position of being able to pick out the best qualities of each nation. I like the sense of duty and orderliness of the Germans – which makes living in Germany pleasant. I like the ability of the Belgians to find a way out of every situation and I admire the "Savoir vivre" of the French, their art of engineering and their clear and logical way of thinking.
Philip: Which languages do you speak?
Ernest: French and German mother tongue, English and Dutch very well and a bit of Spanish – which I am currently trying to improve through a language course.
Paul: You work 2 days per week; what do you do with the rest of your time?
Ernest: Sadly, I can read very little because my eyesight is not so good, but in the meantime I have built up a very good library of audible books. I work in my garden, go for nice long walks or visit my new lady friend who lives a good 40 kilometres away. I also have a rather wonderful hobby: I have a model railway with the most fabulous trains – they run through 3 rooms.
Philip: You enjoy good food – do you have a favourite dish?
Ernest: It's true, I do eat a lot and I do eat well – one of my favourite culinary memories is of a restaurant in Brussels

5

Mit Bedauern geben wir bekannt, dass der Leitende Wissenschaftler Dr. Ernest de Ruiter anlässlich seines 80. Geburtstages im Jahr 2005 in den Ruhestand getreten ist. Wir danken Herrn Dr. de Ruiter für seine sehr wertvolle Arbeit der letzten 24 Jahre und freuen uns auf eine lang anhaltende zukünftige Zusammenarbeit mit ihm in seiner neuen Rolle als Berater für besondere Projektentwicklungen. We regret to announce the early retirement of Dr. Ernest de Ruiter, Senior Scientist, on the occasion of his 80th birthday in 2005. We thank Dr. de Ruiter for his valuable work over the last 24 years and look forward to a long lasting cooperation in the future with him in his new role as consultant for special developemt projects.

Let's do it! Let's go! *Broschüre*

Brochure

Unternehmen — Company Hasso von Blücher — Mettmanner Straße 25 — 40699 Erkrath — Germany — www.bluecher.com

Design Hesse Design GmbH — Klaus Hesse, Sandra Klemt, René Clohse — www.hesse-design.com

Die *agenda edition* „Strom + Mensch“ besteht aus zwei Teilen, die in zwei Blöcken als Schweizer Broschur gegeneinander versetzt gebunden sind. Ein literarisch-journalistisches Textbuch, rein grafisch gestaltet, und ein Fotobuch von Meisterklassen verschiedener Kunsthochschulen. In der Dunkelheit leuchtet durch die fluoreszierende Beschichtung die negativ weiße, kaum wahrnehmbare Typo schwarz vor einer grünlichen Oberfläche. The *agenda edition* entitled “Power + People” is made up of two parts: a literary-journalistic textbook with no illustrations and purely graphic design, and a photo book. The two books were bound back-to-back so that agenda edition “Power + People” can be opened from either end. The fluorescent pale-green colour coating of the book cover unfolds its full impact in the dark, where it glows brightly.

agenda edition *Unternehmenspublikation*

Corporate book

Unternehmen — Company RWE AG — Opernplatz 1 — 42128 Essen — Germany
Design HOFFMANN UND CAMPE Corporate Publishing — Dirk Linke, Sabine Keller, Jessica Winter — www.hoca-cp.de

results ist das Corporate and Investment Bank Magazin der *Deutschen Bank AG.* *results* setzt die umfangreiche wirtschaftliche Kompetenz des Geldinstituts in Magazinform um und macht das Expertenwissen für den unternehmerischen Alltag nutzbar. Ein klarer Heftaufbau, ein großzügiges Layout und eine eigenständige Bildsprache begleiten die Beiträge und unterstreichen den hohen journalistischen Anspruch. *results* is the corporate and investment bank magazine of *Deutsche Bank AG.* *results* translates the financial institution's wide-ranging business expertise into a magazine format and conveys specialist knowledge for use in day-to-day business. A clear magazine structure, generous layout and independent visuals complement the contributions and underline the magazine's high standards of journalism.

results *Unternehmenspublikation*

B-to-B Magazine

Unternehmen — Company Deutsche Bank AG — Taunusanlage 12 — 60325 Frankfurt am Main — Germany — www.deutsche-bank.de

Design HOFFMANN UND CAMPE Corporate Publishing — Wolf Dammann — www.hoca-cp.de

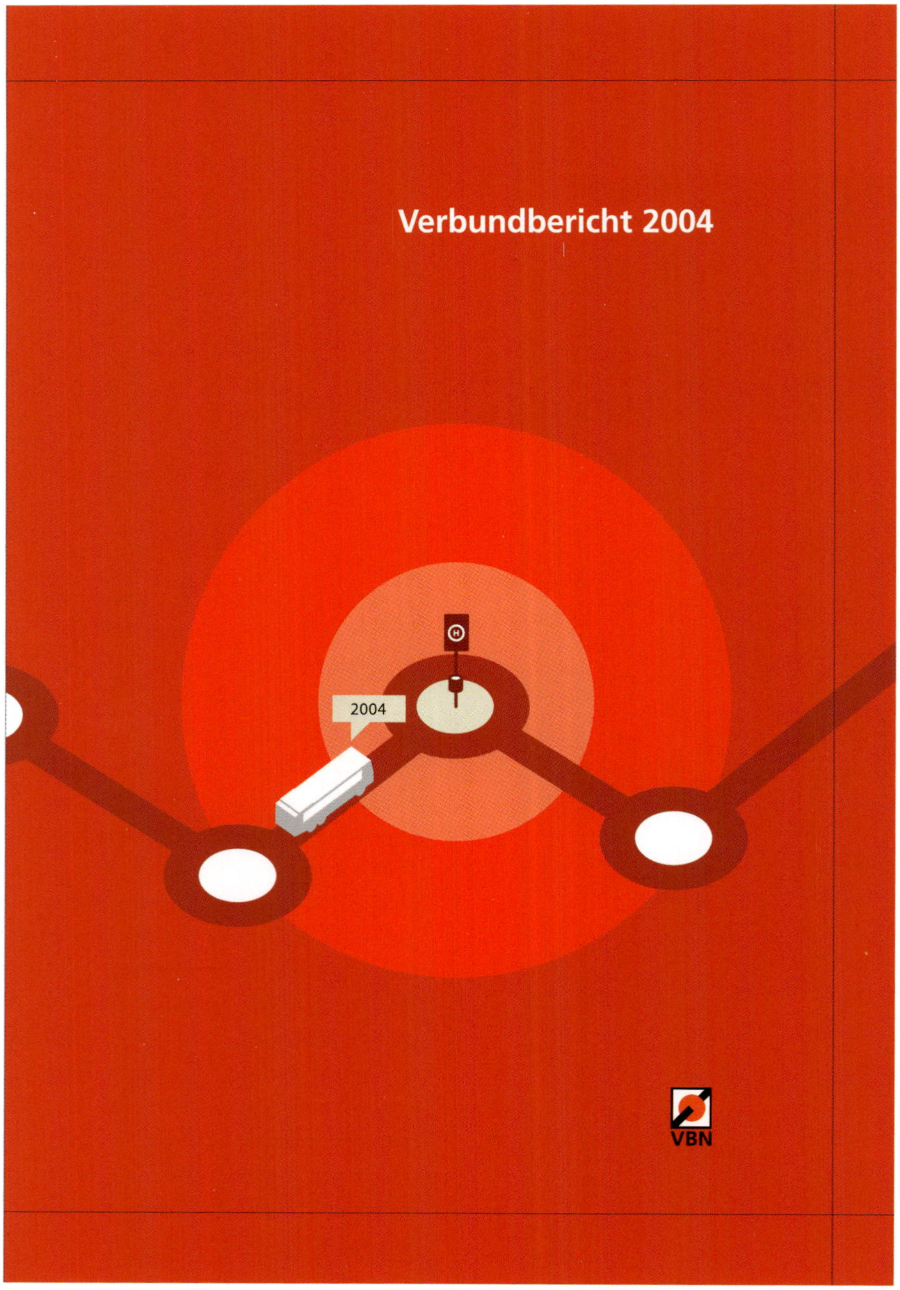

Aufgabe war der Rückblick auf die Schwerpunktthemen des Geschäftsjahrs 2004 mit dem Ziel, den *Verkehrsverbund* als kompetenten Ansprechpartner für öffentlichen Personennahverkehr (ÖPNV) im Großraum Bremen / Oldenburg darzustellen. In diesem Sinne übernimmt der *Verbundbericht* auch die Funktion einer Imagebroschüre. Gleichzeitig sollen bei der Zielgruppe Interesse an der aktuell anstehenden Weiterentwicklung des Verkehrsverbundes geweckt und erste konkrete Aussagen zur Ausgestaltung der neuen Organisation getroffen werden. The task was to take a look back at the key issues of the 2004 financial year, the aim being to present the integrated public transport company as a competent organisation to contact with regard to local public transport in the Bremen / Oldenburg region. In this sense the *Annual Report* also assumes the role of an image brochure. At the same time the aim was also to arouse interest in the upcoming further development of the transport company among people in the target group and make initial definitive statements about the design of the new organisation.

VBN Verbundbericht 2004
VBN Annual Report 2004

Geschäftsbericht
Annual report

Unternehmen — Company Verkehrsverbund Bremen / Niedersachsen (VBN) — Willy-Brandt-Platz 7 — 28357 Bremen — Germany — www.vbn.de
Design construktiv GmbH — Miriam Edelmann, Kai Tietjen, Svenja Miller, Stefan Dziallas — www.construktiv.de

In Zusammenarbeit mit dem Architekten und Designer Adalberto Dal Lago wurden 131 Farben definiert und in die neue Kollektion aufgenommen. Die Verbindung zwischen Tradition und Moderne sowie der Bezug zur Materialität bestimmen die Produkt- und Farbauswahl. Symbolisches Leitmotiv der Kollektion ist das bekannte Linoleum-Motiv „Kreis und Quadrat" von Peter Behrens aus dem Jahr 1905 in einer farblichen Verfremdung. Das Musterbuch *Armstrong DLW Linoleum – World of Colours* zeigt alle 131 Farben strukturübergreifend nach Farbverlauf. Together with renowned architect and designer Adalberto Dal Lago, we defined 131 colours for use in the new Collection. The choice of products and colours centres on blending traditional and modern ideas while referencing the materials' qualities. The Collection takes its symbolic cue from the famous Linoleum motif Peter Behrens devised in 1905, namely that of "Circle and Square", given a coloured twist here. *The Armstrong DLW Linoleum-World of Colours* specimen book presents all colours and the respective textures, by colour.

Armstrong DLW Linoleum – World of Colour Collection 2005

Musterbuch
Specimen book

Unternehmen — Company Armstrong DLW AG — Stuttgarter Straße 75 — 74321 Bietigheim-Bissingen — Germany — www.armstrong.eu

Design Dal Lago Associati — Adalberto Dal Lago, Laura Cesari — www.dallagoassociati.com // Werksdesign/Inhouse — Markus Deimling, Anneliese Müller, Emmi Mieskes, Helen Blank

Das *Magazin m&it* ist als Instrument der Kundenansprache eine eigene Marke von *sd&m* geworden. Die Beiträge über Software und Management blicken über den Tellerrand des eigenen Unternehmens hinaus. Wechselnde Themenschwerpunkte, die den Softwareingenieuren und IT-Beratern bei ihrer Tätigkeit, aber auch im Kundendialog begegnen, runden die *m&it* ab. Inhaltliche Qualität gepaart mit anspruchsvollem Layout sollen die Leser für das Magazin interessieren. Designed as an instrument for customer communications, the *magazine m&it* has become part of the *sd&m* brand. In the articles on software and management, the magazine investigates the world beyond the own four walls. A range of changing focal topics, which the software engineers and IT consultants are faced with in their work as and when speaking with their customers round off *m&it.* High-value content coupled with an appealing design are intended to generate reader interest in the magazine.

Magazin m&it *Magazin zur Kundenansprache*

Magazine for customers

Unternehmen — Company sd&m AG — software design & management — Carl-Wery-Straße 42 — 81739 München — Germany — www.sdm.de
Design Horst Metzner GmbH — Horst Metzner, Thomas Scheuplein — www.horstmetzner.de //
Werksdesign/Inhouse — Edmund Küpper, Rüdiger Azone, Dr. Uwe Dunslaff

Das Magazin des Baudienstleisters HOCHTIEF stellt die vielfältigen und komplexen Projekte in den Bereichen Entwicklung, Bau, Dienstleistung, Konzessionen und Betrieb vor. Faszinierende Reportagen und hochwertige Wirtschaftsberichte nehmen den Leser mit auf eine Reise in das globale Netzwerk eines internationalen Konzerns mit seinen Projekten, Gesichtern und Geschichten. Auch die klare und moderne Optik von *concepts* spiegelt den Qualitätsanspruch von HOCHTIEF wider.
The magazine of the HOCHTIEF construction group presents its varied and complex projects in the fields of development, construction, services, franchising and operations. Fascinating reportage and high-quality business reports take the reader on a journey into the global network of an international group with its manifold projects, faces and stories. The clear and modern visuals of *concepts* magazine reflect the world of HOCHTIEF and its commitment to quality.

concepts by HOCHTIEF *Unternehmenspublikation B-to-B Magazine*

Unternehmen — Company HOCHTIEF Aktiengesellschaft — Opernplatz 2 — 45128 Essen — Germany — www.hochtief.de
Design HOFFMANN UND CAMPE Corporate Publishing — Annette Arnheim, Dirk Linke — www.hoca-cp.de

Klein anfangen und Großes erreichen – lautet das Motto des 102seitigen *Greenpeace Magazins.* Bei den 57 Tipps für eine bessere Welt erscheint selbst Schweres leicht. Der „Aktionsfaktor" von 1 bis 5 lässt Laien und Profis auf einen Blick erkennen, wie viel Engagement ein Tipp erfordert. 15 Textseiten liefern Hintergrundinformationen, weiterführende Kontakte und Internet-Links. Start small and have a big impact – the slogan of the *Greenpeace magazine* reads. The 57 tips for a better world on 102 pages make even more serious topics appear in a light fashion. The "action factor" (from 1 to 5) shows at a glance how much commitment is necessary for each tip. 15 additional text pages offer background information as well as helpful contacts and internet links.

„Tu was!" Greenpeace Magazin 01/06

Magazin
Magazine

Unternehmen — Company Greenpeace Media GmbH — Große Elbstraße 145d — 22767 Hamburg — Germany — www.greenpeace-magazin.de
Design Büro Hamburg JK. PW. Gesellschaft für Kommunikationsdesign mbH — Jürgen Kaffer, Kerstin Leesch, Bettina Rosenow, Annette Gassner — www.buero-hamburg.de

Dieses Poster soll für eine Ausstellung auf der *Stockholmer Möbelmesse* werben. In einem Teil der Ausstellung werden von schwedischen Designern entworfene Produkte gezeigt, die im Ausland produziert werden; im anderen Teil von ausländischen Designern entworfene Produkte, die von schwedischen Firmen produziert werden. Auf diese Weise wird die Frage nach der Bedeutung oder Bedeutungslosigkeit des Konzepts von Nationalität im Design reflektiert. Dieses Siebdruckplakat arbeitet mit einfachsten Mitteln: drei Farben, Schwarz, Gelb und Silber, sowie einer Schriftart, der *Berthold Akzidenz Grotesk*. Die auf dem Kopf stehende Typographie lässt das Thema Möbel und Design assoziieren und stellt zugleich die Frage nach Identität und Nationalität im Design.

This poster is designed for promoting an exhibition at *The Stockholm Furniture Fair*. One part of the exhibition shows products designed by Swedish designers but produced by foreign producers. The other part shows products designed by foreign designers and produced by Swedish companies. In this way reflecting upon the importance or the unimportance of the nationality in design. This screen printed poster works only with the simplest things, three colours, black, yellow and silver and one typeface; the *Berthold Akzidenz Grotesk*. The upside-down typography suggests the association to the subject furniture and design but at the same time asking a question about identity, nationality in design.

„Made in Sweden" *Plakat*

Poster

Unternehmen — Company Svensk Form — Holmamiralens väg 2 — 11149 Stockholm — Sweden

Design Gabor Palotai Design — Gabor Palotai — Sweden — www.gaborpalotai.com

Die niederländische Werbewelt ist müde. Die VEA will diese mentale Trägheit durchbrechen. Widerstand gegen Veränderungen umwandeln in die treibende Kraft der Erneuerung. Dazu sollen Werbeleute zwischen 28 und 35 Jahren wach gerüttelt werden: Mit explosiven Sessions unter dem Titel *Re:set.* Poster in knalligen, explosiven Farben machen auf einen Blick klar, worum es geht: Wenn Sie sich selbst „resetten“ oder wach rütteln wollen, dann besuchen Sie die Sessions der VEA. Dutch advertising has drifted off to sleep. The VEA wants to break through this mental laziness and convert resistance to change into a drive towards innovation. This will be achieved by stimulating advertising men and women between the ages of 28 and 35 in explosive sessions under the title *Re:set.* Brightly coloured, dazzling posters make everything instantly clear: if you want to “reset” or reinvent yourself, then come to the VEA sessions.

VEA RE:set Reclame *Poster und Faltblätter*

Posters and leaflets

Unternehmen — Company PPGHJWT Artists — Rietlandpark 301 — 1019 DW Amsterdam — Netherlands — www.ppghjwt-artists.nl
Design Werksdesign/Inhouse — John de Vries

In der Kampagne VITA *Zähne* wurden künstliche Zähne zum Leben erweckt, indem sie humorvolle Aussagen treffen, die doppeldeutig auf ihre Produktvorteile übertragbar sind. Die Zähne wurden schräg auf eine Zahngarnitur gestellt oder bewegten sich sogar aus ihr heraus. Dies erregte bei Zahntechnikern hohe Aufmerksamkeit, da sie aufgrund ihrer perfektionistischen Arbeitsweise geneigt sind, die Zähne auf einer Garnitur gerade anzuordnen. In the VITA *Teeth campaign*, artificial teeth were brought to life by offering humorous statements that can be ambiguously applied to their product advantages. The teeth were placed on a tooth set at an angle or moved within such a set. This caught the attention of the dental technicians, as given their perfectionist approach to their work always place the teeth in a clean line in the set.

VITA Zähne
VITA Teeth

Anzeigenkampagne

Advertising campaign

Unternehmen — Company VITA Zahnfabrik H. Rauter GmbH & Co. KG — Spitalgasse 3 — 79713 Bad Säckingen — Germany — www.vita-zahnfabrik.com

Design Studio Laeis, Design & Werbung — www.laeis.com

Um von seinen Kunden im B2B-Bereich besser wahrgenommen zu werden, entschloss sich der Hamburger Musikverlag *Selected Sound* (ein Tochterunternehmen der EMI *Music Publishing*) zu einem neuen, zeitgemäßen Corporate Design. Ein neues Logo, eine auffällige Farbgebung und moderne Bildwelten helfen seit 2005 dem Musikarchiv, sich international erfolgreich von seinen Mitbewerbern abzugrenzen und als Kompetenz im Produktionsmusik-Markt zu profilieren. To attract more attention from its customers in the B2B field, the Hamburg music publisher *Selected Sound* (a subsidiary of EMI *Music Publishing*), chose a new, more contemporary Corporate Design. Since 2005 and with global success, a new logo, eye-catching colours and modern images have helped the music archive to stand out from its competitors and profile its expertise in the production music market.

Selected Sound

Corporate Design

Corporate Design

Unternehmen — Company SELECTED SOUND MUSIKVERLAG GMBH & Co. KG — Alsterufer 1 — 20354 Hamburg — Germany — www.selected-sound.de
Design form one - visuelle kommunikation — Niels Bünemann — www.form-one.de

Porsche Design ist fundamental, nicht dekorativ. „Engineered", nicht einfach nur perfekt gestaltet. Die Funktion aller Produkte diktiert die Form. Die Image-Kampagne übersetzt die einzigartige Engineering-Kompetenz der Marke in großen, ungesehenen Bildwelten, spektakulären Landschaften, urbanen Szenarien. Die Markendimensionen *passion, precision, purism, performance* werden symbolisch reflektiert. Die Produkt-Kampagne visualisiert einzelne Produkt-Benefits. *Porsche Design* is truly unique: Fundamental, not decorative. Engineered, not just perfectly designed. Each object is reduced to its essential function. The image campaign translates the brand's engineering competence in unseen picture worlds, spectacular landscapes, urban sceneries. The brand's dimensions *passion, precision, purism, performance* are reflected symbolically. The product campaign visualise specific product benefits.

Porsche Design Corporate Design *Kampagne, Broschüren*

Campaign, brochures

Unternehmen — Company Porsche Lizenz-und Handelsgesellschaft mbH & Co. KG — Porschestraße 1 — 74321 Bietigheim-Bissingen — Germany — www.porsche-design.com

Design KW43 BRANDDESIGN — Margit Tabert — www.kw43.de // KMS Team GmbH — Knut Maierhofer, Patrick Märki — www.kms-team.de

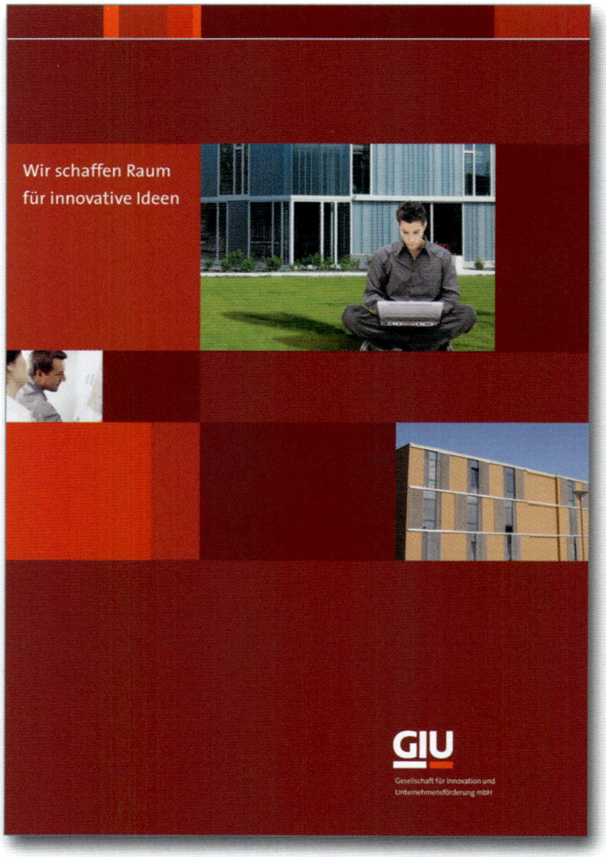

Entwicklung eines ganzheitlichen Corporate Designs für die GIU *GmbH* unter Berücksichtigung aller Kommunikationsmedien inklusive eines anwenderfreundlichen CD-Manuals mit einfachen und präzisen Gestaltungsrichtlinien. Ein variabler Baukasten, auf den jeder Anwender mit viel eigener Kreativität zurückgreifen kann. *Development of a universal Corporate Design for* GIU *GmbH* that covers all communications media with a user-friendly CD manual and simple and detailed design guidelines as a variable module that any really creative user can make good use of.

Entwicklung eines neuen Corporate Designs der GIU mbH
Development of a universal Corporate Design for GIU mbH

Corporate Design

Corporate Design

Unternehmen — Company GIU — Gesellschaft für Innovation und Unternehmensförderung mbH — Nell-Breuning-Allee 8 — 66115 Saarbrücken — Germany — www.giu.de

Design FBO Gesellschaft für Marketing-Kommunikation mbH — Martin Oberhauser, Dominik Friedrich — www.fbo.de

Guy Tillim ist nicht nur der Titel des Kataloges, sondern auch der Name des Fotografen. Mit seiner Serie dokumentarischer Kriegsfotografien wurde er mit dem *DaimlerChrysler Award* für südafrikanische Fotografie ausgezeichnet. *Guy Tillim* is not only the title of the catalogue, but also the name of the photographer. With his documentary photographs of war he was awarded with the *DaimlerChrysler Award* for South African photography.

Guy Tillim *Katalog für DaimlerChrylser Award, Südafrika*
Catalogue for DaimlerChrylser Award, South-Africa

Unternehmen — Company hackenschuh com. design — Lindenspürstraße 22 — 70176 Stuttgart — Germany — www.hackenschuh.com
Design Werksdesign / Inhouse — Christina Hackenschuh

Auftritt der kompletten Kollektion in mehreren Akten. Die Hauptdarsteller sind kapriziöse Diven, die ihre Griffe, Riemen und Schlaufen mit einer spielerischen Bewegung präsentieren. Sie tanzen schon mal aus der Reihe und durchbrechen die grafische Ordnung. Ansonsten halten sie sich diszipliniert an die Choreographie des schattenlosen Lichts. Der Rest der Kompanie folgt Strukturen. Präzise und detailgetreu verharren sie in Reihe oder solo. The collection takes centre-stage, in several acts. The stars of the show are capricious divas, presenting their handles, straps and loops with playful gestures. Now and then they dance out of line, disrupting the order of the design. Otherwise they stay neatly in place, following the choreography of shadowless light. The rest of the company keeps to clear structures, frozen in place – solo or corps de ballet – with every minute detail in sharp focus.

Katalog Bree Collection
Catalogue Bree Collection

Katalog
Catalogue

Unternehmen — Company BREE Collection GmbH & Co. KG — Gerberstraße 3 — 30916 Isernhagen — Germany — www.bree.de
Design büro uebele visuelle kommunikation — Gerd Häußler, Prof. Andreas Uebele — www.uebele.com

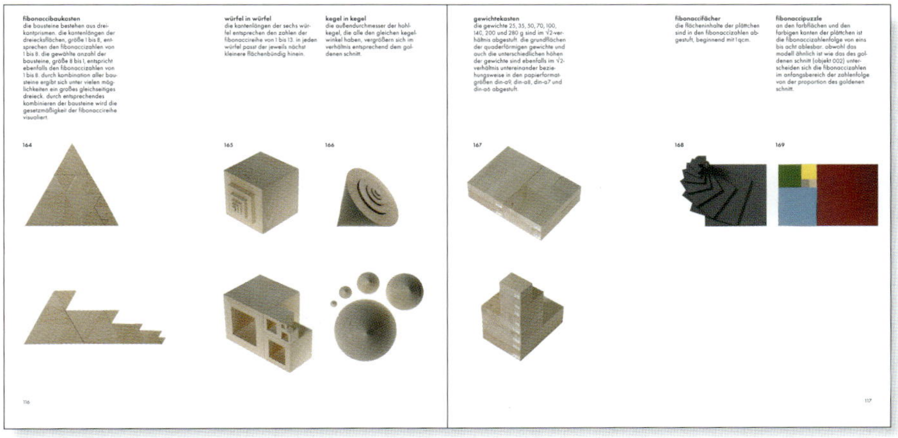

Ein Buch über die Freude am Gestalten. Über die optimale Form. Über das Wunder der Natur. Ein Buch, das Mathematik anschaulich macht, das überraschende Verbindungen zwischen geometrischen Formen und Formen der Natur aufdeckt. Ein Buch mit Bildern und Zeichnungen, die die Strukturen und Gesetzmäßigkeiten zeigen, die diesen Formen zugrunde liegen. Ein Lehrbuch und Nachschlagewerk für Produktgestalter, für Architekten und für Kommunikationsdesigner. A book about the joys of the designer's art, about the ideal form, about the wonders of nature. A book that makes mathematics accessible and reveals surprising links between geometrical and natural forms. A book full of images and drawings, which illustrate the structures and laws on which these forms are based. A textbook and a reference book for product designers, for architects and for communications designers.

„formvollendet“
“The perfect form”

Ausstellungskatalog
Exhibition catalogue

Unternehmen — Company Fachhochschule Düsseldorf — Fachbereich Design — Georg-Glock-Straße 15 — 40474 Düsseldorf — Germany — www.fh.duesseldorf.de

Design Martin Hess, Andreas Uebele, Svenja Knödler, Jessica Krier, Yvonne Niewerth, Silke Sabow, Sarah Lenz

Die Nominierungen von A bis Z
The Nominees from A to Z

a

Unternehmen — Company
A & W Architektur & Wohnen
Poßmoorweg 2
22301 Hamburg
Germany
www.awmagazin.de
Produkt — Product
Die Geschichte von der Fliege ... —
The tale of the fly...
Veröffentlichung im Magazin A & W Architektur & Wohnen — Publication in the magazine A & W Architektur & Wohnen
Design
Werksdesign / Inhouse
Barbara Friedrich, Thomas Elmenhorst, Jan van Rossem, Peter Cederling, Monika Aichele

Unternehmen — Company
A & W Architektur & Wohnen
Poßmoorweg 2
22301 Hamburg
Germany
www.awmagazin.de
Produkt — Product
AUF DIE PLÄTZE!
Veröffentlichung im Magazin A & W Architektur & Wohnen — Publication in the magazine A & W Architektur & Wohnen
Design
Werksdesign / Inhouse
Barbara Friedrich, Thomas Elmenhorst, Juliane Bennien, Jan-Peter Westermann

Unternehmen — Company
A & W Architektur & Wohnen
Poßmoorweg 2
22301 Hamburg
Germany
www.awmagazin.de
Produkt — Product
A & W-DESIGN-MERKSPIEL —
A & W-Design-Memo-Game
DESIGN-MERKSPIEL — Design-memo
Design
Factor Design
www.factordesign.com
337

Unternehmen — Company
A & W Architektur & Wohnen
Poßmoorweg 2
22301 Hamburg
Germany
www.awmagazin.de
Produkt — Product
A & W-KALENDER 2006 — A & W-Calendar 2006
Wöchentlicher Wandkalender 2006 —
Weekly wall calendar 2006
Design
Factor Design
www.factordesign.com

Unternehmen — Company
A & W Architektur & Wohnen
Poßmoorweg 2
22301 Hamburg
Germany
www.awmagazin.de
Produkt — Product
A & W-DESIGNER DES JAHRES 2005 RICHARD SAPPER — A & W-DESIGNER OF THE YEAR 2005 RICHARD SAPPER
Kommunikationsmaterialien —
Communication tools
Design
Factor Design
Johannes Erler, Jan Kruse
www.factordesign.com
//
Werksdesign / Inhouse
Barbara Friedrich, Thomas Elmenhorst, Giovanni Castell, Jan van Rossem

Unternehmen — Company
A. W. Faber-Castell Vertrieb GmbH
Nürnberger Straße 2
90546 Stein
Germany
www.faber-castell.de
Produkt — Product
e-motion Drehkugelschreiber und Drehbleistift aus Edelharz — e-motion Twist ballpoint pen and twist mechanical pencil made in precious resin
Schreibgerät — Writing instrument
Design
Industrial Design Team
Heinrich Stukenkemper
240

Unternehmen — Company
A. W. Faber-Castell Vertrieb GmbH
Nürnberger Straße 2
90546 Stein
Germany
www.faber-castell.de
Produkt — Product
P'3140
Kugelschreiber — Ballpoint pen
Design
Porsche Design GmbH
Roland Heiler
www.porsche-design.com
241

Unternehmen — Company
AB Berg & Berg
Flyttblocksvägen 2
37552 Kallinge
Sweden
www.berg-berg.com
Produkt — Product
Berg & Berg
Visuelle Identität — Visual identity
Design
Heine Warnecke Design GmbH
Dirk Heine, Cord Warnecke
www.heinewarnecke.com
//
Eberhard Franke Fotografie
Eberhard Franke
www.eberhardfranke.de

Unternehmen — Company
ABC-TEAM Spielplatzgeräte GmbH
Eisensteinstraße (Industriegebiet Rohr)
56235 Ransbach-Baumbach
Germany
www.abc-team.de
Produkt — Product
Kletterschnecke — Play sculputure „Climbing Snail"
Spielplatzgerät, Spielskulptur für öffentliche Spielplätze — Playground equipment, play sculpture for public places
Design
designa
Marie-Luise Peres-Krepele

Unternehmen — Company
ABUS August Bremicker Söhne KG
Heimstraße 32
56479 Rehe
Germany
Produkt — Product
Bordo 6000
Faltschloss — Foldable lock
Design
Werksdesign / Inhouse
Martin Hommel, Winfried Buseman

Unternehmen — Company
Ade Hauser Lacour
Westendstraße 84
60325 Frankfurt
Germany
www.adehauserlacour.de
Produkt — Product
„Und keiner hat für uns Kaddisch gesagt"
Ausstellungskatalog — Exhibition catalogue
Design
Ade Hauser Lacour
Kommunikationsgesellschaft GmbH
Lauren Lacour
www.adehauserlacour.de

Unternehmen — Company
Ade Hauser Lacour
Westendstraße 84
60325 Frankfurt
Germany
www.adehauserlacour.de
Produkt — Product
mg technologies ag „annual report 2004"
Geschäftsbericht — Annual report
Design
Ade Hauser Lacour
Kommunikationsgestaltung GmbH
Lauren Lacour
www.adehauserlacour.de

Unternehmen — Company
adidas AG
Adi-Dassler-Straße 1
91074 Herzogenaurach
Germany
www.adidas.com
Produkt — Product
+F50 TUNIT
Fußballschuh — Football shoe
Design
Werksdesign / Inhouse
adidas Global Design
Matthias Bieling, Senior Designer Football
108 Silber

Unternehmen — Company
adidas AG
Adi-Dassler-Straße 1
91074 Herzogenaurach
Germany
www.adidas.com
Produkt — Product
+Predator Twin Flex
Schienbeinschoner-Fußball —
Shin guard-football
Design
Werksdesign / Inhouse
adidas Design-Department

Unternehmen — Company
adidas AG
Adi-Dassler-Straße 1
91074 Herzogenaurach
Germany
www.adidas.com
Produkt — Product
+Teamgeist
Fußball — Football
Design
Werksdesign / Inhouse
adidas-Design Department

Unternehmen — Company
adidas AG
Adi-Dassler-Straße 1
91074 Herzogenaurach
Germany
www.adidas.com
Produkt — Product
AdiStar Ultra
Radschuh — Road cycling shoe
Design
Mavic
Sylvain Richards
www.mavic.com
//
Werksdesign / Inhouse
adidas-Design Department

Unternehmen — Company
adidas AG
Adi-Dassler-Straße 1
91074 Herzogenaurach
Germany
www.adidas.com
Produkt — Product
Fingersave Allround
Torwarthandschuh — Gloves
Design
Werksdesign / Inhouse
adidas-Design Department

Unternehmen — Company
adidas AG
Adi-Dassler-Straße 1
91074 Herzogenaurach
Germany
www.adidas.com
Produkt — Product
T-Mac 5
Basketball-Schuh — Basketball-shoe
Design
Werksdesign / Inhouse
adidas-Design Department

Unternehmen — Company
ADO Gardinenwerke GmbH & Co. KG
Hüntestraße 68
26871 Aschendorf
Germany
www.ado-international.com
Produkt — Product
ADO Dessin 9700 / 9705 / ADO Cordon®
Fadenvorhang — String curtain
Design
Werksdesign / Inhouse
Andreas Wulf

Unternehmen — Company
Adolf Würth GmbH & Co. KG
Reinhold-Würth-Straße 12.17
74653 Künzelsau
Germany
www.wuerth.com
Produkt — Product
Universal Montage- und Verschraubungswerkzeug — Universal assembly and screw fastening tool
Werkzeug — Tool
Design
Werksdesign / Inhouse
Herbert Reuther

Unternehmen — Company
ADOMA GmbH
Pettermandstraße 4
88239 Wangen / Allgäu
Germany
www.adoma.de
Produkt — Product
KristallFrost***
Flaschenkühler — Bottle cooler
Design
Werksdesign / Inhouse
261

Unternehmen — Company
AEG Hausgeräte GmbH
Electrolux Dienstleistungs GmbH
Muggenhoferstraße 135
90429 Nürnberg
Germany
www.aeg-electrolux.de
Produkt — Product
AEG-Electrolux 41029 G-m
Range Cooker Produktreihe —
Range cooker product range
Design
Werksdesign / Inhouse

Unternehmen — Company
AEG Hausgeräte GmbH
Electrolux Dienstleistungs GmbH
Muggenhoferstraße 135
90429 Nürnberg
Germany
www.aeg-electrolux.de
Produkt — Product
AEG-Electrolux 79900 G-m
Gas-Kochmulde auf Glas-Touch-Control —
Gas cooking bowl on glas touch control hob
Design
Werksdesign / Inhouse

Unternehmen — Company
AEG Hausgeräte GmbH
Electrolux Dienstleistungs GmbH
Muggenhoferstraße 135
90429 Nürnberg
Germany
www.aeg-electrolux.de
Produkt — Product
AEG-Electrolux Neue Klasse PREFERENCE
Avantgarde Compact Küchengeräte-Einbaureihe — Avantgarde compact built in range
Design
Werksdesign / Inhouse

Unternehmen — Company
AEG Hausgeräte GmbH
Electrolux Dienstleistungs GmbH
Muggenhoferstraße 135
90429 Nürnberg
Germany
www.aeg-electrolux.de
Produkt — Product
AEG-Electrolux SANTO W 98820
Einbau-Weinklimaschrank —
Built-in winecooler
Design
Werksdesign / Inhouse

Unternehmen — Company
AEG Hausgeräte GmbH
Electrolux Dienstleistungs GmbH
Muggenhoferstraße 135
90429 Nürnberg
Germany
www.aeg-electrolux.de
Produkt — Product
DD 8696
Ceran-Wand-Dunsthaube — Ceran-wall-hood
Design
Werksdesign / Inhouse

Unternehmen — Company
Aerocrine AB
P.O. Box 1024
171 21 Solna
Sweden
www.aerocrine.com / www.nioxmino.com
Produkt — Product
NIOX MINO
Monitoring von Atemwegsentzündungen —
Airway inflammation monitor
Design
Ergonomidesign
David Crafoord
www.ergonomidesign.com

Unternehmen — Company
AFG Arbonia Forster Riesa GmbH
Heinrich-Schönberg-Straße 3
01591 Riesa
www.afg.ch
Germany
Produkt — Product
BAGNOTHERM BTOV
Bad-Designraumwärmer —
Bathroom and designer room heaters
Design
Struppler Associates Design GmbH
Andreas Struppler

Unternehmen — Company
agape
Via Po Barna 69
46031 Bagnolo San Vito
Italy
www.agapedesign.it
Produkt — Product
KAA
Duscharmaturen — Shower taps
Design
Giulio Gianturco
www.giuliogianturco.com
80 Silber

Unternehmen — Company
AGCO GmbH
Fendt
Johann-Georg-Fendt-Straße 4
87616 Marktoberdorf
Germany
www.fendt.com
Produkt — Product
Fendt 936 Vario
Traktor — Tractor
Design
Werksdesign / Inhouse
Johann Epple, Tom Maier
92 Silber

Unternehmen — Company
akf bank / petra hattab büro für design
Friedrichstraße 51
42105 Wuppertal
Germany
www.akf.de
Produkt — Product
akf bank Geschäftsbericht 2003 / 2004
akf bank annual report 2003 / 2004
Geschäftsbericht — Annual report
Design
herzogenrathsaxler kommunikationsdesign
Margarethe Saxler, Matthias Herzogenrath
www.herzogenrathsaxler.de

Unternehmen — Company
Alape GmbH
Am Gräbicht 1–9
38644 Goslar
Germany
www.alape.com
Produkt — Product
Der Messestand als urbaner Raum —
The exhibition stand as urban area
Messestand — Fair stand
Design
Dipl. Ing. Innenarchitektur
Andrea Jürgens

Unternehmen — Company
ALFEX SA
Centro Galleria 3, Via Cantronale
6928 Manno
Switzerland
www.alfex.de
Produkt — Product
FLATLINE
Armbanduhr — Wrist watch
Design
Plum-Design
Georg Plum
www.plum-design.de

Unternehmen — Company
Alfred Kärcher GmbH & Co. KG
Alfred-Kärcher-Straße 28–40
71364 Winnenden
Germany
www.kaercher.de
Produkt — Product
CB-Line
Pkw-Portalwaschanlagen — Gantry car washes
Design
Werksdesign / Inhouse
Denis Dammköhler

Unternehmen — Company
Alfred Kärcher GmbH & Co. KG
Alfred-Kärcher-Straße 28–40
71364 Winnenden
Germany
www.kaercher.de
Produkt — Product
DDC 50
Bohrstaubtool — Drill dust catcher
Design
Werksdesign / Inhouse
Michael Meyer

Unternehmen — Company
Alfred Kärcher GmbH & Co. KG
Alfred-Kärcher-Straße 28–40
71364 Winnenden
Germany
www.kaercher.de
Produkt — Product
BR 55 / 40 W
Scheuersaugmaschine — Scrubber dryer
Design
Pearl Creative
Christian Rummel
www.pearlcreative.com
//
Werksdesign / Inhouse
Denis Dammköhler

Unternehmen — Company
Alfred Kärcher GmbH & Co. KG
Alfred-Kärcher-Straße 28–40
71364 Winnenden
Germany
www.kaercher.de
Produkt — Product
HD 7 / 18-4 Mx Plus
Hochdruckreiniger — High pressure cleaner
Design
Pearl Creative
Christian Rummel
www.pearlcreative.com
//
Werksdesign / Inhouse
Denis Dammköhler

Unternehmen — Company
Alfred Kärcher GmbH & Co. KG
Alfred-Kärcher-Straße 28–40
71364 Winnenden
Germany
www.kaercher.de
Produkt — Product
K 6.91
Hochdruckreiniger — High pressure cleaner
Design
Teams Design
Ulrich Schweig
www.teams-design.de
//
Werksdesign / Inhouse
Michael Meyer

Unternehmen — Company
Alfred Kärcher GmbH & Co. KG
Alfred-Kärcher-Straße 28–40
71364 Winnenden
Germany
www.kaercher.de
Produkt — Product
VC 6300
Staubsauger — Vacuum cleaner
Design
Teams Design
Ulrich Schweig
www.teams-design.de
//
Werksdesign / Inhouse
Michael Meyer

Unternehmen — Company
Alfred Sternjakob GmbH & Co. KG
Frankenstraße 47–55
67227 Frankenthal
Germany
www.scout1975.de
Produkt — Product
Scout 1975
Freizeit- / Business-Taschen-Serie —
Bag series
Design
Estragon GmbH
www.estragon.ch
▤ 250

Unternehmen — Company
Alfred Sternjakob GmbH & Co. KG
Frankenstraße 47–55
67227 Frankenthal
Germany
www.logstoff.com
Produkt — Product
logstoff.com
Taschenkollektion — Bag collection
Design
ID4 Industriedesign
Axel Ricker
www.id4.de
▤ 248

Unternehmen — Company
Allerhand Limited
Unit 2401–02, 24 floor,
Wai Wah Commercial Centre,
6 Wilmer Street, Sheung Wan
Hong Kong
www.allerhand.com
Produkt — Product
Allerhand – Trendy
Taschenkollektion — Bag collection
Design
Werksdesign / Inhouse
Stefanie Jung

Unternehmen — Company
Aloys F. Dornbracht GmbH & Co. KG
Köbbingser Mühle 6
58640 Iserlohn
Germany
www.dornbracht.com
Produkt — Product
ISERLOHN the Dornbracht
Culture Projects Vol.1
Kunst und Kulturmagazin —
Art and culture magazine
Design
Neo Noto Mike Meiré, Cologne
Mike Meiré, Florian Lambl, Tanja Wintels
www.m2web.de

Unternehmen — Company
Aloys F. Dornbracht GmbH & Co. KG
Köbbingser Mühle 6
58640 Iserlohn
Germany
www.dornbracht.com
Produkt — Product
ISH Messestand 2005 — ISH Booth 2005
Messestand — Fair stand
Design
Meiré und Meiré
Mike Meiré
www.m2web.de

Unternehmen — Company
Aloys F. Dornbracht GmbH & Co. KG
Köbbingser Mühle 6
58640 Iserlohn
Germany
www.dornbracht.com
Produkt — Product
TARA. Armatur und Archetypus. Eine Huldigung. — TARA. Homage to a sanitary fitting and an archetype.
Publikation — Publication
Design
Meiré und Meiré
Mike Meiré, Katja Fössel
www.m2web.de

Unternehmen — Company
ALTANA Pharma AG
Byk-Gulden-Straße 2
78467 Konstanz
Germany
Produkt — Product
Pantozol®
Medikamentenverpackung —
Pharmaceutical packaging
Design
Designconnection
Andreas Klatt, Klaus Hinderegger
www.designconnection.eu
▤ 281

Unternehmen — Company
Amann Girrbach GmbH
Herrschaftswiesen 1
6842 Koblach
Austria
www.amanngirrbach.com
Produkt — Product
Artex® Carbon Kieferbewegungssimulator für die Zahnmedizintechnik — Artex® Carbon Sensitive jaw movement simulator for dental technology
Artikulator — Articulator
Design
Ingenieurbüro Wäger GmbH & Co. KEG
Karl-Heinz Wäger, Michael Schrom
www.ib-waeger.at
297

Unternehmen — Company
ANKER-TEPPICHBODEN
Gebr. Schoeller GmbH + Co. KG
Zollhausstraße 112
52353 Düren
Germany
www.anker-teppichboden.de
Produkt — Product
TAURUS
Objekt-Teppichboden — Contract carpet
Design
Werksdesign / Inhouse
Ines Binder
193

Unternehmen — Company
ANKER Teppichboden
Gebr. Schoeller GmbH & Co. KG
Zollhausstraße 112
52353 Düren
Germany
www.anker-dueren.de
//
rohi stoffe GmbH
Schönlinderstraße 1
82538 Geretsried
Germany
www.rohi.com
Produkt — Product
rohi / ANKER Aircraft Collection 2006
Flugzeugtextilien (Sitzbezugstoffe, Vorhänge, Kopflätzchen, Teppichböden) — Aircraft textiles (seat-cover fabrics, curtains, headrests, carpets)
Design
Werksdesign / Inhouse
rohi designteam / ANKER Designteam

Unternehmen — Company
Anthologie Quartett
Schloss Hünnefeld
49152 Bad Essen
Germany
www.anthologie-quartett.de
Produkt — Product
Faruno
Hängeleuchte — Pendant
Design
Gabriel & Evenhuis
Alex Gabriel, Willeke Evenhuis
www.rp-lamps.com

Unternehmen — Company
ARDEX GmbH
Friedrich-Ebert-Straße 45
58453 Witten
Germany
www.ardex.de
Produkt — Product
PANDOMO W1
Dekorspachtelmasse — Decorative finishing render
Design
Werksdesign / Inhouse

Unternehmen — Company
ARDEX GmbH
Friedrich-Ebert-Straße 45
58453 Witten
Germany
www.ardex.de
Produkt — Product
PANDOMO TerrazzoPlus
Hochwertiger Bodenbelag in Terrazzo-Optik mit einem neuartigen ARDEX Schnellzement und mineralischen Schmuckkörnungen. — High-class floor in terrazzo-look out of a new kind of ARDEX rapid cement screed and mineral grains.
Design
Werksdesign / Inhouse

Unternehmen — Company
Arjowiggins Feinpapier GmbH
Abraham-Lincoln-Straße 16
65189 Wiesbaden
Germany
www.arjowiggins.com
Produkt — Product
Angeber-Set — Blow-your-own-trumpet set
Papier-Promotion — Paper promotion kit
Design
Q Kreativgesellschaft mbH
Thilo von Debschitz, Kirsten Fabinski, Zuni Höfer
www.q-home.de

Unternehmen — Company
Armstrong DLW AG
Stuttgarter Straße 75
74321 Bietigheim-Bissingen
Germany
www.armstrong.eu
Produkt — Product
Armstrong DLW Linoleum – World of Colour Collection 2005
Musterbuch — Specimen book
Design
Dal Lago Associati
Adalberto Dal Lago, Laura Cesari
www.dallagoassociati.com
//
Werksdesign / Inhouse
Markus Deimling, Anneliese Müller,
Emmi Mieskes, Helen Blank
357

Unternehmen — Company
AEC
Ars Electronica Futurelab
Hauptstraße 2–4
4040 Linz
Austria
www.futurelab.aec.at
Produkt — Product
Humphrey II Installation
Flugsimulator — Flight simulator
Design
Werksdesign / Inhouse

Unternehmen — Company
Arte G.E.I.E.
4 quai du Chanoine Winterer
67080 Strassbourg Cedex
France
www.arte-tv.com
Produkt — Product
Arte Kultur
Arte Kultur Magazin — Arte culture magazine
Design
Velvet Medien Design
Mathias Zentner
www.velvet.de

Unternehmen — Company
Artemide Group S.p.A.
Via Bergamo 18
20010 Pregnana Milanese
Italy
www.artemide.com
Produkt — Product
One-Line Morsetto
Tischleuchte in Klemmausführung
Table lamp
Design
Ora Ito

Unternehmen — Company
Artemide Group S.p.A.
Via Bergamo 18
20010 Pregnana Milanese
Italy
www.artemide.com
Produkt — Product
Mesmeri
Wandleuchte — Wall lamp
Design
Eric Sole

Unternehmen — Company
Artequa Ltd.
Hermiankatu 6–8 H
33720 Tampere
Finland
www.artequa.com
Produkt — Product
Arctic Sun
Leuchte — Luminaire
Design
Werksdesign / Inhouse
Mauri Drufva

Unternehmen — Company
atelier brückner gmbh
Quellenstraße 7
70376 Stuttgart
Germany
www.atelier-brueckner.de
Produkt — Product
Am Ball der Zeit – Deutschland und die Fußballweltmeisterschaften seit 1954 —
„Am Ball der Zeit" – Germany and the football world championships since 1954
Temporäre Ausstellung im Historischen Museum der Pfalz, Speyer —
Temporary exhibition at the Historical Museum of the Palatinate, Speyer
Design
atelier brückner
www.atelier-brueckner.de
//
Attraktive Grautöne
www.attraktivegrautoene.de
//
JangledNerves
www.janglednerves.com

Unternehmen — Company
Ateljé Lyktan
Fyrvaktaregatan 7
29681 Åhus
Sweden
www.atelje-lyktan.se
Produkt — Product
Tundra
Leuchtenserie — Luminaire series
Design
Kristian Andréason & Kristin Leibel
www.pigdesign.se

Unternehmen — Company
AUDI AG
Online Kommunikation
85045 Ingolstadt
Germany
www.audi.de
Produkt — Product
AUDI – DRIVING EXPERIENCE
Microsite / Internetseite — Microsite / Website
Design
Elephant Seven GmbH Nord
Oliver Viets, Sven Mentel
www.e-7.com

Unternehmen — Company
Audi AG
Ettinger Straße
85049 Ingolstadt
Germany
www.audi.de
Produkt — Product
AUDI – ALLROAD QUATTRO CONCEPT
Microsite / Internetseite — Microsite / Website
Design
Elephant Seven GmbH Nord
Markus Weiss, Oliver Viets, Sven Mentel
www.e-7.com

Unternehmen — Company
Audi AG
Ettingerstraße
85045 Ingolstadt
Germany
www.audi.de
Produkt — Product
AUDI HIGHLINES
Kinetische Installation — Kinetic installation
Design
Mutabor Design GmbH
Johannes Plass, Heinrich Paravicini,
Simone Campe, Kai Riemland
www.mutabor.de
//
Werksdesign / Inhouse
Bernhard Neumann

Unternehmen — Company
Auerhahn Bestecke GmbH
Im oberen Tal 9
72213 Altensteig
Germany
www.auerhahn-bestecke.de
Produkt — Product
A / Design
Tischschale aus Aluminium —
Bowl in aluminium
Design
designafairs GmbH
Christoph Böninger
www.designafairs.com

Unternehmen — Company
Auerhahn Bestecke GmbH
Im oberen Tal 9
72213 Altensteig
Germany
www.auerhahn-bestecke.de
Produkt — Product
A / Design
Dekanter-Set — Decanter-set
Design
Helix-Akantus Formgestaltung GmbH & Co. KG
www.akantus.de

Unternehmen — Company
Authentics GmbH
Am Ölbach 28
33334 Gütersloh
Germany
www.authentics.de
Produkt — Product
BRAVIT
Kerze — Candle
Design
Van Bömmel Studio
www.vanboemmel.com

Unternehmen — Company
Authentics GmbH
Am Ölbach 28
33334 Gütersloh
Germany
www.authentics.de
Produkt — Product
Table Stories
Porzellan- und Glas-Kollektion —
Porcelain and glass series
Design
Tord Boontje
www.tordboontje.com

Unternehmen — Company
Authentics GmbH
Am Ölbach 28
33334 Gütersloh
Germany
www.authentics.de
Produkt — Product
Eiko
Eierkocher — Egg boiler
Design
Christina Schäfer

Unternehmen — Company
Automobili Lamborghini S.p.A.
Via Modena 12
40019 Sant'Agata Bolognese
Italy
www.lamborghini.com
Produkt — Product
Automobili Lamborghini | Messeauftritt Genf
2004 — Exhibition Stand Geneva 2004
Messestand — Fair stand
Design
Werksdesign / Inhouse

Unternehmen — Company
Automobilmanufaktur Dresden
Lennéstraße 1
01069 Dresden
Germany
Produkt — Product
Intentionen — Intentions
Kalender — Calendar
Design
Gingco Werbeagentur GmbH & Co. KG
www.gingco.net
//
Werksdesign / Inhouse
Barbara Geising

Unternehmen — Company
Axel Springer AG
Axel-Springer-Straße 65
10888 Berlin
Germany
www.wams.de
Produkt — Product
Welt am Sonntag
Tageszeitung — Daily newspaper
Design
KircherBurkhardt
Editorial & Corporate Communication GmbH
Lukas Kircher
www.kircher-burkhardt.com

Unternehmen — Company
Axelmeiselicht
Wiener Platz 7 Rbg
81667 München
Germany
www.occhio-sento.de
Produkt — Product
OCCHIO SENTO
Leuchtensystem — Lighting system
Design
Werksdesign / Inhouse
Axel Meise, Christoph Kügler

b

Unternehmen — Company
B. T. Dibbern GmbH & Co. KG
Heinrich-Hertz-Straße 1
22941 Bargteheide
Germany
www.dibbern.de
Produkt — Product
Pure
Porzellanserie aus Fine Bone China — Porcelain series made of fine bone china porcellain
Design
Werksdesign / Inhouse

Unternehmen — Company
B+W stuhl+tisch zentrum GmbH & Co. KG
Mergenthaler Allee 23–25
65760 Eschborn
Germany
www.bisterfeldundweiss.de
Produkt — Product
P1
Tischsystem — Table programme
Design
greutmann bolzern AG für Gestaltung
www.gbdesign.ch

Unternehmen — Company
B+W stuhl+tisch zentrum GmbH & Co. KG
Mergenthaler Allee 23–25
65760 Eschborn
Germany
www.bisterfeldundweiss.de
Produkt — Product
S20 B+W Kollektion
Multifunktionaler Objektstuhl — Multifunctional contract chairs
Design
greutmann bolzern AG für Gestaltung
www.gbdesign.ch

Unternehmen — Company
Bank 1 Saar
Kaiserstraße 20
66111 Saarbrücken
Germany
www.bank1saar.de
Produkt — Product
Leitbild der BANK 1 Saar
Broschüre — Brochure
Design
Maksunovic & Partners
www.maksunovic.de

Unternehmen — Company
Barbara Flügel Porzellan
Unterweißenbacherweg 6
95100 Selb
Germany
www.fluegel-porzellan.de
Produkt — Product
Aladin und Drache — Aladin and Dragon
Porzellanserie — Porcelain series
Design
Werksdesign / Inhouse
Barbara Flügel
271

Unternehmen — Company
Barbara Schneider Human development
Bismarckstraße 2
87700 Memmingen
Germany
www.schneider-human.de
Produkt — Product
Wo die Menschen wachsen, wachsen die Unternehmen — Where the persons grow, the businesses grow
Image Broschüre — Image brochure
Design
Jan Bazing
//
Bruce B. GmbH
corporate communication
Thomas Elser, Thomas Waschke
www.bruce-b.com

Unternehmen — Company
BARK CLOTH Europe
Talhauser Straße 18
79285 Ebringen
Germany
www.barktex.com
Produkt — Product
BARKTEX® - Low friction 0577
Polymerverstärktes abrieboptimiertes Rindentuch® (Unterrinde von Ficus natalensis) — Polymer reinforced abrasion-optimized BARK CLOTH® (Sub-bark of Ficus natalensis)
Design
CHT R. Beitlich GmbH
Andreas Troscheit
www.cht.com
//
Werksdesign / Inhouse
Oliver Heintz

Unternehmen — Company
Bassier, Bergmann & Kindler
Digital Sales and Brand Specialists GmbH
Rheinlandstraße 11
71636 Ludwigsburg
Germany
www.bb-k.com
Produkt — Product
Bewegungs-Freiheit 911. Das neue 911 Carrera Cabriolet und 911 Carrera S Cabriolet. — Moving. The new 911 Carrera Cabriolet and the 911 Carrera S Cabriolet.
Internetseite — Website
Design
Werksdesign / Inhouse
Alex Müller, Anja Krumpholz

Unternehmen — Company
Bassier, Bergmann & Kindler
Digital Sales and Brand Specialists GmbH
Rheinlandstraße 11
71636 Ludwigsburg
Germany
www.bb-k.com
Produkt — Product
Der neue Cayman S. Konsequent Porsche. — The new Cayman S. Instantly Porsche.
Internetseite — Website
Design
Werksdesign / Inhouse
Alex Müller, Anja Krumpholz

Unternehmen — Company
Bauerfeind AG
Triebeser Straße 16
07937 Zeulenroda-Triebes
Germany
www.bauerfeind.com
Produkt — Product
GenuTrain A3
Train Aktivbandage — Train active supports
Design
Rokitta Produkt & Markenästhetik
Hermann Rokitta
www.rokitta.de
276

Unternehmen — Company
Bau-Fritz GmbH & Co.KG, seit 1896
Alpenstraße 25
87746 Erkheim
Germany
www.baufritz.de
Produkt — Product
Editionshaus 478
Holzfertighaus — Timber houses
Design
Gatermann & Schossig
Elmar Schossig
www.gatermann-schossig.de
162

Unternehmen — Company
BBDO Düsseldorf GmbH
Königsallee 92
40212 Düsseldorf
Germany
www.bbdo.de
Produkt — Product
Ascii
AIDS-Präventionskampagne — AIDS Prevention campaign
Design
Werksdesign / Inhouse
Marco Obermann, Jean-Pierre Gregor, Concetta Milione, Michael Ohanian, Raphael Milczarek, Hans-Holger Pollack
128 Silber

Unternehmen — Company
BBDO Düsseldorf GmbH
Königsallee 92
40212 Düsseldorf
Germany
www.bbdo.de
Produkt — Product
Pepsi light
Nichtalkoholische Getränke — Non alcoholic drinks
Design
Werksdesign / Inhouse
Michael Ohanian, Fabian Kirner, Felix Lemcke

Unternehmen — Company
BEKO Deutschland GmbH
Hermannstraße 54–56
63263 Neu Isenburg
Germany
www.beko-deutschland.de
//
Arcelik A.S.
Karaagac Caddesi No: 2–6
34445 Sütlüce / Istanbul
Turkey
www.blomberginternational.com
Produkt — Product
GSN 1581 X
Geschirrspüler — Dishwasher
Design
Frog Design
Hartmut Esslinger
www.frogdesign.com

Unternehmen — Company
BEKO Deutschland GmbH
Hermannstraße 54–56
63263 Neu-Isenburg
Germany
www.beko-deutschland.de
//
Arcelik A.S.
Karaagac Caddesi No: 2–6
34445 Sütlüce / Istanbul
Turkey
www.blomberginternational.com
Produkt — Product
TKF 1350 S
Trockner — Dryer
Design
Frog Design
Hartmut Esslinger
www.frogdesign.com

Unternehmen — Company
Bene AG
Schwarzwiesenstraße 3
3340 Waidhofen / Ybbs
Austria
http://bene.com
Produkt — Product
AL Conferencing
Konferenz-Programm —
Conference programme
Design
Werksdesign / Inhouse
Kai Stania
227

Unternehmen — Company
BenQ Corporation
18 Jihu Road, Neihu
114 Taipei Taiwan
Chinese Taipei
www.BenQ.com
Produkt — Product
FP72V
LCD-Monitor — LCD-Monitor
Design
Werksdesign / Inhouse
BenQ Lifestyle Design Center

Unternehmen — Company
BenQ Corporation
18 Jihu Road, Neihu
114 Taipei Taiwan
Chinese Taipei
www.BenQ.com
Produkt — Product
FP93V
LCD-Monitor — LCD-Monitor
Design
Werksdesign / Inhouse
BenQ Lifestyle Design Center

Unternehmen — Company
BenQ Corporation
18 Jihu Road, Neihu
114 Taipei Taiwan
Chinese Taipei
www.BenQ.com
Produkt — Product
7350CT
Scanner — Scanner
Design
Werksdesign / Inhouse
BenQ Lifestyle Design Center

Unternehmen — Company
BenQ Deutschland GmbH
Große Elbstraße 39
22767 Hamburg
Germany
www.BenQ.de
Produkt — Product
EU HHB-700 Bluetooth Headset
Kommunikationselektronik —
Communication electronics
Design
B / F Design-Team
www.bf-design.de

Unternehmen — Company
BenQ Mobile GmbH & Co. OHG
Haidenauplatz 1
81667 München
Germany
www.benq.com
Produkt — Product
SL75
Mobiltelefon — Mobile phone
Design
Phoenix Design
www.phoenixdesign.de

Unternehmen — Company
Berendsohn AG
Alte Sülldorfer Landstraße 400
22559 Hamburg
Germany
www.berendsohn.com
Produkt — Product
DR. FEEL GOOD
Massagegerät — Massaging device
Design
Nicole Syben
www.artworklab.de
275

Unternehmen — Company
Berker GmbH & Co. KG
Klagebach 38
58579 Schalksmühle
Germany
www.berker.de
Produkt — Product
Berker K.5
Flächenprogramm — Surface range
Design
OCO-Design
Theo Wolf
www.oco-design.de

Unternehmen — Company
Berker GmbH & Co. KG
Klagebach 38
58579 Schalksmühle
Germany
www.berker.de
Produkt — Product
Berker B.IQ
Tastsensor — Tactile sensor system
Design
OCO-Design
Theo Wolf
www.oco-design.de

Unternehmen — Company
Berker GmbH & Co.KG
Klagebach 38
58579 Schalksmühle
Germany
www.berker.de
Produkt — Product
Berker B.IG FIVE
Kampagne — Campaign
Design
Thomas Biswanger Design
Thomas Biswanger
www.thomasbiswanger.de

Unternehmen — Company
Berliner Seilfabrik GmbH & Co.
Lengeder Straße 4
13407 Berlin
Germany
www.berliner-seilfabrik.com
Produkt — Product
Mega-Fussball
Kletter-Seilspielgerät —
Rope Play Climbing Structure
Design
Werksdesign / Inhouse
Karl Köhler
157

Unternehmen — Company
Bette GmbH & Co. KG
Heinrich-Bette-Straße 1
33129 Delbrück
Germany
www.bette.de
Produkt — Product
BetteContura
Duschwanne — Shower tray
Design
Schmiddem Design
www.schmiddem-design.de

Unternehmen — Company
Betty TV AG
Rüdesheimer Straße 11
80686 München
Germany
www.betty-tv.com
Produkt — Product
Betty
TV-Fernbedienung — TV remote control
Design
designafairs GmbH
Dov Rattan, Philipp Kleinlein, Annke Osthues
www.designafairs.com

Unternehmen — Company
Beurer GmbH & Co. KG
Söflinger Straße 218
89077 Ulm
Germany
www.beurer.de
Produkt — Product
BF 18
Diagnosewaage — Diagnostic scale
Design
Uli Schade Industriedesign
Uli Schade
www.schadedesign.de

Unternehmen — Company
Bielefelder Philharmoniker
Brunnenstraße 3–9
33602 Bielefeld
Germany
www.bielefelder-philharmoniker.de
Produkt — Product
Spielzeit 2004_5
Saisonheft — Season brochure
Design
nodesign
www.nodesign.com

Unternehmen — Company
Bikatec Metall- und Textilverarbeitungs GmbH
Ten-Brink Straße 14a
78269 Volkertshausen
Germany
www.bikatec.de
Produkt — Product
Structurelab S1 Schirmsystem —
Structurelab S1 sunshade system
Verschattungssystem — Sunshade system
Design
StructureLab GmbH
Jürgen Schubert, Jürgen Gendriesch,
Alexander Prang
www.structurelab.com
158

Unternehmen — Company
BINZ GmbH & Co.
Draisstraße 11
76307 Karlsbad
Germany
www.binz-bestattungsfahrzeuge.de
Produkt — Product
BINZ Omniel
Bestattungsfahrzeug — Hearse
Design
Werksdesign / Inhouse
Volker Sieber, Gerhard Kurr

Unternehmen — Company
Blå Station AB
Box 100
29622 Åhus
Sweden
www.blastation.se
Produkt — Product
Clip
Klapptisch — Folding table
Design
Firma Jonas Forsman
Jonas Forsman
www.jonasforsman.se

Unternehmen — Company
Black & Decker GmbH
DeWALT Professionelle Elektrowerkzeuge
Black und Decker Straße 40
65510 Idstein
Germany
www.blackanddecker.com
Produkt — Product
DeWalt D 25701K
Hochleistungs-Kombihammer 7 kg-Klasse —
High performance combination hammer,
7 kg class
Design
Werksdesign / Inhouse
Michael Stirm

Unternehmen — Company
Black & Decker GmbH
DeWALT Professionelle Elektrowerkzeuge
Black und Decker Straße 40
65510 Idstein
Germany
www.blackanddecker.com
Produkt — Product
DeWalt D 25730K
Hochleistungs-Kombihammer 9 kg-Klasse —
High performance combination hammer,
9 kg class
Design
Werksdesign / Inhouse
Michael Stirm

Unternehmen — Company
Black & Decker GmbH
DeWALT Professionelle Elektrowerkzeuge
Black und Decker Straße 40
65510 Idstein
Germany
www.blackanddecker.com
Produkt — Product
DeWalt D 25840K
Hochleistungs-Meißelhammer 7 kg-Klasse —
High performance chiseling hammer 7 kg class
Design
Werksdesign / Inhouse
Michael Stirm

Unternehmen — Company
Black & Decker GmbH
Black und Decker Straße 40
65510 Idstein
Germany
www.blackanddecker.de
Produkt — Product
GK1000 – ALLIGATOR
Elektro-Astschere — Powered lopper
Design
Werksdesign / Inhouse
Lawrie Cunningham

Unternehmen — Company
Black Diamond Equipment AG
Christoph Merian Ring 7
4153 Reinach
Switzerland
www.BlackDiamondEquipment.com
Produkt — Product
Covert mit AvaLung(tm)
Rucksack — Backpack
Design
Werksdesign / Inhouse
Peater Wilkening, David Schipper, Paul Terry

Unternehmen — Company
Blindkilde
Gunnar Clausens Vej 58
8260 Viby J
Denmark
www.byno.dk
Produkt — Product
byNo. #4
Gartenfackel — Garden torch
Design
DOK 54
Jacob M. Lund, Thomas Brinch-Møller
www.dok54.dk
159

Unternehmen — Company
BMW AG
Petuelring 130
80788 München
Germany
www.bmw.de
Produkt — Product
Der BMW 6er — The BMW 6 Series
Unternehmenspublikation — Corporate Book
Design
HOFFMANN UND CAMPE Corporate Publishing
Dirk Linke, Anna Clea Skoluda
www.pc.hoca.de

Unternehmen — Company
BMW AG,
MINI Brand Management
Heidemannstraße 164
80788 München
Germany
www.mini.de
Produkt — Product
MINIInternational
Unternehmenspublikation — B-to-C Magazine
Design
Meiré und Meiré
Mike Meiré
www.m2web.de
//
HOFFMANN UND CAMPE Corporate Publishing
Stefan Pietsch
www.cp.hoca.de

Unternehmen — Company
BMW Group
Knorrstraße 147
80788 München
Germany
www.bmw.com
Produkt — Product
BMW K1200 R
Motorrad — Motorcycle
Design
Werksdesign / Inhouse
BMW Group

Unternehmen — Company
BMW Group
Knorrstraße 147
80788 München
Germany
www.bmw.com
Produkt — Product
BMW R 1200 RT
Motorrad — Motorcycle
Design
Werksdesign / Inhouse
BMW Group

Unternehmen — Company
BMW Group
Knorrstraße 147
80788 München
Germany
www.bmw.com
Produkt — Product
BMW 3er Limousine — BMW 3er Series Saloon
Pkw — Passenger car
Design
Werksdesign / Inhouse
BMW Group Design

Unternehmen — Company
BMW Group
Knorrstraße 147
80788 München
Germany
www.bmw.com
Produkt — Product
BMW Cruising Bike
Fahrrad — Bicycle
Design
BMW Group Design
Designworks USA
Studio München
104 Silber

Unternehmen — Company
BMW Group
Knorrstraße 147
80788 München
Germany
www.bmw.com
Produkt — Product
NBG
Fahrradhelm — Bicycle helmet
Design
Werksdesign / Inhouse
BMW Group Design

Unternehmen — Company
BMW Group
Knorrstraße 147
80788 München
Germany
www.bmw.com
Produkt — Product
BMW Kidsbike
Kinderfahrrad — Juvenile bicycle
Design
Werksdesign / Inhouse
BMW Group Design

Unternehmen — Company
BMW Group
Knorrstraße 147
80788 München
Germany
www.bmw.com
Produkt — Product
BMW Lifestyle Taschenkollektion —
BMW Lifestyle Bag Collection
Taschenkollektion — Bag collection
Design
Werksdesign / Inhouse
BMW Group Design

Unternehmen — Company
BMW Group
Knorrstraße 147
80788 München
Germany
www.bmw.com
Produkt — Product
MINI Brillenkollektion 2005 —
MINI Collection Eyewear
Sonnen- und optische Brillen — Eyewear
Design
Werksdesign / Inhouse
BMW Group Design

Unternehmen — Company
BMW Group
Knorrstraße 147
80788 München
Germany
www.bmw.com
Produkt — Product
BMW Motorrad R 1200 S
Motorrad — Motorcycle
Design
Werksdesign / Inhouse
BMW Group Design

Unternehmen — Company
BODUM AG
Bodum Design Group
Kantonsstrasse 100
6234 Triengen
Switzerland
www.bodum.com
Produkt — Product
SHIN BISTRO
Kork Linie — Cork line
Design
Werksdesign / Inhouse
Bodum Design Group

Unternehmen — Company
BODUM AG
Kantonsstrasse 100
6234 Triengen
Switzerland
www.bodum.com
Produkt — Product
BODUM CHEF
Messer Linie — Knife line
Design
BODUM AG Bodum Design Group
www.bodum.com

Unternehmen — Company
Boffi spa
Via Oberdan 70
20030 Lentate sul Seveso (MI)
Italy
www.boffi.com
Produkt — Product
folio
Waschtisch — Wash basin
Design
Neunzig ° Design
www.neunzig-grad.com

Unternehmen — Company
Bosch Rexroth AG
Maria-Theresien-Straße 23
97816 Lohr am Main
Germany
www.boschrexroth.com
Produkt — Product
AS2
System zur Druckluftaufbereitung —
Air preparation unit
Design
Henssler & Schultheiss Fullservice
Productdesign GmbH
Martin Schultheiss
www.henssler-schultheiss.de
298

Unternehmen — Company
Bosch Transferzentrum Venture Capital
Ingersheimer Straße 16
70499 Stuttgart
Germany
Produkt — Product
Idee. Transferzentrum. Unternehmen. —
Ideas. Transfer centre. Business.
Broschüre — Brochure
Design
Beaufort 8
Christine Elsässer, Philipp Heimsch,
Annette Petry, Sabine Schmidt
www.beaufort8.com

Unternehmen — Company
Bosch und Siemens Hausgeräte GmbH
Markendesign Neff / Constructa
Carl-Wery-Straße 34
81739 München
Germany
www.neff.de
Produkt — Product
DKG 59
Wandesse
Wall mounted hood
Design
Werksdesign / Inhouse
Ralf Grobleben, Gerhard Nüssler
252

Unternehmen — Company
Bosch und Siemens Hausgeräte GmbH
Markendesign Neff / Constructa
Carl-Wery-Straße 34
81739 München
Germany
www.neff.de
Produkt — Product
SH 4664 A
Elektro-Einbaubackofen — Built in electric oven
Design
Werksdesign / Inhouse
Ralf Grobleben, Gerhard Nüssler
253

Unternehmen — Company
BOSE
Max-Planck-Straße 36
61381 Friedrichsdorf
Germany
www.bose.de
Produkt — Product
3 2 1 GS DVD Home Entertainment System
DVD-Anlagen — DVD home entertainment system
Design
Werksdesign / Inhouse

Unternehmen — Company
Bowa electronic GmbH
Thomas Morgan Straße 2
72810 Gomaringen
Germany
www.bowa.de
Produkt — Product
Tissue Seal
Chirurgisches Instrument — Surgical instrument
Design
Held & Team
Burkhard Peters
www.heldundteam.de

Unternehmen — Company
BRANDIT
Marketing Kommunikation
Mittelstraße 12–14
50672 Köln
Germany
www.brandit.de
Produkt — Product
Midnight Ride ... oder die Macht herausragender Kommunikation — Midnight Ride ... or the power of outstanding communication
Broschüre zur Eigenkommunikation — A house advertising brochure
Design
Werksdesign / Inhouse
Peter Specht, Angela Strecker
349

Unternehmen — Company
BREE Collection GmbH & Co. KG
Gerberstraße 3
30916 Isernhagen
Germany
www.bree.de
Produkt — Product
Katalog Bree Collection — Catalogue Bree Collection
Katalog — Catalogue
Design
büro uebele
visuelle kommunikation
Gerd Häußler, Prof. Andreas Uebele
www.uebele.com
372

Unternehmen — Company
BREE Collection GmbH & Co. KG
Gerberstraße 3
30916 Isernhagen
Germany
www.bree.com
Produkt — Product
GUM
Damentasche — Handbag
Design
Werksdesign / Inhouse
BREE Design-Team

Unternehmen — Company
BREE Collection GmbH & Co. KG
Gerberstraße 3
30916 Isernhagen
Germany
www.bree.com
Produkt — Product
Dallas
Reisegepäck — Luggage
Design
Werksdesign / Inhouse
BREE Design-Team

Unternehmen — Company
Brill Gartengeräte GmbH
Därmannsbusch 7
58456 Witten
Germany
www.brill.de
Produkt — Product
Evolution 36 EF
Elektro-Rasenmäher — Electric lawnmover
Design
Weinberg & Ruf Produktgestaltung
H. Weinberg, H. Ruf
www.weinberg-ruf.de

Unternehmen — Company
Bruandet S.A.
13, rue de Chamois
39700 La Barre
France
www.bruandet.com
//
TENTE-ROLLEN GmbH
Herrlinghausen 75
42929 Wermelskirchen
Germany
www.tente.com
Produkt — Product
SMILES Serie 5520
Möbelrolle — Castor for furniture
Design
Emmanuel Dietrich
www.emmanueldietrich.com
//
Werksdesign / Inhouse
Christian Marchand
229

Unternehmen — Company
Brühl & Sippold GmbH
Alter Bad Stebener Weg 1
95138 Bad Steben
Germany
www.bruehl.com
Produkt — Product
Mya
Bettsofa – Polstermöbel — Sofabed – Upholstered furniture
Design
Designbüro Laprell / Classen

Unternehmen — Company
Brühl & Sippold GmbH
Alter Bad Stebener Weg 1
95138 Bad Steben
Germany
www.bruehl.com
Produkt — Product
SUNRISE – Sofa-3 und Openend –
SUNRISE – Openend
Polstermöbel — Upholstered furniture
Design
Werksdesign / Inhouse
Kati Meyer-Brühl

Unternehmen — Company
Brunner GmbH
Stühle I Tische I Objektmöbel
Im Salmenkopf 10
77866 Rheinau
Germany
www.brunner-group.com
Produkt — Product
ST. PETER Modell 1080
Stapelstuhl — Stacking chair
Design
f / p design gmbh
www.fp-design-gmbh.com
223

Unternehmen — Company
BSH Bosch und Siemens Hausgeräte GmbH
Carl-Wery-Straße 34
81739 München
Germany
www.bsh-group.de
Produkt — Product
MSM 6A50 mixxo
Stabmixer — Hand blender
Design
Daniels & Koitzsch
www.daniels-koitzsch.de
//
Werksdesign / Inhouse
Tobias Krüger, Helmut Kaiser

Unternehmen — Company
BSH Bosch und Siemens Hausgeräte GmbH
Carl-Wery-Straße 34
81739 München
Germany
www.bsh-group.de
Produkt — Product
TFB 1620
Friteuse — Deep fryer
Design
Daniels & Koitzsch
www.daniels-koitzsch.de
//
Werksdesign / Inhouse
Gregor Luippold, Helmut Kaiser

Unternehmen — Company
Buffetissimo GmbH
Wielandstraße 34
28203 Bremen
Germany
www. buffetissimo.com
Produkt — Product
flix Buffet-System — Typ: flix Cool
Präsentationssystem für Catering & Hotellerie zur Präsentation von Buffets — Presentation system for catering & hotel business to present buffets
Design
Box
Gernot Krause
//
Werksdesign / Inhouse
Matthias Biehler

Unternehmen — Company
Bundesministerium der Finanzen
Wilhelmstraße 97
10117 Berlin
Germany
www.bmf.bund.de
Produkt — Product
Flugboot DO-X
Briefmarke — Postage stamp
Design
Klein & Neumann
www.knndn.de

Unternehmen — Company
Burkhardt Leitner constructiv GmbH & Co. KG
Breitwiesenstraße 17
70565 Stuttgart
Germany
www.burkhardtleitner.de
Produkt — Product
constructiv PON
Messe-, Ausstellungs-, Display-, Shop- und Office-System — Trade fair-, exhibition-, display-, shop- and office system
Design
Werksdesign / Inhouse
324

Unternehmen — Company
Burkhardt Leitner constructiv GmbH & Co. KG
Breitwiesenstraße 17
70565 Stuttgart
Germany
www.burkhardtleitner.de
Produkt — Product
constructiv TELVIS I und II
Präsentations- und Displaysystem — Presentation and display system
Design
Werksdesign / Inhouse
323

Unternehmen — Company
büro für gestaltung diane ziegler
Paulusstraße 18
70197 Stuttgart
Germany
www.dianeziegler.de
Produkt — Product
leporello büro für gestaltung diane ziegler — fanfold flyer
Leporello für Eigenwerbung — Fanfold flyer, corporate brochure
Design
büro uebele
visuelle kommunikation
Prof. Andreas Uebele, Beate Gerner
www.uebele.com

Unternehmen — Company
Büro für Gestaltung Wangler & Abele
Hohenzollernstraße 89
80796 München
Germany
www.bfgest.de
Produkt — Product
„gestalten drei" und „gestalten vier"
Büroberichte — Office reports
Design
Werksdesign / Inhouse
Ursula Wangler, Frank Abele, Amelie von Büdingen, Kerstin Ullrich
340

Unternehmen — Company
büro uebele
visuelle kommunikation
Heusteigstraße 94a
70180 Stuttgart
Germany
www.uebele.com
Produkt — Product
Weg Zeichen / my type of place
Buch — Book
Design
Werksdesign / Inhouse
Prof. Andreas Uebele
www.uebele.com

Unternehmen — Company
Büro X Design GmbH
Salzgries 18
1010 Wien
Austria
www.buerox.at
Produkt — Product
baumaxx – Der Do-it-yourself Geschäftsbericht 2003
Geschäftsbericht 2003 — Annual Report 2003
Design
Werksdesign / Inhouse
Andreas Miedaner, Florian Ribisch, Michael Pichler

Unternehmen — Company
BURO LUBBERS
Oude Dieze 17
5200 BN 's-Hertogenbosch
Netherlands
www.burolubbers.nl
//
Gemeinde Sluis, Projektleiter Tiny Maenhout
//
Aquavia Sas van Gent
Produkt — Product
Aussichtssteg & Panoramaterrasse Breskens
Aussichtspunkte — Viewpoints
Design
BURO LUBBERS landschapsarchitectuur & stedelijk ontwerp
Peter Lubbers, Marian de Vries, Caroline de Feijter, Roy Wouters
156

Unternehmen — Company
Büttenpapier Gmund GmbH & Co. KG
Mangfallstraße 5
83703 Gmund
Germany
www.gmund.com
Produkt — Product
Die Geschichte der Büttenpapierfabrik Gmund — The history of Gmund
Geschichtsbuch und Kassette mit Papiermustern — History book with paper samples in box
Design
Feine Werbung
www.feinewerbung.de

C

Unternehmen — Company
C. & E. Fein GmbH
Hans-Fein-Straße 81
73529 Schwäbisch Gmünd
Germany
www.fein.de
Produkt — Product
ASM 9-2
Akku-Schrauber — Battery screwdiver
Design
Henssler und Schultheiss Fullservice Productdesign GmbH
www.henssler-schultheiss.de
301

Unternehmen — Company
C. Hafner GmbH & Co. KG
Bleichstraße 13–17
75173 Pforzheim
Germany
www.c-hafner.de
Produkt — Product
CeHa White ECS
Zahntechnisches Laborgerät — Dental lab unit
Design
Formgestaltung
Alexander Schnell-Waltenberger
www.formgestaltung.de
//
Werksdesign / Inhouse
Heiko Grusche, Wilhelm Heim, Uwe Böhm, Sven Gude

Unternehmen — Company
C. Josef Lamy GmbH
Grenzhöfer Weg 32
69123 Heidelberg
Germany
www.lamy.de
Produkt — Product
LAMY studio
Drehkugelschreiber und Füllhalter mit Stahlfeder — Rotating ballpoint and steel-nibbed fountain pen
Design
Werksdesign / Inhouse

Unternehmen — Company
Caliboor Design Studio
Theresienstraße 104
80333 München
Germany
www.caliboor.com
Produkt — Product
Caliboor Design Studio
Internetseite — Website
Design
Werksdesign / Inhouse
Jan Schlösser

Unternehmen — Company
Calimax GmbH
Bundesstraße 102
6830 Rankweil
Austria
www.calimax.com
Produkt — Product
Bellina 5 kW
Pelletofen — Pellet stove
Design
Werksdesign / Inhouse
170

Unternehmen — Company
CAPAROL
Roßdörfer Straße 50
64372 Ober-Ramstadt
Germany
www.caparol.de
Produkt — Product
Metallorcyl
Metallisch wirkende Effektbeschichtung auf Dispersionsbasis — Metallic-effect coating, emulsion based
Design
Werksdesign / Inhouse

Unternehmen — Company
CARL MERTENS
CMS-GRASOLI Besteckfabrik GmbH + Co. KG
Krahenhöher Weg 8
42659 Solingen
Germany
www.carl-mertens.com
Produkt — Product
Corporate Design CARL MERTENS
Logo, Typografie, Anzeigen, Verpackung, Display, Katalog, Imageprospekt — Logo, typography, advertisement, packaging, display, catalogue, teaser
Design
BOROS
Christian Boros, Ingo Maak
www.boros.de

Unternehmen — Company
Carl Zeiss Sports Optics GmbH
Gloelstraße 3–5
35576 Wetzlar
Germany
www.zeiss.de / sportsoptics
Produkt — Product
Victory 42 FL
Fernglas — Binocular
Design
zed AG
Thomas Steuri
www.zednetwork.com
246

Unternehmen — Company
Carpet Concept Objekt Teppichboden GmbH
Bunzlauer Straße 7
33719 Bielefeld
Germany
www.carpet-concept.de
Produkt — Product
Mod
Gewebter PVC + Glasfasern — Woven pvc + glas filaments
Design
Carsten Gollnick
www.gollnick-design.com

Unternehmen — Company
Carpet Concept Objekt Teppichboden GmbH
Bunzlauer Straße 7
33719 Bielefeld
Germany
www.carpet-concept.de
Produkt — Product
Sqr
Webware-Teppichboden Kollektion — Woven wilton floorcovering collection
Design
Carsten Gollnick
www.gollnick-design.com
//
Peter Maly
www.peter-maly.de

Unternehmen — Company
Carsten Kraemer
Am Kieselhumes 34
66123 Saarbrücken
Germany
www.schritt-macher.com
Produkt — Product
„Schrittmacher Berlin"
Stadtführer — City guide
Design
Werksdesign / Inhouse

Unternehmen — Company
Carsten Kraemer
Am Kieselhumes 34
66123 Saarbrücken
Germany
Produkt — Product
Katalog „Kunst macht Schule"
Ausstellungskatalog — Art catalogue
Design
Werksdesign / Inhouse

Unternehmen — Company
Cascando Products b.v.
Het Vergun 5
6931 KC Westervoort
Netherlands
www.cascando.nl
Produkt — Product
Round20
Accessoires für Wohn- und Büroeinrichtung —
Accessoiries for home and office
Design
Smool Designstudio
Robert Bronwasser
www.smool.nl

Unternehmen — Company
Cassina S.p.A.
Via L. Busnelli, 1
20036 Meda (MI)
Italy
www.cassina.it
Produkt — Product
AUCKLAND
Drehsessel — Swivel armchair
Design
studio massaud
Jean-Marie Massaud
www.massaud.com

Unternehmen — Company
Cassina S.p.A.
Via L. Busnelli, 1
20036 Meda (MI)
Italy
www.cassina.it
Produkt — Product
Hola Sprint
Stuhl — Chair
Design
zed AG
Hannes Wettstein
www.zednetwork.com

Unternehmen — Company
cat-interiors.de
Bahnhofstraße 5
96215 Lichtenfels
Germany
www.cat-interiors.de
Produkt — Product
RONDO STAND Filz
Katzenmöbel — Cat interior
Design
Meyer Design
Otto Meyer, Christel Meyer
www.meyer-design.net
215

Unternehmen — Company
Central-druck Trost GmbH & Co.
Industriestraße 2
63150 Heusenstamm
Germany
Produkt — Product
DD Magazin
Magazin — Magazine
Design
4gzl designkontext
Markus Frenzl, Alexander Michaelopoulos,
Jörg Niehage, Jörg Baumann
//
Pict
Bastian, Eicke, Sturm GbR
Sven Bastian, Andrea Eicke, Uta Sturm
www.pict.de

Unternehmen — Company
Christian Büning Informationsgestalter
Hafenstraße 64
48153 Münster
Germany
www.christianbuening.de
Produkt — Product
Vincent van Gogh – alle Bilder —
Vincent van Gogh – the complete paintings
Schautafel — Poster
Design
Werksdesign / Inhouse
Christian Büning

Unternehmen — Company
Christian Tuermer
Minimal Lighting
Oberländer Ufer 186
50968 Köln
Germany
www.christiantuermer.de
Produkt — Product
minimal compact series
Aufbaustrahler — Surface-mounted spotlights
Design
Werksdesign / Inhouse
Christian Tuermer

Unternehmen — Company
Clauss Markisen Projekt GmbH
Bissinger Straße 9
73266 Bissingen-Ochsenwang
Germany
www.clauss-markisen.de
Produkt — Product
s_onro®
Rollladen und Sonnenschutz —
Shutters and sun shade
Design
Werksdesign / Inhouse
Ulrich Clauss, Hans Friedrich Schur,
Klaus Westenberger, Pasquale Cianci

Unternehmen — Company
Colmio Oy
Keulapuomintie 7
90500 Oulu
Finland
www.colmio.com
Produkt — Product
Woodpecker
Schaukelstuhl — Rocking all-purpose chair
Design
Werksdesign / Inhouse
Päivi Mikola

Unternehmen — Company
Compo GmbH & Co. KG
Gildenstraße 38
48157 Münster
Germany
www.compo.de
Produkt — Product
Compo Flüssigdünger — Compo liquid fertilizer
Flaschenserie für Flüssigdünger —
Bottle range for liquid fertilizer
Design
SOLUTIONS Branding & Design Companies AG
Karel Golta, Volker Schumann
www.solutions.de

Unternehmen — Company
CONCORD Gmbh
Industriestraße 25
D-95346 Stadtsteinach
Germany
www.concord.de
Produkt — Product
ION
Babyliegeschale — Baby seat
Design
FOURHEAD Productdesign
www.fourhead.de
151

Unternehmen — Company
ContiTech Elastomer-Beschichtungen GmbH
Breslauer Straße 14
37154 Northeim
Germany
www.contitech.de
Produkt — Product
ContiMetalflex
Flexibler elastomerer Verbundwerkstoff mit beschichteten Metall- oder Effektpigmenten — Flexible elastomer composite material with coated metal or special-effect pigments
Design
Werksdesign / Inhouse
Dr. Thomas Perkovic, Thomas Weber

Unternehmen — Company
COR Sitzmöbel
Helmut Lübke GmbH & Co. KG
Nonenstraße 12
33378 Rheda-Wiedenbrück
Germany
www.cor.de
Produkt — Product
COR Gesamtkatalog — COR product catalogue
Produktkatalog — Print catalogue
Design
Factor Design AG
Jan Kruse
www.factordesign.com

Unternehmen — Company
COR Sitzmöbel
Helmut Lübke GmbH & Co. KG
Nonenstraße 12
33378 Rheda-Wiedenbrück
Germany
www.cor.de
Produkt — Product
COR Jahrbuch — COR jubilee book
Buch — Book
Design
Factor Design AG
www.factordesign.com

Unternehmen — Company
COR Sitzmöbel Helmut Lübke GmbH & Co KG
Nonenstraße 12
33378 Rheda-Wiedenbrück
Germany
www.cor.de
Produkt — Product
COR Lesereise — COR lecture tour
Plakatserie — Poster series
Design
Factor Design AG
Jan Kruse, Olaf Stein
www.factordesign.com

Unternehmen — Company
cosmic Werbeagentur Bern AG
Wabernstrasse 40
3000 Bern 23
Switzerland
www.cosmic.ch
Produkt — Product
DBMB „Daddy broke my bike"
Buch — Book
Design
Werksdesign / Inhouse
Thom Pfister, Roland Zenger, Philipp Lüthi, Julien Junghäni

Unternehmen — Company
COUNTRY
Poßmoorweg 2
22301 Hamburg
Germany
www.country-online.de
Produkt — Product
DA GEHT WAS AUF! —
Something is Going to Prove!
Veröffentlichung im Magazin COUNTRY —
Feature in COUNTRY magazine
Design
Werksdesign / Inhouse
Barbara Friedrich, Nicole Vansteenkiste, Oliver Schwarzwald, Maria Grossmann, Volker Hobl

Unternehmen — Company
Creation Club (CC) GmbH
Agentur für integrierte Kommunikation
Medienallee 19
85774 Unterföhring
Germany
www.creation-club.de
Produkt — Product
Creation Club: Plug TV - START-UP
START-UP ist eine Serie über die Neugründung eines Unternehmens — START-UP is a series about the formation of a company
Design
Werksdesign/Inhouse
Cora Riegel, Michael Engelhardt

Unternehmen — Company
Creation Club (CC) GmbH
Agentur für integrierte Kommunikation
Medienallee 19
85774 Unterföhring
Germany
www.creation-club.de
Produkt — Product
Creation Club: Premiere Sport - 4 Minuten im Mai - 4 Jahre danach — Creation Club: Premiere Sport - 4 minutes in May - 4 years after
TV-Dokumentation — TV-documentary
Design
Werksdesign/Inhouse
Volker Bahr, Philip Wolfarth, Bernd Mutscheller, Michael Engelhardt

Unternehmen — Company
Creation Club (CC) GmbH
Agentur für integrierte Kommunikation
Medienallee 19
85774 Unterföhring
Germany
www.creation-club.de
Produkt — Product
Creation Club: www.creation-club.de
Internetauftritt Creation Club — Internet presence Creation Club
Design
Werksdesign/Inhouse
Lilly Wong de Weis, Michael Engelhardt, Martin Mueller, Markus Goles

Unternehmen — Company
Creation Club (CC) GmbH
Agentur für integrierte Kommunikation
Medienallee 19
85774 Unterföhring
Germany
www.creation-club.de
Produkt — Product
Creation Club: Plug TV - The Simple Life 2
TV-Serie — TV-series
Design
Werksdesign/Inhouse
Markus Goles, Cora Riegel, Philip Wolfarth, Michael Engelhardt

Unternehmen — Company
Creation Club (CC) GmbH
Agentur für integrierte Kommunikation
Medienallee 19
85774 Unterföhring
Germany
www.creation-club.de
Produkt — Product
Creation Club: Creation Club Launch
Logo, Corporate Idendity, Geschäftsausstattung, U.R.L., Demo Tape — Logo, Corporate Idendity, office equipment, U.R.L., demo tape
Design
Werksdesign/Inhouse
Lila Bagherzadeh, Lilly Wong de Weiss, Arman Kavrailoglu, Michael Engelhardt

Unternehmen — Company
Creation Club (CC) GmbH
Agentur für integrierte Kommunikation
Medienallee 19
85774 Unterföhring
Germany
www.creation-club.de
Produkt — Product
Re-Design Deutscher Kamerapreis
Corporate Design — Corporate design
Design
Werksdesign/Inhouse
Michael Engelhardt, Lila Bagherzadeh, Markus Goles, Michaela Hoess

Unternehmen — Company
CZERNY_taschen
Obere Hauptstraße 41
09244 Lichtenau/Chemnitz
Germany
www.lederdesign.de
Produkt — Product
Virage
Handtasche — Handbag
Design
Werksdesign/Inhouse
Ulrich Czerny

d

Unternehmen — Company
DAB DISEÑO ACTUAL BARCELONA, S.L.
Av/de la Cerdanya, nave 40, P.I. Pomar de Dalt
08915 Badalona, Barcelona
Spain
www.dab.es
Produkt — Product
JUDD
Licht, Hängelampe, serie: JUDD S 160/250 — Lighting, suspension, series: JUDD S 160/250
Design
David Abad
www.davidabad.com

Unternehmen — Company
DAB DISEÑO ACTUAL BARCELONA, S.L.
Av/ de la Cerdanya, nave 40, P.I. Pomar de Dalt
08915 Badalona, Barcelona
Spain
www.dab.es
Produkt — Product
ROOF
Licht, Hängelampe, Deckenleuchte oder Wandleuchte — Lighting, hanging, ceiling and wall lamp
Design
David Abad
www.davidabad.con

Unternehmen — Company
DaimlerChrysler AG
Epplestraße 225
70546 Stuttgart
Germany
www.daimlerchrysler.com
Produkt — Product
„Faszination Motorsport"
Ausstellung — Exhibition
Design
design hoch drei mit Ludwig architekten
Wolfram Schäffer, Almut Weinecke-Ludwig
www.design-hoch-drei.de
321

Unternehmen — Company
DaimlerChrysler AG
Abteilung MKP/K, HPC 0944
70546 Stuttgart
Germany
www.daimlerchrysler.com
Produkt — Product
The Porter – Webspecial zum CLS-Produktfilm mit Bryan Ferry
Website
Design
Scholz & Volkmer, Wiesbaden
Heike Brockmann
www.s-v.de

Unternehmen — Company
DaimlerChrysler AG
Epplestraße 225
70546 Stuttgart
Germany
www.daimlerchrysler.com
Produkt — Product
Energy for the Future – DaimlerChrysler-Konzernpräsentation auf der Mondial de l'Automobile, Paris 2004 — Energy for the Future – DaimlerChrysler Corporate Presentation at the 2004 Mondial de l'Automobile, Paris 2004
Messeauftritt — Fair stand
Design
Atelier Markgraph
www.markgraph.de
//
Kauffmann Theilig & Partner
Freie Architekten BDA
www.ktp-architekten.de
//
Design Hoch Drei GmbH & Co.KG
www.design-hoch-drei.de
//
TLD Planungsgruppe
www.tldplan.com
//
Freisteel
www.freisteel.de
//
Ernst F. Ambrosius & Sohn
www.ambrosius.de
325

Unternehmen — Company
DaimlerChrysler AG
Abteilung Brand Communications.
HPK 000/G328
70546 Stuttgart
Germany
www.mercedes-benz.com/museum
Produkt — Product
Mercedes-Benz Museum
Internetseite Mercedes-Benz Museum — Website Mercedes-Benz museum
Design
Scholz & Volkmer
Heike Brockmann, Katja Rickert, Thorsten Kraus, Rainer Eidemüller, Philippe Just, Carla Meinhard, Peter Reichard, Ron Kellermann
www.s-v.de
334

Unternehmen — Company
DaimlerChrysler AG, Zentrale
Epplestraße 225
70546 Stuttgart
Germany
www.mercedes-benz.de
Produkt — Product
Mercedes-Benz CLS-Klasse – Viertüriges Coupé — Mercedes-Benz CLS-Class – 4-doors-Coupé
PKW — Passenger car
Design
Werksdesign/Inhouse
Werksdesign Sindelfingen
144

Unternehmen — Company
DaimlerChrysler AG, Zentrale
Epplestraße 225
70546 Stuttgart
Germany
www.mercedes-benz.de
Produkt — Product
Mercedes-Benz R-Klasse
PKW — Passenger car
Design
Werksdesign/Inhouse
142

Unternehmen — Company
DaimlerChrysler AG, Zentrale
Epplestraße 225
70546 Stuttgart
Germany
www.mercedes-benz.de
Produkt — Product
Mercedes-Benz S-Klasse S350L
PKW — Passenger car
Design
Werksdesign / Inhouse
143

Unternehmen — Company
Dark
Vliegplein 43
9991 Adegem
Belgium
www.dark.be
Produkt — Product
POC
Leuchte — Lighting
Design
Helen Högberg

Unternehmen — Company
D'ART DESIGN GRUPPE
Am Zollhafen 5
41460 Neuss
Germany
www.d-art-design.de
Produkt — Product
D'ART DESIGN „Sample"
Messeauftrit auf der Euroshop 2005, Düsseldorf — Trade fair appearance at the Euroshop 2005, Düsseldorf
Design
Werksdesign / Inhouse
317

Unternehmen — Company
De Longhi Deutschland GmbH
Am Reitpfad 20
63500 Seligenstadt
Germany
www.delonghi.de
Produkt — Product
EN 90 G Essenza
Espressomaschine mit Nespresso-Kapselsystem — Espressomachine portion closed with Nespresso System
Design
Werksdesign / Inhouse
Antoine Cahen

Unternehmen — Company
DEDON GmbH
Zeppelinstraße 22
21337 Lüneburg
Germany
www.dedon.de
Produkt — Product
LEAF
Gartenmöbel — Outdoor furniture
Design
Frank Ligthart
211

Unternehmen — Company
Dekodur GmbH & Co.KG
Langenthaler Straße 4
69434 Hirschhorn
Germany
www.dekodur.de
Produkt — Product
dekowood bark cloth
Dekorative Hochdruckschichtstoffplatte (HPL) — Decorative high pressure laminate (HPL)
Design
BARK CLOTH
Oliver Heintz
www.barktex.com
//
Werksdesign / Inhouse
Volkmar André

Unternehmen — Company
Dell Inc.
Dell House, The Boulevard, Cain Road
RG12 1LF Bracknell, Berkshire,
Great Bitain
www1.euro.dell.com
Produkt — Product
Dell All-In-One Photo Printer 964
Drucker — Printer
Design
HLB Inc.
www.hlb.com
//
Werksdesign / Inhouse
Dell ID & Usability Group

Unternehmen — Company
Dell Inc.
Dell House, The Boulevard, Cain Road
RG12 1LF Bracknell, Berkshire,
Great Bitain
www1.euro.dell.com
Produkt — Product
New keyboard family — Dell USB Keyboard, Multimedia
Tastatur, Multimedia-Tastaturen — Computer keyboard, multimedia
Design
NPK Industrial Design
www.npk.nl
//
Werksdesign / Inhouse
Dell ID & Usability Group

Unternehmen — Company
Dell Inc.
One Dell Way
78682 Round Rock, Texas
USA
www.dell.com
Produkt — Product
Dell 30 inch UltraSharpTM Widescreen Flat Screen Monitor.
Breitformat-Flachbildschirm — PC Monitor, 30-inch flat panel display
Design
Werksdesign / Inhouse
Dell ID & Usability Group

Unternehmen — Company
DER BLAUE HUMMER
Große Elbstraße 212
22767 Hamburg
Germany
www.der-blaue-hummer.de
Produkt — Product
Caterer DER BLAUE HUMMER
Internetseite — Website
Design
weissraum.de(sign)
Lucas Buchholz, Bernd Brink, Irena Celecki,
Patric Trautwein
www.weissraum.de
333

Unternehmen — Company
DESEO GmbH, Agentur für Gestaltung
Koppstraße 2/RGB
81379 München
Germany
www.deseo.de
Produkt — Product
BMW Aktion 4Z - Zylindersimulation —
BMW Aktion 4Z - cylinder light simulation
Lichtdesign — Light design
Design
Werksdesign/Inhouse
Manfred Beck, Philipp Stegmüller
311

Unternehmen — Company
design hoch drei_GmbH & Co. KG
Hallstraße 25a
70376 Stuttgart
Germany
www.design-hoch-drei.de
Produkt — Product
Springerle - Du bist, was du isst —
Springerle - You are what you eat
Springerle Gebäck — Biscuit
Design
Werksdesign/Inhouse
Wolfram Schäffer, Marcus Wichmann

Unternehmen — Company
design hoch drei_GmbH & Co.KG
Hallstraße 25a
70376 Stuttgart
Germany
www.design-hoch-drei.de
Produkt — Product
MAUL-TASCHE
Papier-Tragetasche — Paper bags
Design
Werksdesign/Inhouse
Wolfram Schäffer, Susanne Wacker,
Marcus Wichmann

Unternehmen — Company
deSIGN Markus T. GmbH
Isselhorster Straße 374
33334 Gütersloh
Germany
www.markus-t.com
Produkt — Product
me Markus T
Brille — Spectacles
Design
Werksdesign/Inhouse
Markus Temming
116 Silber

Unternehmen — Company
Deuter Sport GmbH
Siemensstraße 1
86368 Gersthofen
Germany
www.deuter.com
Produkt — Product
ATTACK
Rückenprotektorrucksack —
Backprotector backpack
Design
Werksdesign/Inhouse
Steve Buffinton, Thomas Hilger

Unternehmen — Company
Deutsche Bahn AG
Potsdamer Platz
10785 Berlin
Germany
www.bahn.de
//
URW ++
Poppenbütteler Bogen 36
22399 Hamburg
Germany
www.urwpp.de
Produkt — Product
DB Type
Exklusives Schriftsystem für die Deutsche Bahn
— Exclusive system of corporate typefaces for
Deutsche Bahn
Design
SpiekermannPartners
Erik Spiekermann, Christian Schwartz
www.spiekermannpartners.com
54 Gold

Unternehmen — Company
Deutsche Bahn AG
Potsdamer Platz 2
10785 Berlin
Germany
www.db.de
Produkt — Product
Konzernportal Deutsche Bahn AG - www.db.de
Corporate Website
Internetseite — Website
Design
IBM Deutschland GmbH
IBM e-business Innovation Center
Kristjan Kristjanson, Renate Mittelstrass,
Christianne Mares
www.ibm.com/de

Unternehmen — Company
Deutsche Bank AG
Taunusanlage 12
60325 Frankfurt am Main
Germany
www.deutsche-bank.de
Produkt — Product
results
Unternehmenspublikation — B-to-B Magazine
Design
HOFFMANN UND CAMPE Corporate Publishing
Wolf Dammann
www.hoca-cp.de
355

Unternehmen — Company
Deutsche BP AG
Wittener Straße 45
44789 Bochum
Germany
www.deutschebp.de
Produkt — Product
ROSIplus
Das Training ROSIplus Interactive training CD-ROM — The Training ROSIplus Interactive training CD-ROM
Design
designverign gmbh
Christoph Scheufeld
www.designverign.de

Unternehmen — Company
Deutsche Energie-Agentur GmbH
dena
Chausseestraße 128a
10115 Berlin
Germany
www.dena.de
Produkt — Product
Powerscout
Online-Computerspiel — Online computer game
Design
unit-berlin, Dedecke Dertmann GbR
Hinnerk Dedecke, Thilo Fuchs, Torsten Sperling, Pierre le Gonidec
www.unit-berlin.de

Unternehmen — Company
Deutsche Post AG
Charles-de-Gaulle-Straße 20
53113 Bonn
Germany
www.deutschepost.de
Produkt — Product
Postkartenkalender „Schöne Grüße!" — The postcard calendar „Schöne Grüße!"
Kalender — Calendar
Design
lyhssauer GbR
Sabrina Lyhs-Webelhaus, Jane Sauer
www.lyhs-sauer.de

Unternehmen — Company
Deutsche See GmbH
Maifischstraße 3–9
27572 Bremerhaven
Germany
www.deutschesee.de
Produkt — Product
„Deutsche See" Wellenverpackung — „Deutsche See" wavepack
Frischfisch-Verpackung — Freshness packaging
Design
feldmann+schultchen design studios
www.fsdesign.de
260

Unternehmen — Company
Deutsche Steinzeug Keramik GmbH
Postfach 49
92519 Schwarzenfeld
Germany
www.deutsche-Steinzeug.de
Produkt — Product
Deutsche Steinzeug auf der BAU 2005 — Deutsche Steinzeug BAU 2005
Messestand — Fair stand
Design
Zeeh Bahls & Partner Design
Thomas Honegg, Frank Hermann
www.zeeba.de

Unternehmen — Company
Deutsche Telekom AG
Friedrich-Ebert-Allee 140
53113 Bonn
Germany
www.telekom.de
Produkt — Product
Zukunftscamp http://zukunftscamp.telekom.de/ — Future camp
Internetseite — Website
Design
people interactive GmbH
www.people-interactive.de
335

Unternehmen — Company
Deutsche Telekom AG
Hochstadenring 50
53119 Hamburg
Germany
www.telekom.de
Produkt — Product
T-Com Dual Phone TC 300
GSM-/IP-Telefon — GSM-/IP-Telefone
Design
Design 3 Produktdesign
Jan-Michael von Lewinski, Alexander Ksoll
www.design3.de

Unternehmen — Company
Deutsche Telekom AG, T-Com
Hochstadenring 50
53119 Bonn
Germany
www.telekom.de
Produkt — Product
T-Com Sinus TC 300
Basistation — Base station
Design
Design 3 Produktdesign
Jan-Michael von Lewinski
www.design3.de
//
Werksdesign/Inhouse

Unternehmen — Company
Deutsche Tinnitus-Liga e.V.
Am Lohsiepen 18
42652 Wuppertal-Ronsdorf
Germany
www.tinnitus-liga.de
Produkt — Product
Plakat – Die Unerträgliche —
Plakat – The unbearable
Plakat — Poster
Design
Heye & Partner GmbH
Gudrun Muschalla, Barbara Eisenhauer,
Anja Stiffel
www.heye.de

Unternehmen — Company
DHO-DON HIERRO Sdad. Ltda.
Crta. Tavernes-Alzira Km.1
46760 Tavernes Valldigna
Spain
www.donhierro.com
Produkt — Product
Delica
Auswahl von Tabletts — Natural wooden trays
Design
ZOOCREATIVE
www.zoocreative.com

Unternehmen — Company
Die ArtDiktatoren
Wilhelmshöher Straße 149
60389 Frankfurt
Germany
www.artdiktatoren.de
Produkt — Product
Die ArtDiktatoren
Geschäftsausstattung — Business stationery
Design
Werksdesign / Inhouse
Die ArtDiktatoren
Michael Emmel, Anke Stache

Unternehmen — Company
Die Firma GmbH
Marcus Wenig
Schwalbacher Straße 74
65183 Wiesbaden
Germany
www.diefirma.de
Produkt — Product
cubicon immobilien service /
www.cubicon-immobilien.de
Internetseite — Website
Design
Werksdesign / Inhouse

Unternehmen — Company
Die Firma GmbH
Schwalbacher Straße 74
65183 Wiesbaden
Germany
www.diefirma.de
Produkt — Product
www.schmidtholzinger.de
Internetseite — Website
Design
Werksdesign / Inhouse

Unternehmen — Company
Die Neue Sammlung –
Staatliches Museum für angewandte Kunst |
Design in der Pinakothek der Moderne
Türkenstraße 15
80333 München
Germany
www.die-neue-sammlung.de
Produkt — Product
Karim Rashid – Change
Ausstellungsinstallation —
Exhibition installation
Design
bangertprojects
Dr. Albrecht Bangert
www.bangertverlag.com
//
Karim Rashid
319

Unternehmen — Company
Die Neue Sammlung.
Staatliches Museum für angewandte Kunst |
Design in der Pinakothek der Moderne
Türkenstraße 15
80333 München
Germany
www.die-neue-sammlung.de
Produkt — Product
„Mythen. Automobili Lamborghini in der
Pinakothek der Moderne" — "Myths. Automobili
Lamborghini at the Pinakothek der Moderne"
Ausstellung — Exhibition
Design
KMS Team GmbH
Michael Keller, Birgit Vogel, Wahan
Mechitarian, Dirk Koy
www.kms-team.de

Unternehmen — Company
Die Wethje GmbH Kunststofftechnik
Donaustraße 35–41
94491 Hengersberg
Germany
www.wethje-gmbh.com
Produkt — Product
Sattelauflieger aus Hochleistungsverbundstoff
— Trailer made of high performance composite
material
LKW Sattelaufleger — Trailer
Design
TTT The Team Technology GmbH und Wethje
Entwicklungs GmbH
www.theteam-hamburg.de

Unternehmen — Company
Dionex
Dornierstraße 4
82110 Germering
Germany
www.softron.de
//
Whipsaw, Inc.
434 South First Street
95113 California San Jose
USA
www.whipsawinc.com
Produkt — Product
Dionex ICS-3000
Ionenchromatographie Maschine —
Ion chromatography machine
Design
Whipsaw, Inc.
Dan Harden, Ari Turgel
www.whipsawinc.com

Unternehmen — Company
Doepke Schaltgeräte GmbH & Co. KG
Stellmacherstraße 11
26506 Norden
Germany
www.doepke.de
Produkt — Product
smartFIS
Biometrische Zutrittskontrolle —
Biometry access control
Design
Riemann Design
Herbert Riemann, Dörte Waldöstl
www.riemanndesign.de
//
Werksdesign / Inhouse
Johann Meints, Gerold Roolfs
290

Unternehmen — Company
Doerner Institut Bayerische
Staatsgemäldesammlungen
Barer Straße 29
80799 München
Germany
www.doernerinstitut.de
Produkt — Product
Geschäftspapierausstattung des Doerner
Instituts — Stationery of the Doerner Institut
Geschäftspapier — Stationery
Design
KMS Team GmbH
Michael Keller, Marion Fink, Angela Keesman,
Melanie Sauer
www.kms-team.de

Unternehmen — Company
Doorzand
Stuurmankade 144
1019 WB Amsterdam
Netherlands
www.doorzand.com
Produkt — Product
Airdrive
Einwegtransportsystem für Spenderorgane —
Donor organ transport system
Design
Indes
www.indes.nl

Unternehmen — Company
doppler und saner
Industrieareal Ried 6
4222 Zwingen
Switzerland
www.dopplerundsaner.ch
Produkt — Product
CIRRUS
Offenes Regal — Open shelf
Design
Werksdesign / Inhouse
Rainer Saner

Unternehmen — Company
Dorine de Vos
Westzeedijk 126 F
3016 AJ Rotterdam
Netherlands
Produkt — Product
Harry Wich Een liefdesverklaring
Buch — Book
Design
Karelse & den Besten
John den Besten, Jan Kees Karelse
www.karelse-denbesten.nl

Unternehmen — Company
DORMA GmbH + Co. KG
Breckerfelderstraße 42–48
58256 Ennepetal
Germany
www.dorma.de
Produkt — Product
PHA 2500
Panikbeschlag — Panic handle
Design
Kaluza Design – Büro für Industrial Design
Georg Kaluza
www.kaluza-design.de

Unternehmen — Company
DORMA GmbH + Co. KG
Breckerfelder Straße 42 -48
58256 Ennepetal
Germany
www.dorma.com
Produkt — Product
TEGO Glass
Ganzglas-Türsystemlösung —
All-glass door system solution
Design
Kaluza Design
Büro für Industrial Design
Georg Kaluza
www.kaluza-design.de
//
Werksdesign / Inhouse
Michael Glanz
167

Unternehmen — Company
DORMA Hüppe
Raumtrennsysteme GmbH + Co. KG
Industriestraße 5
26655 Westerstede / Ocholt
Germany
www.dorma-hueppe.com
Produkt — Product
Dorma Moveo
Raumtrennsystem — Operable partition system
Design
Kaluza-Design - Büro für Industrial Design
Georg Kaluza
www.kaluza-design.de
//
Werksdesign / Inhouse
Jan-Gerd Behrens

Unternehmen — Company
Dr. August Oetker KG
Lutterstraße 14
33617 Bielefeld
Germany
www.oetker.de
Produkt — Product
Dr. August Oetker KG | Markenausstellung 2005:
„Dr. Oetker Welt" —
Dr. August Oetker KG | Brandworld 2005:
"Dr. Oetker World"
Ausstellung — Exhibition
Design
Triad Berlin Projektgesellschaft mbH
www.triad.de
316

Unternehmen — Company
Dr. Ing. h.c. F. Porsche AG
Porschestraße 15-19
71634 Ludwigsburg
Germany
www.porsche.com
Produkt — Product
Driving Identities
Porsche Engineering — Porsche engineering
Design
Klaus E. Küster Werbeagentur GmbH
Lars Vollert
www.kek.de

Unternehmen — Company
Dr. Ing. h.c. F. Porsche AG
Porschestraße 15-19
71634 Ludwigsburg
Germany
www.porsche.com
Produkt — Product
Trotzdem — Nevertheless
Imagefilm
Design
Klaus E. Küster Werbeagentur GmbH
Klaus Erich Küster
www.kek.de

Unternehmen — Company
dr. momo
mobility solutions
Siederstraße 40
78054 VS-Schwenningen
Germany
www.doktormomo.de
Produkt — Product
„khargo" Einkaufs-und Gehhilfe —
Shopping and walking aid
Mobilitätshilfe — Mobility aid
Design
DIRK MENSENINDUSTRIAL DESIGN
Dirk Mensen
www.dirkmensen.de

Unternehmen — Company
Draenert Studio GmbH
Steigwiesen 3
88080 Immenstaad
Germany
www.draenert.de
Produkt — Product
LOOP
Esszimmertisch — Dining table
Design
Studio Werner Aisslinger
Werner Aisslinger
www.aisslinger.de

Unternehmen — Company
Draenert Studio GmbH
Steigwiesen 3
88090 Immenstaad
Germany
www.draenert.de
Produkt — Product
TWIST
Couchtisch / Loungetisch —
Coffee table / Lounge table
Design
Büro für Gestaltungskonzeptionen
Prof. Wulf Schneider und Partner
www.profwulfschneider.de

Unternehmen — Company
Dräger Medical AG & Co. KG
Moislinger Allee 53-55
23542 Lübeck
Germany
www.draeger-medical.com
Produkt — Product
Stella
Operationsleuchte — Surgical light
Design
Barskidesign
Olaf Barski
www.barskidesign.com
174

Unternehmen — Company
Druckerei Schmitt GmbH
Lotschstraße 3
76829 Landau
Germany
www.sldruck.de
Produkt — Product
Jumus wird ein Ritter
Bilder-, Spiel- und Bastelmappe —
Folder for pictures, to play and crafts materials
Design
Kluge Gestaltung
www.klugegestaltung.de

Unternehmen — Company
DuPont de Nemours (Deutschland) GmbH
Du Pont-Straße 1
61352 Bad Homburg
Germany
www.zodiaq.com
Produkt — Product
Zodiaq® Concrete Block
Technischer Stein — Engineered stone
Design
Werksdesign / Inhouse
187

Unternehmen — Company
Duravit
Werderstraße 36
78123 Hornberg
Germany
www.duravit.de
Produkt — Product
Starck X Corporate Design
Broschüre und Internetseite — Brochure and website
Design
Philippe Starck
www.philippe-starck.com
//
Werbung etc.
Hans-Jürgen Fein, Andreas Sauer
www.werbungetc.de
//
Kaiser + more
Joachim Kaiser
www.km-net.de

Unternehmen — Company
Duravit AG
Werderstraße 36
78132 Hornberg
Germany
www.duravit.de
Produkt — Product
Paiova
Badewanne mit Möbelverkleidung und Rollcontainer in Amerikanisch Nussbaum — Bathtub with furniture casing and mobile storage in american walnut
Design
EOOS Design GmbH
Martin Bergmann, Gernot Bohmann, Harald Gründl
www.eoos.com
200

Unternehmen — Company
Duravit AG
Werderstraße 36
78132 Hornberg
Germany
www.duravit.de
Produkt — Product
Starck X
Badewanne freistehend mit Holzaufleger für Wannengraben — Bathtub free standing with wooden inlays for tub trench
Design
Philippe Starck
www.philippe-starck.com
198

Unternehmen — Company
Düring AG
Brunnenwiesenstrasse 14
8108 Dällikon
Switzerland
www.dueringag.ch
Produkt — Product
WC-ENTE AirDeo Kampagne „Blumen-WCs" — Toilet Duck AirDeo campaign "flower toilets"
Lufterfrischer — Airfreshner
Design
walker Werbeagentur
Pius Walker, Sophie Broadbridge, Glenn Gibbins
www.walker.ag

Unternehmen — Company
DW-TV
DEUTSCHE WELLE
Voltastraße 6
13355 Berlin
Germany
www.dw-world.de
Produkt — Product
Made in Germany
Identität für TV Magazin — Identity for TV magazine
Design
Velvet Mediendesign GmbH
Holger Geisler, Andi List, Torsten Lippstock, Jochen Kraus
www.velvet.de
//
Werksdesign / Inhouse
Holger Zeh
329

Unternehmen — Company
Dyckerhoff AG
Biebricher Straße 69
65203 Wiesbaden
Germany
www.dyckerhoff.com
Produkt — Product
Dyckerhoff TERRAPLAN
Betonboden — Concrete floor
Design
Werksdesign / Inhouse
Andre Conrady, Dr. Reinhard Winzer
194

Unternehmen — Company
Dyckerhoff AG
Biebricher Straße 69
65203 Wiesbaden
Germany
www.dyckerhoff.com
Produkt — Product
GESCHÄFTSBERICHT 2004 DYCKERHOFF AG
Geschäftsbericht — Annual report
Design
Heisters & Partner, Büro für Kommunikations-design
Valentin Heisters, Anja Griesler, Amöne Schmidt
www.heisters-partner.com

Unternehmen — Company
Dynatherm Medical, Inc.
46716 Lakeview Blvd
94538 Fremont
USA
www.dynathermmedical.com
Produkt — Product
Dynatherm Vital Heat
Therapeutisches Gerät — Therapeutic device
Design
Whipsaw, Inc.
Dan Harden, Ari Turgel, Hirotomi Teranishi,
David Benavidez, Tom Keegan, Gerson Goldberg
www.whipsawinc.com

Unternehmen — Company
Dyson GmbH
Lichtstraße 43b
50825 Köln
Germany
www.dyson.de
Produkt — Product
DC15 The Ball
Bürstensauger — Vacuum cleaner upright
Design
Dyson Team Forschung/Design/Entwicklung
251

e

Unternehmen — Company
edition GALERIE VEVAIS
Verlagsgesellschaft mbH
An der Dornbuschmühle 7
16269 Vevais
Germany
www.minotaurusverlag.de
Produkt — Product
William Blake, Die Hochzeit von Himmel und Hölle — William Blake, The Marriage of Heaven and Hell
Buchobjekt — Book object
Design
die Typonauten®
Krepinsky Krömer GbR
Stefan Krömer, Ingo Krepinsky
www.typonauten.de
348

Unternehmen — Company
Edition Panorama
G7,17
68159 Mannheim
Germany
www.editionpanorama.de
Produkt — Product
Alpen 2005 — The Alps 2005
Panorama-Kalender — Panorama calendar
Design
Wolff Kommunikation
Carsten Wolff, Wim Schafspelz, Thomas Rott,
Christiane Wolff
www.wolff-kommunikation.de

Unternehmen — Company
Edition Panorama
G7,17
68159 Mannheim
Germany
www.editionpanorama.de
Produkt — Product
Edition Panorama Bibliothek
Buchreihe im Panoramaformat — Panoramic photo book series
Design
Wolff Kommunikation
Carsten Wolff, Christiane Wolff, Thomas Rot,
Wim Schafspelz
www.wolff-kommunikation.de

Unternehmen — Company
Edition Panorama
G7,17
68159 Mannheim
Germany
www.editionpanorama.de
Produkt — Product
www.editionpanorama.de
Internetseite und Webshop — Website and webshop
Design
Wolff Kommunikation
Carsten Wolff, Wim Schafspelz, Thomas Rott,
Arve D. Fühler
www.wolff-kommunikation.de

Unternehmen — Company
ELCO Shared Services GmbH
Hohenzollernstraße 31
72379 Hechingen
Germany
www.elco.net
Produkt — Product
AURON
Vakuumröhrenkollektor —
Vacuum tube solar collector
Design
Weinberg & Ruf
Andreas Weinberg
www.weinberg-ruf.de
//
Werksdesign / Inhouse
Markus Hödl, Thomas Weidemann
309

Unternehmen — Company
Elephant Seven AG
Gerhofstraße 1–3
20354 Hamburg
Germany
www.e-7.com
Produkt — Product
25 Jahre quattro — 25 years of quattro
Internetseite — Website
Design
Werksdesign / Inhouse
Oliver Viets, Jan Hellberg, Elke Maasdorff,
Tina Kläring

Unternehmen — Company
Elephant Seven AG
Gerhofstraße 1–3
20354 Hamburg
Germany
www.e-7.com
Produkt — Product
B-Salat — Chewed up
Banner
Design
Werksdesign / Inhouse
Dirk Ollmann, Kai Becker, Lavinia Malkowski

Unternehmen — Company
Elephant Seven AG
Gerhofstraße 1–3
20354 Hamburg
Germany
www.e-7.com
Produkt — Product
Der Berg ruft. — The mountains are calling.
Banner
Design
Werksdesign / Inhouse
Dirk Ollmann, Sven Giese, Kai Becker,
Till Hinrichs

Unternehmen — Company
Elephant Seven AG
Gerhofstraße 1–3
20354 Hamburg
Germany
www.e-7.com
Produkt — Product
Punto Blanco update
Internetseite — Website
Design
Werksdesign / Inhouse
Joniel Franco da Silva, Kathrin Wiegandt,
Natalia Castaño

Unternehmen — Company
elmarflötotto GmbH
Am Ölbach 28
33334 Gütersloh
Germany
www.elmarfloetotto.de
Produkt — Product
LE BOCK
Tischgestell / Tischbock — Trestle
Design
Rupert Kopp

Unternehmen — Company
EMCO Bad GmbH & Co. KG
Breslauer Straße 34–38
49808 Lingen
Germany
www.emco.de
Produkt — Product
EMCO ASIS
Einbau-und Aufbaumodule für das Bad.
Lichtspiegelschränke — Build in-and wall
mounted modules for the bathroom.
Illuminated mirror cabinets
Design
Nexus Product Design
Ulli Finkeldey, Kai Uetrecht
www.nexusproductdesign.de

Unternehmen — Company
Eoos
Zelinkagasse 2 / 6
1010 Wien
Austria
www.eoos.com
Produkt — Product
a1 lounge – Flagshipstore
Flagshipstore für Mobilnetzbetreiber —
Flagshipstore for mobile phone provider
Design
Werksdesign / Inhouse

Unternehmen — Company
Eppendorf AG
Barkhausenweg 1
22339 Hamburg
Germany
www.eppendorf.com
Produkt — Product
Multipette® Xstream
Elektronischer Dispenser —
Electronic dispenser
Design
Koop Industrial Design
Norbert Koop
www.koopdesign.de

Unternehmen — Company
ERBE Elektromedizin GmbH
Waldhörnlestraße 17
72072 Tübingen
Germany
www.erbe-med.de
Produkt — Product
VIO 300 S
Hochfrequenz-Chirugiesystem — Electrosurgical system
Design
W & R Produktgestaltung
Andreas Weinberg
www.weinberg-ruf.de

Unternehmen — Company
ERLUS AG
Hauptstraße 106
84088 Neufahrn
Germany
www.erlus.com
Produkt — Product
Ergoldsbacher Karat XXL
Tondachziegel — Clay roof tile
Design
Hans-Wilhelm Seitz

Unternehmen — Company
Erwin Halder KG
Erwin-Halder-Straße 5–9
88480 Achstetten-Bronnen
Germany
www.halder.de
Produkt — Product
EH 2237. / EH 2238.
Kugelsperrbolzen mit elastischem Griff — Single-Acting Ball Lock Pin with elastic grip
Design
einmaleins – Büro für Gestaltung
Rupert Bläsi
www.einmaleins.net

Unternehmen — Company
Eschenbach Optik GmbH + Co
Schopenhauerstraße 10
90409 Nürnberg
Germany
www.eschenbach-optik.de / www.freizeitoptik.de
Produkt — Product
farlux selector V 8–15 x 35 B
Fernglas — Binocular
Design
Prof. Ralph Michel
//
Werksdesign / Inhouse
Cornelia Bäuml

Unternehmen — Company
Eschenbach Optik GmbH + Co
Schopenhauerstraße 10
90409 Nürnberg
Germany
www.eschenbach-optik.de / www.freizeitoptik.de
Produkt — Product
trophy AS / S 15–45 x 60 B
Spektiv — Spotting scope
Design
Prof. Ralph Michel
//
Werksdesign / Inhouse
Cornelia Bäuml

Unternehmen — Company
Esso Schweiz GmbH
Uraniastrasse 40
8001 Zürich
Switzerland
www.esso.ch
Produkt — Product
Geschäftsbericht 2004 der Esso Schweiz GmbH — The "Esso Schweiz GmbH" annual report 2004
Broschüre — Brochure
Design
Hans Peter Dubacher
//
Klauser Weibel Design
Tobias Klauser, Corinne Weibel
www.kwdesign.ch

Unternehmen — Company
Eternit Swiss AG
Niederurnen
8867 Niederurnen
Switzerland
www.eternit.ch
Produkt — Product
Twista
Pflanzengefäß Serie — Flowerpot series
Design
Martin Mostböck d.i. arch.
Martin Mostböck
www.martin-mostboeck.com

Unternehmen — Company
etkon AG
Lochhamer Schlag 6
82166 Gräfelfing
Germany
www.etkon.de
Produkt — Product
ES1
Dentalscanner — Dental scanner
Design
Oxxid
Stefan Reichl, Andreas Rößner, Verena Kehrer
www.oxxid.de

Unternehmen — Company
Europcar Autovermietung GmbH
Tangstedter Landstraße 81
22415 Hamburg
Germany
www.europcar.de
Produkt — Product
Europcar Point of Service
Corporate Interior und Branding — Corporate interior and branding
Design
feldmann+schultchen design studios
www.fsdesign.de
318

Unternehmen — Company
Eva Denmark A/S
Maaloev Teknikerby 18–20
2760 Maaloev
Denmark
www.evadenmark.com
Produkt — Product
CafeSolo
Kaffeezubereiter — Coffee maker
Design
Tools Design
Henrik Holbaek
www.toolsdesign.com
262

Unternehmen — Company
Eva Denmark A/S
Maaloev Teknikerby 18–20
2760 Maaloev
Denmark
www.evadenmark.com
Produkt — Product
Eva Solo
Blumentopf mir Wasserreservoir — Self-watering flowerpot
Design
Tools Design
Henrik Holbaek
www.toolsdesign.com
267

Unternehmen — Company
Eva Denmark A/S
Maaloev Teknikerby 18–20
2760 Maaloev
Denmark
www.evadenmark.com
Produkt — Product
Eva Solo
Grillfass — Barrel grill
Design
Tools Design
Henrik Holbaek
www.toolsdesign.com
266

Unternehmen — Company
Eva Denmark A/S
Maaloev Teknikerby 18–20
2760 Maaloev
Denmark
www.evadenmark.com
Produkt — Product
Eva Solo
Tropffreie Öl- und Essigkaraffen & Tropffreier Dressing-Shaker — Drip-free oil and vinegar carafes & Drip-free salad dressing shaker
Design
Tools Design
Henrik Holbaek
www.toolsdesign.com
265

Unternehmen — Company
Eva Denmark A/S
Maaloev Teknikerby 18–20
2760 Maaloev
Denmark
www.evadenmark.com
Produkt — Product
Eva Solo
Zucker-Sahne-Set — Milk and sugar set
Design
Tools Design
Henrik Holbaek
www.tooldesign.com
263

Unternehmen — Company
Eva Denmark A/S
Maaloev Teknikerby 18–20
2760 Maaloev
Denmark
www.evadenmark.com
Produkt — Product
Eva Solo
Teezubereiter — Tea maker
Design
Tools Design
Henrik Holbaek
www.toolsdesign.com
264

Unternehmen — Company
Expotechnik
Heinz Soschinski GmbH
Aarstraße 176
65232 Taunusstein
Germany
www.expotechnik.com
Produkt — Product
Delta
Theken- und Präsentationssystem — Counter and presentation system
Design
NOSE AG Design Intelligence
Ruedi August Müller, Christian Harbeke
www.nose.ch
//
Werksdesign/Inhouse
Heinz H. Soschinski
231

f

Unternehmen — Company
F. X. Nachtmann Bleikristallwerke GmbH
Zacharias-Frank-Straße 7
92660 Neustadt
Germany
www.nachtmann.com
Produkt — Product
Amalfi
Vase, Schale — Vase, bowl
Design
Christian Haas
www.haasdesign.de

Unternehmen — Company
F. W. Oventrop GmbH & Co. KG
Paul-Oventrop-Straße 1
59939 Olsberg
Germany
www.oventrop.de
Produkt — Product
Uni SH
Heizkörperthermostat — Radiator valve
Design
Design & Identität
Ulrich Prof. Hirsch
//
Werksdesign / Inhouse
Entwicklungs-und Konstruktionsabteilung

Unternehmen — Company
Fachhochschule Düsseldorf
Fachbereich Design
Georg-Glock-Straße 15
40474 Düsseldorf
Germany
www.fh.duesseldorf.de
Produkt — Product
„formvollendet" — "The perfect form"
Ausstellungskatalog — Exhibition catalogue
Design
Martin Hess, Andreas Uebele, Svenja Knödler, Jessica Krier, Yvonne Niewerth, Silke Sabow, Sarah Lenz
373

Unternehmen — Company
Fachhochschule Düsseldorf
Georg-Glock-Straße 15
40474 Düsseldorf
Germany
www.fh-duesseldorf.de
Produkt — Product
9xBrochure+Leporello+Factsheet
Broschüre — Brochure
Design
Werksdesign / Inhouse
Projektgruppe Kommunikationsmaschine
Projektleitung: Prof. Wilfried Korfmacher, Prof. Victor Malsy, Prof. Philipp Teufel

Unternehmen — Company
Fachhochschule Wiesbaden
Fachbereich Design / Informatik / Medien
Unter den Eichen 5
65195 Wiesbaden
Germany
www.kalenderkult-retrospektive.de
Produkt — Product
Das Schauspiel ´06 – Theater muss wie Fußball sein — The Schauspiel ´06 – Theatre needs to be like football
Kalender — Calendar
Design
Frotscher Druck GmbH
Wolfgang Frotscher
www.frotscher-druck.de
//
Werksdesign / Inhouse
David Fernandez del Campo, Karolin Kutter, Duc Nguyen, Melanie Tu

Unternehmen — Company
facts+fiction GmbH
Bayenstraße 4
50678 Köln
Germany
www.factsfiction.de
Produkt — Product
Aktion Mensch Messe
re:spect our future – die andere Jugendmesse
Jugendmesse — Youth fair
Design
sinkenarchitekten
Martin Sinken
www.sinken.de
//
Werksdesign / Inhouse
Mark Daniels, Robert Müller

Unternehmen — Company
facts+fiction GmbH
Bayenstraße 4
50678 Köln
Germany
www.factsfiction.de
Produkt — Product
Vivendi Universal Games Messestand
„Game Crazy – Spielverrückt"
Messestand — Fair stand
Design
sinkenarchitekten
Martin Sinken
www.sinken.de
//
Werksdesign / Inhouse
Dietmar Jähn, Martina Haag

Unternehmen — Company
FALKE KG
Oststraße 5
57392 Schmallenberg
Germany
www.FALKE.com
Produkt — Product
FALKE Second Layer
Pullover
Design
N3twork
Dennis Hinze
www.n3twork.de
112 Silber

Unternehmen — Company
FC Bayern Sport-Werbe GmbH
Säbener Straße 51
81547 München
Germany
www.fcbayern.de
Produkt — Product
TV Spot: Fan/FC Bayern München
Imagespot — Imagespot
Design
Heye & Partner GmbH
Norbert Herold, Jan Okusluk
www.heye.de

Unternehmen — Company
FELUWA Pumpen GmbH
Beulertweg
54570 Mürlenbach
Germany
www.feluwa.com
Produkt — Product
MULTISAFE Doppel-Schlauchmembranpumpe —
MULTISAFE Double Hose-Diaphragm Pump
Prozesspumpe — Process pump
Design
Fachhochschule Trier
Jonas Kartenbeck
www.fh-trier.de
//
Werksdesign/Inhouse
Heinz M. Nägel
299

Unternehmen — Company
Festo AG & Co. KG
Ruiter Straße 82
73734 Esslingen
Germany
www.festo.com
Produkt — Product
MS 12
Druckluftaufbereitungseinheit — Air supply unit
Design
Werksdesign/Inhouse
Jan Kleffmann

Unternehmen — Company
Festo Didactic GmbH & Co. KG
Rechbergstraße 82
73770 Denkendorf
Germany
www.festo.com
Produkt — Product
Robotino®
Mobiler Roboter — Training robot
Design
Werksdesign/Inhouse
Karoline Schmidt, Jan Kleffmann
300

Unternehmen — Company
FH Augsburg
Henisiusstraße 1
86152 Augsburg
Germany
www.fh-augsburg.de/gestaltung
Produkt — Product
citycrossing
Animationsfilm — Animated videoclip
Design
Werksdesign/Inhouse
Christian Leberfinger, Christoph Beckmann,
Mathias Bauer

Unternehmen — Company
Fiberline Composites A/S
Nr. Bjertvej 88
6000 Kolding
Denmark
www.fiberline.de
Produkt — Product
Fiberline
Fassadensytem — Facade system
Design
Werksdesign/Inhouse

Unternehmen — Company
Fiedler
Weißenburgstraße 32
70180 Stuttgart
Germany
Produkt — Product
Claro
Stuhl — Chair
Design
Werksdesign/Inhouse
Hugo Fiedler

Unternehmen — Company
Fischer GmbH
Fischerstraße 8
4910 Ried im Innkreis
Austria
www.fischer-ski.com
Produkt — Product
Fischer AMC/All Mountain Cruise
Alpinski — Alpine ski
Design
Spirit Design/Innovation and Branding
Daniel Huber, Guillaume Korompay,
Claudia Leeb
www.spiritdesign.at

Unternehmen — Company
Fischereiverband Bayern e.V.
Pechdeller Straße 16
81545 München
Germany
www.lfvbayern.de
Produkt — Product
Landesfischereiverband Bayern e. V. „Fischen in Bayern" — Fishing association of bavaria "Fishing in bavaria"
Plakat — Poster
Design
Heye & Partner GmbH
Jan Okusluk, Oliver Diehr
www.heye.de

Unternehmen — Company
Fissler GmbH
Harald Fissler Straße 1
55743 Idar-Oberstein
Germany
www.fissler.de
Produkt — Product
Intensa
Intensa Kochsystem — Intensa cooking system
Design
milani d & c
Britta Pukall
www.milani.ch
//
Werksdesign / Inhouse
Arno Thelen

Unternehmen — Company
Flechtwerk-Gestaltung
Laufer Straße 3
96199 Zapfendorf
Germany
Produkt — Product
Flechtleuchten — Woven lamps
Bodenleuchte, 32-teilig, Weide geschält — Floorlamp, 32 pieces, willow-white
Design
Werksdesign / Inhouse

Unternehmen — Company
Florado by Sylvia Gohlke
Blumenversand
Goethestraße 42
64285 Darmstadt
Germany
www.florado.de
Produkt — Product
Kabelstrauß — Cable flowers
Anzeigenentwurf — Advertising script
Design
Jung von Matt - Team Elbe
Carolin Bastian
www.jvm.de

Unternehmen — Company
Flötotto GmbH
Senner Straße 32
33335 Gütersloh
Germany
www.floetotto.de
Produkt — Product
system kvaro
Kastenmöbelsystem — Furniture system
Design
KOLBERGdesign
Justus Kolberg
www.kolbergdesign.com
212

Unternehmen — Company
formabina schmuckgestaltung
Mühlenstrasse 2a
8753 Mollis
Switzerland
Produkt — Product
trema
Armreif — Bangle
Design
Werksdesign / Inhouse
Bettina Geistlich

Unternehmen — Company
Forschungsinstitut und Naturmuseum Senckenberg
Senckenberganlage 25
60325 Frankfurt am Main
Germany
www.senckenberg.de
Produkt — Product
ILLUMINATION: EVOLUTION -
Eine Inszenierung zu Licht und Leben —
ILLUMINATION: EVOLUTION -
A light-and-life-production
Licht-und Medieninszenierung —
Light and media spectacle
Design
Atelier Markgraph GmbH
www.markgraph.de
//
FOUR TO ONE: scale design
Institut für Licht-und Bühnensimulation GmbH
www.4-to-1.com
//
SHOWTEC Beleuchtung- und Beschallungs GmbH
www.showtec.de
//
Praviz Mir-Ali
www.sl15.com
//
235 Media GmbH
www.235media.com
312

Unternehmen — Company
frame eleven
Schiffbaustrasse 10
8005 Zürich
Switzerland
www.frame-eleven.com
Produkt — Product
SF Schweizer Fernsehen, Werbetrenner/Serienlabel — SF Swiss Television, commercial separators/series labels
Broadcast Design — Broadcast design
Design
SF Schweizer Fernsehen
Alex Hefter, Thomas Gloor
www.sf.tv
//
Werksdesign/Inhouse
Marc Haas, Mike Frei

Unternehmen — Company
frame eleven
Schiffbaustrasse 10
8005 Zürich
Switzerland
www.frame-eleven.com
Produkt — Product
Musicnight-Gesamtauftritt — Broadcast Show Appearance
Broadcast Design — Broadcast Design
Design
SF Schweizer Fernsehen
Thomas Gloor, Alex Hefter
www.sf.tv
//
Werksdesign/Inhouse
Marc Haas, Jürg Egli, Adrian Sute, Mike Frei

Unternehmen — Company
Frame Publishers
Lijnbaansgracht 87
1015 GZ Amsterdam
Netherlands
www.framemag.com
Produkt — Product
Forefront: The Culture of Shop Window Design
Buch über 10 Schaufenstergestalter — Book on 10 shop window designers
Design
Thomas Buxo
www.buxo.nl

Unternehmen — Company
Francotyp-Postalia GmbH
Triftweg 21–26
16547 Birkenwerder
Germany
www.francotyp.de
Produkt — Product
optimail 30
Frankiermaschine — Franking machine
Design
CONSTIN GmbH
Lutz Friedrich
www.constin-design.de

Unternehmen — Company
FRANK-CONCEPT International Trading GmbH
Nikolaus-Otto-Straße 2
40721 Hilden
Germany
www.frank-concept.com
Produkt — Product
Cabrio — Cabriolet
Segelzelt — Sail tent, sun shade
Design
Annegret Weimer Kommunikation & Design
www.design-produkte.de

Unternehmen — Company
Franke Kaffeemaschinen AG
Franke-Strasse 9
4663 Aarburg
Switzerland
www.franke-cs.com
Produkt — Product
Spectra S
Kaffeemaschine — Coffee machine
Design
N. & P. Industrial Design GmbH.
Andreas Bergsträsser
www.neumeister-partner.de

Unternehmen — Company
Franz Fertig GmbH
Schützenstraße 7
74722 Buchen/Odenwald
Germany
//
die Collection
dC-Marketing GmbH
Schützenstraße 7
74722 Buchen/Odenwald
Germany
www.die-collection.de
Produkt — Product
TIARA
Polstermöbel — Upholstered furniture
Design
Designbüro Laprell Classen
Volker Classen, Volker Laprell
//
Werksdesign/Inhouse

Unternehmen — Company
Franz Fertig GmbH
Schützenstraße 7
74722 Buchen / Odenwald
Germany
//
die Collection
dC-Marketing GmbH
Schützenstraße 7
74722 Buchen / Odenwald
Germany
www.die-collection.de
Produkt — Product
SINO
Sessel — Armchair
Design
guggenbichlerdesign ...
Harald Guggenbichler
www.guggenbichler.at

Unternehmen — Company
Franz Kaldewei GmbH & Co. KG
Beckumer Straße 33–35
59229 Ahlen
Germany
www.kaldewei.com
Produkt — Product
COMFORT-LEVEL
Spezial Ab- und Überlaufgarnitur —
Drain and overflow fitting
Design
PHOENIX DESIGN
www.phoenixdesign.de

Unternehmen — Company
Freising_Brillen
Bergstraße 58
54295 Trier
Germany
www.hannesfreising.de
Produkt — Product
freising_brillen - modelle: freising august, freising ferdinand, freising gerhard, freising hannes, freising uwe —
freising_eyewear - models: freising august, freising ferdinand, freising gerhard, freising hannes, freising uwe
Brillen — Spectacles
Design
Hannes Freising

Unternehmen — Company
FROLI Kunststoffwerk Fromme GmbH
Liemker Straße 27
33758 Schloß Holte
Germany
www.froli.com
Produkt — Product
frolexus relax and fold
Klappbett — Foldawaybed
Design
www.UNITEDDESIGNWORKERS.com
Reiner Wallbaum, Andreas Kalweit, Christof Paul
www.uniteddesignworkers.com
//
Werksdesign / Inhouse
Hardy Keil, Reinhardt Göstenkors, Lothar Brockgreitens

Unternehmen — Company
Funke Kunststoffe GmbH
Siegenbeckstraße 15
59071 Hamm
Germany
www.funkegruppe.de
Produkt — Product
HS-VARIOmuffe — HS-VARIOcoupler
Rohrverbindungselement — Pipeconnection-element
Design
Werksdesign / Inhouse

g

Unternehmen — Company
Gaggenau Hausgeräte GmbH
Carl-Wery-Straße 34
81739 München
Germany
www.gaggenau.com
Produkt — Product
AT 400
Tischlüftung — Backdraft system
Design
Werksdesign / Inhouse
H. Reinhard Segers

Unternehmen — Company
Gaggenau Hausgeräte GmbH
Carl-Wery-Straße 34
81739 München
Germany
www.gaggenau.com
Produkt — Product
BL 253
Wandbackofen mit Liftmatic —
Wall mounted oven with Liftmatic
Design
Werksdesign / Inhouse
H. Reinhard Segers

Unternehmen — Company
Gaggenau Hausgeräte GmbH
Carl-Wery-Straße 34
81739 München
Germany
www.gaggenau.com
Produkt — Product
CI 481
Induktionskochfeld mit Twist-Pad-Bedienung —
Induction cooktop with twist-pad control
Design
Werksdesign / Inhouse
H. Reinhard Segers

Unternehmen — Company
Gaggenau Hausgeräte GmbH
Carl-Wery-Straße 34
81739 München
Germany
wwww.gaggenau.com
Produkt — Product
CI 482
Induktionskochfeld mit Twist-Pad-Bedienung —
Induction cooktop with twist-pad control
Design
Werksdesign / Inhouse
H. Reinhard Segers

Unternehmen — Company
Gaggenau Hausgeräte GmbH
Carl-Wery-Straße 34
81739 München
Germany
www.gaggenau.com
Produkt — Product
EM 220
Mikrowellengerät — Microwave oven
Design
Werksdesign / Inhouse
H. Reinhard Segers

Unternehmen — Company
Gaggenau Hausgeräte GmbH
Carl-Wery-Straße 34
81739 München
Germany
www.gaggenau.com
Produkt — Product
AW 560
Wandhaube — Wall hood
Design
Werksdesign / Inhouse
H. Reinhard Segers

Unternehmen — Company
Gaggenau Hausgeräte GmbH
Carl-Wery-Straße 34
81739 München
Germany
www.gaggenau.com
Produkt — Product
CE 490
Glaskeramik-Kochfeld — Glas ceramic hob
Design
Werksdesign / Inhouse
H. Reinhard Segers

Unternehmen — Company
Gaggenau Hausgeräte GmbH
Carl-Wery-Straße 34
81739 München
Germany
www.gaggenau.com
Produkt — Product
Gaggenau Magazin
Produktkatalog — Product catalouge
Design
:Ateliergemeinschaft Gerwin Schmidt
Timo Thurner

Unternehmen — Company
Gaggenau Hausgeräte GmbH
Carl-Wery-Straße 34
81739 München
Germany
www.gaggenau.com
Produkt — Product
KG 491
Gaskochfeld — Gas cooktop
Design
Werksdesign / Inhouse
H. Reinhard Segers

Unternehmen — Company
GAIA & GINO
Abdi Ipekci Cd. No:44 K:4
34368 Nisantasi, Istanbul
Turkey
www.gaiaandgino.com
Produkt — Product
Mistic
Kerzenhalter / Vase — Candleholder / Vase
Design
L DESIGN
Arik Levy
www.ldesign.fr

Unternehmen — Company
Galp Energia
Rua Tomás da Fonseca, Torre C
1600-209 Lisboa
Portugal
www.galpenergia.com
Produkt — Product
Pluma
Butangasflasche — Butane gas cylinder
Design
Brandia Creating
Rui Sampaio de Faria (Design Director), José Carlos Mendes (Graphic Design), Teresa Costa Reis, Rita Norton Sampaio, Tiago Mendes
www.brandia.net
46 Gold

Unternehmen — Company
GARDENA Manufacturing GmbH
Hans-Lorenser-Straße 40
89079 Ulm
Germany
www.gardena.com
Produkt — Product
GARDENA Spindelmäher 380 C — GARDENA Hand Cylinder Lawnmower 380 C
Spindelmäher — Cylinder lawnmower
Design
Henssler & Schultheiss Fullservice Productdesign GmbH
www.henssler-schultheiss.de

Unternehmen — Company
GARDENA Manufacturing GmbH
Hans-Lorenser-Straße 40
89079 Ulm
Germany
www.gardena.com
Produkt — Product
GARDENA Comfort Heckenscheren 570 / 600 — GARDENA Comfort (Gear) Hedge Clippers 570 / 600
Mechanische Heckenschere — Mechanical hedge clipper
Design
attivo creative resource
www.attivocreative.com

Unternehmen — Company
Gebr. Niessing GmbH & Co.
Schmuckmanufaktur
Butenwall 117
48691 Vreden
Germany
www.niessing.com
Produkt — Product
Dacapo
Schmucksystem — Jewelry system
Design
Werksdesign / Inhouse
Susanne Winckler
243

Unternehmen — Company
Gebr. Niessing GmbH & Co.
Schmuckmanufaktur
Butenwall 117
48691 Vreden
Germany
www.niessing.com
Produkt — Product
Sinus
Schmuckringe — Rings
Design
Werksdesign / Inhouse
Susanne Winckler, Timo Küchler
242

Unternehmen — Company
Gemeinschaftskrankenhaus Herdecke
Gerhard Kienle Weg 4
58313 Herdecke
Germany
www.gemeinschaftskrankenhaus.de
Produkt — Product
medizin individuell
Magazin — Magazine
Design
hilbig | strübbe partner
Frank Scheele
www.hilbig-struebbe-partner.de

Unternehmen — Company
geobra Brandstätter GmbH & Co. KG
PLAYMOBIL
Brandstätterstraße 2–10
90513 Zirndorf
Germany
www.playmobil.de
Produkt — Product
PLAYMOBIL Transport & Verkehr Hafen — PLAYMOBIL Cargo Harbor
Spielzeug — Toys
Design
Werksdesign / Inhouse
100 Silber

Unternehmen — Company
Georg Schlegel GmbH & Co. KG
Am Kapellenweg
88525 Dürmentingen
Germany
www.schlegel.biz
Produkt — Product
QRBLUVSE
Abschließbare und beleuchtbare Not-Aus-Taste — Lockable and illuminable emergency stop
Design
Werksdesign / Inhouse
E. Schlegel, R. Dolpp

Unternehmen — Company
Georg Schlegel GmbH & Co. KG
Am Kapellenweg
88525 Dürmentingen
Germany
www.schlegel.biz
Produkt — Product
RONTRON-Q-JUWEL
Befehlsgeräte — Control units
Design
Werksdesign / Inhouse
E. Schlegel, R. Dolpp

Unternehmen — Company
Georg Schlegel GmbH & Co. KG
Am Kapellenweg
88525 Dürmentingen
Germany
www.schlegel.biz
Produkt — Product
RONTRON-R-JUWEL
Befehlsgeräte — Control units
Design
Werksdesign / Inhouse
E. Schlegel, R. Dolpp

Unternehmen — Company
Gingco Werbeagentur GmbH & Co. KG
Gartenstraße 13
38114 Braunschweig
Germany
www.gingco.net
Produkt — Product
Gingco.Net United.
Unternehmensdarstellung —
Company presentation
Design
Werksdesign / Inhouse

Unternehmen — Company
GIRA Giersiepen GmbH & Co. KG
Dahlienstraße 12
42477 Radevormwald
Germany
www.gira.de
Produkt — Product
Gira Architektenbroschüre —
Gira architect brochure
Broschüre — Brochure
Design
Schmitz Visuelle Kommunikation
Hans Günter Schmitz, Thordis Ohler
www.hgschmitz.de

Unternehmen — Company
GIRA Giersiepen GmbH & Co. KG
Dahlienstraße 12
42477 Radevormwald
Germany
www.gira.de
Produkt — Product
GIRA SmartSensor
Kontroll-und Bedieneinheit —
Control and operating unit
Design
Ingenhoven Overdiek Architekten GmbH & Co. KG
www.ingenhoven-overdiek.de

Unternehmen — Company
GIRA Giersiepen GmbH & Co. KG
Dahlienstraße 12
42477 Radevormwald
Germany
www.gira.de
Produkt — Product
„Kunststofftechnik von GIRA" —
"Plastics Engineering by GIRA"
Informationsbroschüre — Brochure
Design
Schmitz Visuelle Kommunikation
Hans Günter Schmitz, Thordis Ohler
www.ghschmitz.de

Unternehmen — Company
GIU
Gesellschaft für Innovation und
Unternehmensförderung mbH
Nell-Breuning-Allee 8
66115 Saarbrücken
Germany
www.giu.de
Produkt — Product
Entwicklung eines neuen Corporate Designs
der GIU mbH — Development of a universal
Corporate Design for GIU mbH
Corporate Design — Corporate Design
Design
FBO Gesellschaft für Marketing-
Kommunikation mbH
Martin Oberhauser, Dominik Friedrich
www.fbo.de
370

Unternehmen — Company
Glashütter Uhrenbetrieb GmbH
„Glashütte Original"
Altenbergerstraße 1
01768 Glashütte / Sa.
Germany
www.glashuette-original.com
Produkt — Product
PanoMaticChrono
Mechanische Armbanduhr —
Mechanical wristwatch
Design
Werksdesign / Inhouse

Unternehmen — Company
Glaswerke Arnold
Alfred-Klingele-Straße 15
73630 Remshalden
Germany
www.glaswerke-arnold.de
Produkt — Product
ORNILUX SB1
Isolierglas gegen Vogelschlag —
Insulation glass against bird strike
Design
Werksdesign / Inhouse
Christian Irmscher
165

Unternehmen — Company
Gorenje Vertriebs-GmbH
Garmischer Straße 4–6
80339 München
Germany
www.gorenje.de
Produkt — Product
RK 67365 E
Kühl-Gefrier-Kombination — Fridge-Freezer
Design
Werksdesign / Inhouse

Unternehmen — Company
GP Acoustics Ltd.
Eccleston Road, Tovil
ME15 6QP Maidstone, Kent
Great Bitain
www.kef.com
Produkt — Product
KHT 6000ACE
3-Wege Satelliten-Lautsprecher mit Bass-extender-Standfuß — Three way satellite louds-peaker with 'Bass Extender' bass enhancing stand
Design
Werksdesign / Inhouse
Andrew Watson, Mark Burdett, Barry Dock

Unternehmen — Company
Greenpeace Media GmbH
Große Elbstraße 145d
22767 Hamburg
Germany
www.greenpeace-magazin.de
Produkt — Product
„Tu was!" Greenpeace Magazin 01 / 06
Magazin — Magazine
Design
Büro Hamburg JK. PW. Gesellschaft für Kommunikationsdesign mbH
Jürgen Kaffer, Kerstin Leesch, Bettina Rosenow, Annette Gassner
www.buero-hamburg.de
360

Unternehmen — Company
Greenpeace Media GmbH
Große Elbstraße 145d
22767 Hamburg
Germany
www.greenpeace-magazin.de
Produkt — Product
Greenpeace Magazin 01 / 04
„Die Welt auf Kriegskurs"
Magazin — Magazine
Design
Büro Hamburg JK. PW. Gesellschaft für Kommunikationsdesign mbH
Bettina Rosenow, Kerstin Leesch, Katja Kleinebrecht
www.buero-hamburg.de

Unternehmen — Company
Grohe AG
Industriepark Edelburg
58675 Hemer
Germany
www.grohe.com
Produkt — Product
K4
Küchenarmaturenlinie — Kitchen faucet line
Design
Werksdesign / Inhouse
Grohe Design Team

Unternehmen — Company
Grohe AG
Industriepark Edelburg
58675 Hemer
Germany
www.grohe.com
Produkt — Product
Lineare
Badarmaturenlinie — Faucet line
Design
Werksdesign / Inhouse
Grohe Design Team

Unternehmen — Company
Grohe AG
Ringstraße 18
32427 Minden
Germany
www.grohe.com
Produkt — Product
Messestand zur SHK Essen 2006 — Trade Fair SHK Essen 2006
Messestand — Fair stand
Design
Schmidhuber + Partner GbR
www.schmidhuber.de

Unternehmen — Company
Grohe AG
Industriepark Edelburg
58675 Hemer
Germany
www.grohe.com
Produkt — Product
Allure
Armaturenlinie — Faucet line
Design
Werksdesign / Inhouse
Grohe Design Team

Unternehmen — Company
Groke Türen und Tore GmbH
Wikingerstraße 10
76189 Karlsruhe
Germany
www.groke.de
Produkt — Product
Groke Designer Collection Charactere 100144
Hauseingangstür aus Aluminium — Aluminium entrance door
Design
Frank König
163

Unternehmen — Company
Groupe SEB Deutschland GmbH
Geschäftsbereich KRUPS
Herrnrainweg 5
63067 Offenbach
Germany
www.krups.de
Produkt — Product
KRUPS Prep Expert / KA 8027
Kompakt-Küchenmaschine — Food processor
Design
Konstantin Grcic Industrial Design
Konstantin Grcic
www.konstantin-grcic.com
254

Unternehmen — Company
Groupe SEB Deutschland GmbH
Geschäftsbereich KRUPS
Herrnrainweg 5
63067 Offenbach
Germany
www.krups.de
Produkt — Product
KRUPS Prep Expert / KB 7207
Standmixer — Blender
Design
Konstantin Grcic Industrial Design
Konstantin Grcic
www.konstantin-grcic.com
255

Unternehmen — Company
gumpo Büromöbel
Gobener Weg 27
84130 Dingolfing
Germany
www.gumpo.de
Produkt — Product
pinatec
Schreibtisch — Desktop
Design
Votteler+Votteler
Prof. Arno Votteler, Matthias Votteler

Unternehmen — Company
Gute Gesellschaft mbH
Grafenberger Allee 126
40237 Düsseldorf
Germany
www.gutegesellschaft.com
Produkt — Product
Virtual Republic – Staatsgründung —
Virtual Republic – foundation
Unternehmensbroschüre Mailings, Internetseite — Corporate folder, mailings, website
Design
Werksdesign / Inhouse
Raimund Schmelzer, Ruben Cuellas Is
350

Unternehmen — Company
Gute Gesellschaft mbH
Strategie, Design und Kommunikation
Grafenberger Allee 126
40237 Düsseldorf
Germany
www.gutegesellschaft.com
Produkt — Product
Die Kleine Marie — Little Mary
Corporate Design
Design
Ruben Cuellas Is, Raimund Schmelzer,
Heiko Schulz, Tim Turiak

h

Unternehmen — Company
H & T Raumdesign AG
Falt-und Schiebewände
Rohrerstrasse 20
5001 Aarau
Switzerland
www.schiebewaende.com
Produkt — Product
H & T SGS
Schalldämmende Glasschiebewand —
Sound-absorbing all-glass wall
Design
Werksdesign / Inhouse
Thomas Zulauf, Christian Horlacher
164

Unternehmen — Company
H. Hiendl GmbH & Co. KG
Marke: Serpa
Industriestraße 5 + 6
94327 Bogen-Furth
Germany
www.serpa.de
Produkt — Product
Serpa Kabelbündler D10 —
Serpa cable wraps D10
Kabelbündler — Cable wraps
Design
Werksdesign / Inhouse

Unternehmen — Company
hackenschuh com. design
Lindenspürstraße 22
70176 Stuttgart
Germany
www.hackenschuh.com
Produkt — Product
Guy Tillim
Katalog für DaimlerChrylser Award, Südafrika — Catalogue for DaimlerChrylser Award, South-Africa
Design
Werksdesign / Inhouse
Christina Hackenschuh
371

Unternehmen — Company
häfelinger + wagner design GmbH
Türkenstraße 55–57
80799 München
Germany
www.hwdesign.de
Produkt — Product
mo–fr, Auszüge einer unvollständigen Chronologie
Kalender — Calendar
Design
Werksdesign / Inhouse
Frank Wagner, Kurt Steinebrunner, Martina Lothes

Unternehmen — Company
häfelinger + wagner design GmbH
Türkenstraße 55–57
80799 München
Germany
www.hwdesign.de
Produkt — Product
Zukunft Technik. ThyssenKrupp — Future Technology. ThyssenKrupp
Geschäftsbericht 2003 / 2004 — Annual report 2003 / 2004
Design
Werksdesign / Inhouse
Annette Häfelinger, Jenny Heinecker

Unternehmen — Company
Hager Electro GmbH
Zum Gunterstal
66440 Blieskastel
Germany
www.hager.de
Produkt — Product
Serie VC — VC series
Installationskleinverteiler — Consumer unit
Design
Werksdesign / Inhouse
Klaus Dissel

Unternehmen — Company
Hager Electro GmbH
Zum Gunterstal
66440 Blieskastel
Germany
www.hager.de
Produkt — Product
Corporate Design der Hager Gruppe — Corporate Design of the Hager group
Visuelle Kommunikation, Geschäftsdrucksachen, produktbezogene Drucksachen, Internetauftritt, Drucksachen zum 50-jährigen Jubliäum — Visual communication, stationery, product-related printed matters, Internet presence, printed materials on the occasion of the 50th anniversary of Hager
Design
Stankowski + Duschek
Karl Duschek
www.st-du.de

Unternehmen — Company
HAMM AG
Hammstraße 1
95643 Tirschenreuth
Germany
www.hamm.eu
Produkt — Product
DV 40, DV 70, DV 90
Asphalt-Tandemwalze Serie DV — Asphalt Tandemroller Series DV
Design
Dialogform GmbH
Ulrich Ewringmann, Boris Eickhoff
www.dialogform.de
//
Werksdesign / Inhouse
Hans-Peter Ackermann

Unternehmen — Company
Hanns Schmid Grafikdesign
Bruggerstrasse 69
5400 Baden
Switzerland
www.schmidgrafikdesign.ch
Produkt — Product
SAAB
Plakatserie für PKW SAAB — Poster series for passenger car SAAB
Design
Werksdesign / Inhouse
Hanns Schmid
147

Unternehmen — Company
Hans Otto Theater Potsdam GmbH
Schiffbauergasse 11
14467 Potsdam
Germany
www.hansottotheater.de
Produkt — Product
Corporate Design Hans Otto Theater Potsdam
Corporate Design (Spielzeithefte, Programmkarten, Plakate, Geschäftsausstattung) — Corporate Design (season preview, programme, posters, business stationery)
Design
formdusche gbr, Berlin
Svenja von Döhlen, Borries Schwesinger, Steffen Wierer
www.formdusche.de
367

Unternehmen — Company
Hansaton Akustik GmbH
Stückenstraße 48
22081 Hamburg
Germany
www.hansaton.de
Produkt — Product
Hansaton FREE SOUNDMANAGER
Digitales Hörsystem — Digital hearing system
Design
Pilotfish GmbH
www.pilotfish.eu

Unternehmen — Company
Hansgrohe AG
Auestraße 5–9
77761 Schiltach
Germany
www.hansgrohe.com
Produkt — Product
Focus S Einhebel Waschtischmischer — Focus S single-lever basin mixer
Einhebel-Waschtischmischer — Single-lever basin mixer
Design
Phoenix Design
www.phoenixdesign.de

Unternehmen — Company
Hansgrohe AG
Auestraße 5–9
77761 Schiltach
Germany
www.hansgrohe.com
Produkt — Product
Hansgrohe Croma S 100 3jet
Handbrause — Hand shower
Design
Phoenix Design
www.phoenixdesign.de

Unternehmen — Company
Hansgrohe AG
Auestraße 5–9
77761 Schiltach
Germany
www.hansgrohe.com
Produkt — Product
Imagekampagne „Hans!" — Image Campaign "Hans!"
Imagekampagne — Image campaign
Design
Werbung etc.
www.werbungetc.de

Unternehmen — Company
Hansgrohe AG
Auestraße 5–9
77761 Schiltach
Germany
www.hansgrohe.com
Produkt — Product
Metris S Elektronik Armatur — Metris S Electronic Tap
Waschtischarmatur — Washbasin mixer
Design
Phoenix Design
www.phoenixdesign.de

Unternehmen — Company
Hansgrohe AG
Auestraße 5–9
77761 Schiltach
Germany
www.hansgrohe.com
Produkt — Product
Pharo ShowerArc
Duschsystem — Shower system
Design
Phoenix Design
www.phoenixdesign.de

Unternehmen — Company
Hansgrohe AG
Auestraße 5–9
77761 Schiltach
Germany
www.hansgrohe.com
Produkt — Product
Raindance S 120 AIR 3jet
Handbrause — Hand shower
Design
Phoenix Design
www.phoenixdesign.de
203

Unternehmen — Company
Hansgrohe AG
Auestraße 5–9
77761 Schiltach
Germany
www.hansgrohe.com
Produkt — Product
Axor Massaud Dreiloch-Waschtischmischer — Axor Massaud three-hole washbasin mixer
Dreiloch-Waschtischmischer — Three-hole washbasin mixer
Design
Jean Marie Massaud
www.massaud.com

Unternehmen — Company
Hansgrohe AG
Auestraße 5–9
77761 Schiltach
Germany
www.hansgrohe.com
Produkt — Product
Axor Massaud
Einhebel-Waschtischmischer —
Single-lever washbasin mixer
Design
Jean Marie Massaud
www.massaud.com
202

Unternehmen — Company
Hansgrohe AG
Auestraße 5–9
77761 Schiltach
Germany
www.hansgrohe.com
Produkt — Product
Axor Citterio Küchenarmatur —
Axor Citterio kitchen mixer
Küchenarmatur — Kitchen mixer
Design
Antonio Citterio
www.antonio-citterio.it

Unternehmen — Company
Harman Deutschland GmbH
Hünderstraße 1
74080 Heilbronn
Germany
www.harmanmultimedia.com
//
JBL, Inc.
8500 Balboa Blvd.
CA 91329 Northridge
USA
www.jbl.com
Produkt — Product
JBL OnStage II
iPod-Dock mit Lautsprechern —
iPod docking station with speakers
Design
Werksdesign / Inhouse
Kurt Solland
284

Unternehmen — Company
Harman Deutschland GmbH
Hünderstraße 1
74080 Heilbronn
Germany
www.harmanmultimedia.com
//
JBL, Inc.
8500 Balboa Blvd.
CA 91329 Northridge
USA
www.jbl.com
Produkt — Product
JBL OnTour
Portables Stereo-Lautsprechersystem —
Portable stereo speaker system
Design
Werksdesign / Inhouse
Kurt Solland
285

Unternehmen — Company
Hase Kaminofenbau GmbH
Niederkircher Straße 14
54294 Trier
Germany
www.hase.de
Produkt — Product
AKABA
Kaminofen — Stove
Design
Werksdesign / Inhouse
Walter Blasius

Unternehmen — Company
Hasso von Blücher
Mettmanner Straße 25
40699 Erkrath
Germany
www.bluecher.com
Produkt — Product
Let's do it! Let's go!
Broschüre — Brochure
Design
Hesse Design GmbH
Klaus Hesse, Sandra Klemt, René Clohse
www.hesse-design.com
353

Unternehmen — Company
Haworth GmbH
Am Deisterbahnhof 6
31848 Bad Münder
Germany
www.haworth-europe.com
Produkt — Product
Zody_System 89
Bürodrehstuhl-Serie — Office chair series
Design
ITO Design Team
www.ito-design.com
225

Unternehmen — Company
Hebbel am Ufer
Hallesches Ufer 32
10963 Berlin
Germany
www.hebbel-am-ufer.de
Produkt — Product
HEBBEL AM UFER – HAU123 – Kampagne —
HEBBEL AM UFER – HAU123 – CAMPAIGN
Plakatkampagne — Poster campaign
Design
Double Standards
Chris Rehberger, Todd Schulz,
Monika Rehberger
www.doublestandards.net
132 Silber

Unternehmen — Company
Heidelberger Druckmaschinen AG
Kurfürsten-Anlage 52–60
69115 Heidelberg
Germany
www.heidelberg.com
Produkt — Product
Heidelberger Druckmaschinen AG „Created with Passion, made for Success" Messestand-design für die drupa 04, Düsseldorf
Messestand — Fair stand
Design
stengele+cie.
Nadine Aschenbrücker, Britta Heiner, Volker Stengele
www.stengele-cie.de
//
Generators-Communication GmbH
Thomas Schulze
www.generators-communication.de

Unternehmen — Company
HEINRICH KIPP WERK
Spanntechnik . Normelemente . Bedienteile
Heubergstraße 2
72172 Sulz
Germany
www.kipp.com
Produkt — Product
KIPP Exzenterhebel
Exzenterhebel mit Innen-und Außengewinde — Ex centrical lever with interior and exterior thread
Design
Werksdesign/Inhouse
Heinrich Kipp

Unternehmen — Company
Helix Vertriebsagentur
Limburger Straße 3
61462 Königstein im Taunus
Germany
www.helix-produkte.de
Produkt — Product
HELIOS
Armbanduhr — Wristwatch
Design
Botta-Design
Klaus Botta, Ingmar Boos
www.botta-design.de

Unternehmen — Company
Helly Hansen
Solgaard Skog 139
218 Moss
Norway
www.hellyhansen.com
Produkt — Product
Shorehike
Schuhe — Footwear
Design
David Solk

Unternehmen — Company
HENK
Smelterij 4
2211 SH Noordwijkerhout
Netherlands
www.henk.com
Produkt — Product
HENK
Exklusive Reisekoffer — Exclusive luggage
Design
Spark design·engineering
Robert Barnhoorn, Robbert Boost, Bartel Timmermans, Rogier van Rossen, Antony Weinbeck, Alex van de Wetering
www.sparkdesign.nl

Unternehmen — Company
Hess Form+Licht GmbH
Schlachthausstraße 19–19/3
78050 Villingen-Schwenningen
Germany
www.hess-form-licht.com
Produkt — Product
Millennio
LED-Straßenleuchte — LED streetlights
Design
Bauwerkstadt Dortmund
Karsten Winkels
www.bauwerkstadt.net

Unternehmen — Company
Hess Form+Licht GmbH
Schlachthausstraße 19–19/3
78050 Villingen-Schwenningen
Germany
www.hess-form-licht.com
Produkt — Product
Night Elements
Lichtsäulensystem — Pole luminaire
Design
Vogt & Partner
www.lichtgestaltung.ch

Unternehmen — Company
Hiller Objektmöbel GmbH & Co. KG
Kippenheimer Straße 6
77971 Kippenheim
Germany
www.hiller-moebel.de
Produkt — Product
system 24
Klapptisch — Folding table
Design
Korb+Korb Architektur-Innenarchitektur-Design
Daniel Korb
www.korb-korb.ch
222

Unternehmen — Company
HMH Engineering-Consulting-Trading GmbH
Im Südpark 196
4030 Linz
Austria
www.rubblemaster.com
Produkt — Product
RM70 Compact Recycler
Bauschutt-Recyclinganlage —
Rubble recycling plant
Design
idukk industrial design union kittler
kurz partner
Reinhard Kittler, Heinrich Kurz,
Gerald Fasching
www.idukk.at
96 Silber

Unternehmen — Company
Hochhaus Werbeagentur GmbH
Brüsseler Straße 92
50672 Köln
Germany
www.dashochhaus.de
Produkt — Product
Kölner Wohnzimmertheater Kampagne „Wohnzimmertapete" — Campaign "Living room theater cologne" / "Living room wallpaper"
Plakatserie — Poster series
Design
Werksdesign / Inhouse
Carsten Brück, Esther Bröhl

Unternehmen — Company
Hochschule für Künste Bremen
Am Speicher XI 8
28215 Bremen
Germany
www.hfk-bremen.de
Produkt — Product
Art, Music, Design, Theorie. Veranstaltungen Wintersemester 04/05. Hochschule für Künste / University of the Arts / Bremen —
Art, Music, Design, Theorie. Veranstaltungen Wintersemester 04/05. Hochschule für Künste / University of the Arts / Bremen
Vorlesungsverzeichnis — University calendar
Design
Stefan Bargstedt, Anne Rösemeier,
Philipp Dörrie, Betie Pankoke

Unternehmen — Company
Hochschule Furtwangen /
Fakultät Digitale Medien / tunespoon.tv
Robert-Gerwig-Platz 1
78120 Furtwangen
Germany
www.tunespoon.tv
Produkt — Product
tunespoon.tv – Online Musikfernsehen —
tunespoon.tv – online music television
Internetseite — Website
Design
Patrick Schwab, Martin Schmidt,
Johannes Wagner
www.sense-art.de
//
Werksdesign / Inhouse
Florian Brandel, Christian Ensslen,
Uwe Erlenwein, Kirsten Franke

Unternehmen — Company
HOCHTIEF Aktiengesellschaft
Opernplatz 2
45128 Essen
Germany
www.hochtief.de
Produkt — Product
concepts by HOCHTIEF
Unternehmenspublikation — B-to-B Magazine
Design
HOFFMANN UND CAMPE Corporate Publishing
Annette Arnheim, Dirk Linke
www.hoca-cp.de
359

Unternehmen — Company
HOFFMEISTER LEUCHTEN GmbH
Am Neuen Haus 4–10
58507 Lüdenscheid
Germany
www.hoffmeister.de
Produkt — Product
elli.bi
Strahler — Projector
Design
Ottenwälder und Ottenwälder
www.ottenwaelder.de

Unternehmen — Company
HOFFRICHTER GmbH
Mettenheimer Straße 12/14
19061 Schwerin
Germany
www.hoffrichter.de
Produkt — Product
CARAT II
Heimbeatmungsgerät — Home care ventilation
Design
Werksdesign / Inhouse
Gerald Hross

Unternehmen — Company
Holzprojekte Härtenberger HKH GmbH
Hauptstraße 39
94336 Hunderdorf
Germany
www.haertenberger.de
Produkt — Product
Jumbo
Küchenmöbel — Kitchen furniture
Design
ALNO – Produktdesign
www.alno.de
//
Werksdesign / Inhouse
Stephan Härtenberger
205

Unternehmen — Company
Hove Møbler
Stordal
6250 Stordal
Norwegen
www.hove-mobler.no
//
fjordfiesta
Fjordfiesta Furniture AS
Fuglsetbakken 41
6416 Molde
Norwegen
www.fjordfiesta.com
Produkt — Product
SCANDIA SENIOR
Lehnstuhl — Easy chair
Design
Hans Brattrud

Unternehmen — Company
Huidi Lauhoff
Wilhelminestraße 4
65193 Wiesbaden
Germany
www.huidi-lauhoff.de
Produkt — Product
Website Huidi Lauhoff
Internetauftritt zur Präsentation der Modekollektion von Huidi Lauhoff — Webspecial to present the fashion collection designed by Huidi Lauhoff
Design
Scholz & Volkmer, Wiesbaden
www.s-v.de

Unternehmen — Company
HWAM Heat Design AS
Nydamsvej 53
8362 Hoerning
Denmark
www.hwam.com
Produkt — Product
HWAM 3320
Kaminofen — Stove
Design
STRAND + HVASS
Niels Hvass, Christina Strand
www.strand-hvass.com
171

Unternehmen — Company
HWG Inductoheat GmbH
Ostweg 5
73262 Reichenbach
Germany
www.hwg-inductoheat.de
Produkt — Product
INDUCTOSCAN-flex
Induktionshärteanlage — Induction hardening system
Design
SK Design
Cordula Schneider, Reinhard Kohn
www.sk-design-bw.de

Unternehmen — Company
HYDAC ELECTRONIC GMBH
Hauptstraße 27
66128 Saarbrücken
Germany
www.hydac.com
Produkt — Product
EDS 3000
Elektronischer Druckschalter — Electronic pressure switch
Design
industrial:design
Prof. Wolfgang Schabbach
//
Werksdesign / Inhouse
Thomas Herrmann

Unternehmen — Company
HYDAC ELECTRONIC GMBH
Hauptstraße 27
66128 Saarbrücken
Germany
www.hydac.com
Produkt — Product
HMG 3000
Mobiles Handmessgerät — Hand-held measuring instrument
Design
industrial:design
Prof. Wolfgang Schabbach
//
Werksdesign / Inhouse
Thomas Herrmann

Unternehmen — Company
HYMER AG
Holzstraße 19
88339 Bad Waldsee
Germany
www.hymer.com
Produkt — Product
Eriba-Feeling
Wohnwagen — Caravan
Design
Werksdesign / Inhouse
Rainer Buck
150

Unternehmen — Company
HYMER AG
Holzstraße 19
88339 Bad Waldsee
Germany
www.hymer.com
Produkt — Product
HYMER Van
Reisemobil — Motorhome
Design
Hymer IDC
Prof. Johann Tomforde
www.hymer-idc.de
//
Werksdesign / Inhouse
Rainer Buck
149

i

Unternehmen — Company
i / i / d Institut für Integriertes Design
Am Seicher XI, Abt. 7, Bd. 3
28217 Bremen
Germany
www.iidbremen.de
Produkt — Product
://creating future / 99–04
Publikation — Publication
Design
i / i / d Institut für Integriertes Design
Detlef Rahe, Michael Wilde

Unternehmen — Company
ic! Berlin Brillen GmbH
Saarbrücker Straße 37
10405 Berlin
Germany
www.ic-berlin.de
Produkt — Product
www.ic-berlin.de
Internetseite — Website
Design
unit-berlin, dedecke, dertmann, gbr
Hinnerk Dedecke, Heike Dertmann,
Thilo Fuchs, Pierre le Gonidec
www.unit-berlin.de

Unternehmen — Company
ic! Berlin Brillen GmbH
Saarbrücker Straße 37
10405 Berlin
Germany
www.ic-berlin.de
Produkt — Product
„natascha"
Sonnenbrillen, Kollektion spring / summer 2005
— Sunglasses, spring / summer 2005
Design
Werksdesign / Inhouse
Fabian Hofmann

Unternehmen — Company
icotec ag
Industriestrasse 12
9450 Altstätten
Switzerland
www.icotec.biz
Produkt — Product
Kimba Cage nach Magerl —
Kimba Cage by Magerl
Wirbelsäulenimplantat aus Composite —
Spinal implant made of composite
Design
Werksdesign / Inhouse
Prof. Friedrich Magerl, Roger Tognini,
Bekim Bajramaj

Unternehmen — Company
IDEAL STANDARD GMBH
Industriekultur
Euskirchener Straße 80
53121 Bonn
Germany
Produkt — Product
Ceraluna
Küchenarmatur — Kitchen fitting
Design
ARTEFAKT
www.artefakt.de

Unternehmen — Company
Ideenwerkstatt Pfaffenberger
Römersbergweg 13
95448 Bayreuth
Germany
www.pfaffenbergeridee.de
Produkt — Product
Eggi
Stuhl — Chair
Design
Werksdesign / Inhouse
Micheal Pfaffenberger

Unternehmen — Company
iDoc-Institut
Institut für medizinische Fachinformation,
Patientenaufklärung und Diagnostik
GmbH & Co. KG
Posthofstraße 8
14467 Potsdam
Germany
www.idoc.de
Produkt — Product
iDoc-Schlafapnoe — iDoc Sleep Apnea
Telemedizinisches Informations-und
Diagnostiksystem — Telemedic information
and diagnostic system
Design
Bildbau AG Neue Medien
www.bildbau.de
//
Werksdesign / Inhouse
Nikolaus Böhning

Unternehmen — Company
IKK-Bundesverband
Friedrich-Ebert-Straße / TechnologiePark
51429 Bergisch Gladbach
Germany
www.ikk.de
Produkt — Product
www.ikk.de
Internetseite — Website
Design
denkwerk gmbh
Uwe Thiel, Nadine Müller, Sandra Griffel,
Jörg Offermann, Christoph Langnickel
www.denkwerk.com

Unternehmen — Company
ims Industrial Micro Systems AG
Zürcherstrasse 262
8406 Winterthur
Switzerland
www.ims-ag.ch
Produkt — Product
haze
LED-Tischleuchte, simled —
LED table lamp, simled
Design
Greutmann Bolzern Designstudio
Carmen Greutmann, Urs Greutmann
www.gbdesign.ch

Unternehmen — Company
in(corporate GmbH
communication + design GmbH
Rosenthaler Straße 38
10178 Berlin
Germany
www.incorporate.de
Produkt — Product
Vermarktungsmedien Ackerlofts —
Brochure Ackerlofts
Verkaufsprospekt — Sales brochure
Design
Werksdesign / Inhouse
366

Unternehmen — Company
Infotech AG
Vogelherdstrasse 4
4500 Solothurn
Switzerland
www.infotech-automation.com
Produkt — Product
Onyx 25
Löt-Maschine — Soldering system
Design
i Design AG
Michael Koch
www.i-design.ch

Unternehmen — Company
Ing. Castaldi Illuminazione Srl
Via Carlo Goldoni 18
20090 Trezzano sul Naviglio
Italy
www.castaldilluminazione.com
Produkt — Product
D46 diablo
Leuchte für Aussenbeleuchtung LED bollard-
typ — Outdoor lighting fixture LED bollard-style
Design
Werksdesign / Inhouse

Unternehmen — Company
Ing. Erich Pfeiffer.de
Oeschlestraße 54–56
78315 Radolfzell
Germany
www.pfeiffer-group.com
Produkt — Product
User Independent Package
Mikrodosiersystem — Micro spray system
Design
yellow design | yellow circle
www.yellowdesign.com
280

Unternehmen — Company
Inga-Lisa Burst
Wörthstraße 36
81667 München
Germany
www.ingalisa.com
Produkt — Product
Subjekt: private -public
Themenspezifisches Magazin in Form einer
Akte — Specialised magazine in file format
Design
Werksdesign / Inhouse
Inga-Lisa Burst

Unternehmen — Company
Ingenhoven Architekten
Plange Mühle 1
40221 Düsseldorf
Germany
www.ingenhovenarchitekten.de
Produkt — Product
Flash-Animation / Video / Website
www.ingenhovenarchitekten.de
Internetseite — Website
Design
schmitz Visuelle Kommunikation
Hans Günter Schmitz, Nicolas Markwald
www.hgschmitz.de

Unternehmen — Company
INSTANT Corporate Culture
Leerbachstraße 57
60322 Frankfurt / Main
Germany
www.e-instant.de
Produkt — Product
INSTANT Nr. 57: Männer am Rande des
Nervenzusammenbruchs — INSTANT Nr. 57:
Men on the brink of nervous breakdown
Broschüre — Brochure
Design
Heye & Partner GmbH
Thomas Feicht, Norbert Herold, Marc Herold,
Jörg Stöckigt
www.heye.de

Unternehmen — Company
Interflex Datensysteme GmbH & Co. KG
Zettachring 16
70567 Stuttgart
Germany
www.interflex.de
Produkt — Product
IF-701
Terminal für Zutrittskontrolle —
Terminal for access controll
Design
Neumeister & Partner
www.neumeister-partner.de
289

Unternehmen — Company
interlübke
Gebr. Lübke GmbH & Co. KG
Ringstraße 145
33378 Rheda-Wiedenbrück
Germany
www.interluebke.com
Produkt — Product
Sipario
Schwebe-/Schiebetürensystem mit Raumteilerfunktion — Partition system with gliding doors (room deviding system)
Design
Schneider GmbH
Klaus Schneider
www.der-ideenschreiner.de

Unternehmen — Company
International School of Stuttgart e.V.
Sigmaringerstraße 257
70597 Stuttgart
Germany
www.international-school-stuttgart.de
Produkt — Product
International School of Stuttgart, „viewpoints"
Buch — Book
Design
Engenhart Visuelle Kommunikation
Marc Engenhart
www.engenhart.com
//
Lothar Bertrams Fotograf DGPh
www.lotharbertrams.com
345

Unternehmen — Company
INTERSPORT Deutschland eG
Wannenäckerstraße 50
74078 Heilbronn
Germany
www.intersport.de
Produkt — Product
INTERSPORT FunWheelGuide 2005
www.intersport-funwheelguide.de
Internetseite — Website
Design
work.id Werbeagentur Gesellschaft für Communication & Marketing mbH
Jürgen Hensler
www.workid.com

Unternehmen — Company
Interstil, Diedrichsen GmbH & Co. KG
Liebigstraße 1–3
33803 Steinhagen
Germany
www.interstil.de
Produkt — Product
outlook 4
Vorhanggarnitur — Curtain set
Design
Design AG
Frank Greiser
www.design-ag.de

j

Unternehmen — Company
J.P. Sauer & Sohn
Maschinenbau GmbH
Brauner Berg 15
24159 Kiel
Germany
www.sauersohn.de
Produkt — Product
Tornado
Hochdruckkompressor —
High pressure compressor
Design
mehnert corporate design gmbh & co. kg
Prof. Kurt Mehnert, Jan Eickhoff,
Alasdair Wright
www.mehnertdesign.de
307

Unternehmen — Company
JAB Josef Anstoetz KG
Potsdamer Straße 160
33719 Bielefeld
Germany
www.jab.de
Produkt — Product
Metropolitan
Produkt- und Bestellkatalog —
Product and order catalogue
Design
Peter Maly
www.peter-maly.de
//
Carsten Gollnick
www.gollnick-design.de
//
Werksdesign / Inhouse
Maria Loer-Reuschel
339

Unternehmen — Company
JAB Josef Anstoetz KG
Potsdamer Straße 160
33719 Bielefeld
Germany
www.jab.de
Produkt — Product
Score 1-7825-126
Dekostoff — Home textiles
Design
Marty Lamers

Unternehmen — Company
JAB Josef Anstoetz KG
Potsdamer Straße 160
33719 Bielefeld
Germany
www.jab.de
Produkt — Product
Swirl 1-7831-124
Dekostoff — Home textiles
Design
Marty Lamers

Unternehmen — Company
JAB TEPPICHE HEINZ ANSTOETZ KG
Dammheider Straße 67
32052 Herford
Germany
www.jab.de
Produkt — Product
LASA 7204/311
Teppich-Kollektion LANA L III —
Rug collection LANA L III
Design
Werksdesign/Inhouse
191

Unternehmen — Company
JAB TEPPICHE HEINZ ANSTOETZ KG
Dammheider Straße 67
32052 Herford
Germany
www.jab.de
Produkt — Product
LEM 7205/520
Teppich-Kollektion LANA L III —
Rug collection LANA L III
Design
Werksdesign/Inhouse
190

Unternehmen — Company
JAB TEPPICHE HEINZ ANSTOETZ KG
Dammheider Straße 67
32052 Herford
Germany
www.jab.de
Produkt — Product
LOA 7200/760
Teppich-Kollektion LANA L III —
Rug collection LANA L III
Design
Werksdesign/Inhouse
192

Unternehmen — Company
Jaxx GmbH
Behringstraße 16a
22765 Hamburg
Germany
www.jaxx.de
Produkt — Product
JAXX Corporate Design
Corporate Design
Design
Factor Design AG
Johannes Erler, Sonja Stroth
www.factordesign.com

Unternehmen — Company
Jean Abt Innenausbau GmbH
Ferdinand-Porsche-Straße 27
60386 Frankfurt am Main
Germany
www.jean-abt.de
//
CC-Concepts LTD
448 Suffolk Road
7011 Nelson
Australia
www.circular-kitchen.com
Produkt — Product
CleverKitchen
Hightech Rundküche — Kitchen
Design
CC-Concepts LTD
Averbeck Alfred
www.circular-kitchen.com

Unternehmen — Company
Jenoptik Laser, Optik, Systeme GmbH
Göschwitzer-Straße 25
07745 Jena
Germany
www.jenoptik-los.de
Produkt — Product
Sinarback eMotion 22
Digitales Rückteil für Mittelformatkameras —
Digital medium format camera back
Design
ART-KON-TOR
Marvin Kock
www.artkontor.de

Unternehmen — Company
Jens Ritter Instruments
Weinstraße
67146 Deidesheim
Germany
www.bassconcept.com
Produkt — Product
ROYA
Bass Gitarre — Bass guitar
Design
Werksdesign/Inhouse

Unternehmen — Company
Joh. Hermann Picard GmbH & Co. KG
Hammerschmiede
Rottsiepen 15
42349 Wuppertal
Germany
Produkt — Product
790
Latthammer — Carpenters roofing hammer
Design
Werksdesign / Inhouse

Unternehmen — Company
Jones Lang Lasalle Holdings GmbH & Co. KG
Mies-van-der-Rohe-Straße 6
80807 München
Germany
www.joneslanglasalle.de
Produkt — Product
Park Hyatt Hotel, Zürich
Buch — Book
Design
UNID – One Voice Branding
Ralph Missy, Martina Donat, Stanislav Erma
www.unid.de
//
Werksdesign / Inhouse
346

Unternehmen — Company
Julius Blum GmbH
Industriestraße 1
6973 Höchst
Austria
www.blum.com
Produkt — Product
Blumotion 973 A
Dämpfungssystem für Türen —
Damping system for doors
Design
form orange
Wolfgang Held
www.form-orange.com

Unternehmen — Company
Jung und Pfeffer: Visuelle Kommunikation
Langenstraße 6–8
28195 Bremen
Germany
www.jungundpfeffer.de
Produkt — Product
:output 07 international yearbook and student award in visual communication 2004
Buch — Book
Design
Jung und Pfeffer: Visuelle Kommunikation
Petra Michel, Ole Kaleschke, Florian Pfeffer
www.jungundpfeffer.de

Unternehmen — Company
JURA Elektroapparate AG
Kaffeeweltstrasse 10
4626 Niederbuchsiten
Switzerland
www.jura.com
Produkt — Product
IMPRESSA Z5 Chrom
Espresso- / Kaffee-Vollautomat —
Fully automatic espresso and coffee machine
Design
Zemp + Partner Design
Werner Zemp
www.zemp-objets.ch

Unternehmen — Company
Jütte-Messedruck Leipzig GmbH
Ostwaldstraße 4
04329 Leipzig
Germany
www.juette-messedruck.de
//
Rosebud, Inc.
Ralf Herms
Pelzetleite 65
90614 Ammerndorf
Germany
www.rosebudmagazine.com
Produkt — Product
+rosebud No.5 „MYSTERY"
Design-Magazin — Design magazine
Design
Werksdesign / Inhouse
Ralf Herms, Fritz T. Magistris, Katja Fössel

k

Unternehmen — Company
KAHLA / Thüringen Porzellan GmbH
Christian-Eckardt-Straße 38
07768 Kahla
Germany
www.kahlaporzellan.com
Produkt — Product
Elixyr
Porzellankollektion — Porcelain collection
Design
Werksdesign / Inhouse
Barbara Schmidt

Unternehmen — Company
KAHLA / Thüringen Porzellan GmbH
Christian-Eckardt-Straße 38
07768 Kahla
Germany
www.kahlaporzellan.com
Produkt — Product
touch!
Porzellan mit samtweicher Oberflächen-gestaltung — Porcelain with a velvety surface
Design
speziell produktgestaltung
Thilo Schwer, Jens Pohlmann,
Sybille Fleckenstein
www.speziell.net
//
Werksdesign / Inhouse
Barbara Schmidt, Cornelia Müller,
Mirjami Rissanen

Unternehmen — Company
Kallenbach – Leipzig
Gohliser Straße 20
04105 Leipzig
Germany
www.kallenbach-leipzig.de
Produkt — Product
S.O.S.
Sessel, Sofa — Seat, sofa
Design
Torsten Klocke
www.desarteur.de

Unternehmen — Company
karl-heinz best . mind the gap! design
Florstädter Straße 37
60385 Frankfurt
Germany
www.mindthegap-design.com
Produkt — Product
Corporate Design karl-heinz best .mind the gap! design
Corporate Design
Design
Werksdesign / Inhouse
259

Unternehmen — Company
KATHREIN Werke KG
Anton Kathrein Straße 1–3
83022 Rosenheim
Germany
www.kathrein.de
Produkt — Product
800 10270
Mobilfunk-Tri-Sector-Antenne —
Mobile communications tri-sector antenna
Design
Werksdesign / Inhouse

Unternehmen — Company
KATHREIN Werke KG
Anton Kathrein Straße 1–3
83022 Rosenheim
Germany
www.kathrein.de
Produkt — Product
BZD 40
DVB-T-Antenne für den Außenbereich —
DVB-T antenna – outdoor
Design
Werksdesign / Inhouse

Unternehmen — Company
KATHREIN Werke KG
Anton Kathrein Straße 1–3
83022 Rosenheim
Germany
www.kathrein.de
Produkt — Product
UFT 171, UFS 301, UFS 702, UFS 821
Digital-Receiver-Familie 2005 —
Digital receivers, product line 2005
Design
Werksdesign / Inhouse

Unternehmen — Company
KBE Engineering GmbH
Robert-Bosch-Straße 5
72172 Sulz a. N.
Germany
www.kbe.de
Produkt — Product
enflo 0060 / 0.5
Windkraftturbine — Wind generator
Design
Werksdesign / Inhouse
Michael Ehrenberger
302

Unternehmen — Company
Keijsers Interieurprojecten
Postbus 6025
5960 AA Horst
Netherlands
www.decoprojekt.nl
//
Deco Project
Aaltenseweg 52
7051 CM Varsseveld
Netherlands
www.keijsers.nl
Produkt — Product
T-Mobile Retail
Flagship store — Flagship store
Design
Storeage
Martijn Hoogendijk
www.store-age.nl

Unternehmen — Company
Kembo
Koningsschot 55
3905 PR Veenendaal
Netherlands
www.kembo.com
Produkt — Product
MI
Büromöbelsystem — Office furniture system
Design
Werksdesign / Inhouse
Erik Munnikhof
226

Unternehmen — Company
KEMMLIT-Bauelemente GmbH
Sanitäreinrichtungen für höchste Ansprüche
Maltschachstraße 37
72144 Dusslingen
Germany
www.kemmlit.de
Produkt — Product
NOXX
Trennwandsystem — Cubicle system
Design
wöhr mieslinger architekten
Wolfram Wöhr, Jörg Mieslinger
//
Werksdesign / Inhouse
Dr. Marc Kemmler, Kurt Binder
230

Unternehmen — Company
Kenwood Ltd.
New Lane, Hampshire
PO9 2NH Havant
Great Bitain
www.kenwoodworld.com
//
De'Longhi Deutschland GmbH
Geschäftsbereich Kenwood
Am Reitpfad 20
63500 Seligenstadt
Germany
www.delonghi.de
Produkt — Product
Kenwood CM 901 Eon-Serie
Kaffeemaschine — Coffee maker
Design
Werksdesign / Inhouse
Mark Morgan

Unternehmen — Company
KeraTür GmbH & Co. KG
Vennekenweg 12
46348 Raesfeld
Germany
www.keratuer.de
Produkt — Product
3010.1 (ExclusivLine modern)
Hauseingangstür — Front door
Design
Werksdesign / Inhouse
Julian Kemming

Unternehmen — Company
KeraTür GmbH & Co. KG
Vennekenweg 12
46348 Raesfeld
Germany
www.keratuer.de
Produkt — Product
63401.2 -4W.3400 (SpecialLine Space)
Hauseingangstür — Front door
Design
Werksdesign / Inhouse
Werner Kemming

Unternehmen — Company
Kermi GmbH
Pankofen-Bahnhof 1
94447 Plattling
Germany
www.kermi.de
Produkt — Product
Walk-in-Glasanlagen — Walk-in-showers
Duschkabine — Shower enclosure
Design
Werksdesign / Inhouse

Unternehmen — Company
Kermi GmbH
Pankofen-Bahnhof 1
94447 Plattling
Germany
www.kermi.de
Produkt — Product
Credo-Half
Badheizkörper — Bathroom radiator
Design
artefakt büro für industriekultur
www.artefakt.de

Unternehmen — Company
Kern GmbH
Elbinger Straße 12
64625 Bensheim
Germany
www.kern.ch
Produkt — Product
Kern 3500
Hochleistungskuvertiersystem 2-Kanal — High-speed inserting system dual-channel
Design
Neufo Design
Guido Vazau
www.neufodesign.ch

Unternehmen — Company
KEUCO GmbH & Co. KG
Oesestraße 36
58675 Hemer
Germany
www.keuco.de
Produkt — Product
Edition 300
Komplettprogramm Badmöbel, Waschtisch, Badaccessoires, Spiegelschrank, Armaturen. — Bathroom furniture, washbasin, bathroom accessoires, mirror cabinet, fittings
Design
Tesseraux + Partner
Dominik Tesseraux
www.tesserauxundpartner.de

Unternehmen — Company
KFF Design
Bahnhofstraße 27
32657 Lemgo
Germany
www.kff.de
Produkt — Product
GLOOH
Stuhlserie — Chair series
Design
Werksdesign / Inhouse
Hansjörg Helweg, Rolf Helweg
221

Unternehmen — Company
King Kong Mietstudios
Brühlstraße 63
71292 Friolzheim
Germany
www.kingkongmietstudios.de
Produkt — Product
King Kong Studios
Corporate Design „Bigger than life"
Geschäftsausstattung / Corporate Design —
Stationery / Corporate Design
Design
BBDO Campaign GmbH Stuttgart
Armin Jochum, Andreas Rell, Maja
Hünemörder, Mimi Schif
www.bbdo.de

Unternehmen — Company
Kinnasand GmbH
Danziger Straße 6
26655 Westerstede
Germany
www.kinnasand.com
Produkt — Product
SPIRIT
(Dekorations-) Stoff —
(Interior decoration) Fabric
Design
Werksdesign / Inhouse
Kinnasand Design Team
188

Unternehmen — Company
KircherBurkhardt GmbH
Oranienburger Straße 66
10117 Berlin
Germany
www.kircher-burkhardt.com
Produkt — Product
Editorial Solutions
Image Broschüre —
Self promotion for KircherBurkhardt
Design
Werksdesign / Inhouse
Katrin Hoffmann, Victoria Sarapina

Unternehmen — Company
Klaus Bartels
Echtholzdesign
Jägerstraße 14
95349 Thurnau
Germany
www.echtholzdesign.de
Produkt — Product
Stuhl „Caresse der Rückenschmeichler" —
Chair "Caress your back"
Stuhlserie — Range of chairs
Design
Werksdesign / Inhouse
Klaus Bartels
219

Unternehmen — Company
Klaus Klein GmbH
Zum Kaiserbusch 18
48165 Münster
Germany
www.kleindesign-muenster.de
Produkt — Product
Element No. 2
Waschtisch — Wash stand
Design
Werksdesign / Inhouse
Jürgen Klein

Unternehmen — Company
KMS Team GmbH
Deroystraße 3–5
80335 München
Germany
www.kms-team.de
Produkt — Product
KMS Showreel
Showreel — Show reel
Design
Werksdesign / Inhouse
Knut Maierhofer, Dirk Koy, Cecil Rustemeyer,
Sarah Graf, Christian Ring

Unternehmen — Company
KNAUS TABBERT GROUP GMBH
Helmut-Knaus-Straße 1
94118 Jandelsbrunn
Germany
www.knaus.de
Produkt — Product
Knaus Sun Ti 650 MF
Reisemobil — Motorhome
Design
DESIGNquadrat GbR
Alexander Christ
www.designquadrat.de
148

Unternehmen — Company
Knürr AG
Mariakirchener Straße 38
94424 Arnstorf
Germany
www.knuerr.com
Produkt — Product
CoolTherm®
Flüssigkeitsgekühlter Serverschrank —
Water-cooled server rack
Design
Werksdesign / Inhouse
Christian Stepputat

Unternehmen — Company
Koziol >>ideas for friends GmbH<<
Werner-von-Siemens-Straße 90
64711 Erbach
Germany
www.koziol.de
Produkt — Product
Josephine
Pendelleuchte — Pendant
Design
Werksdesign / Inhouse
Jürgen Diehl
175

Unternehmen — Company
KRONES AG
Böhmerwaldstraße 5
93073 Neutraubling
Germany
www.krones.com
Produkt — Product
Krones Produktprogramm —
Krones Product Range
Anlagen für Getränkeabfüllung und
-verpackung — Machines and devices for
beverage bottling and packaging
Design
Hartmann & Hartmann
Werner Hartmann, Andreas Kämmerer,
Patrick Bober
www.hartmannundhartmann.com

Unternehmen — Company
KTM Sportmotorcycle AG
Stallhofnerstraße 3
5230 Mattighofen
Austria
www.ktm.at
Produkt — Product
KTM 990 Super Duke
Straßenmotorrad — Streetbike
Design
Kiska GmbH
Gerald Kiska, Bernd Stelzer
www.kiskadesign.com

Unternehmen — Company
Kunstschmiede H.-J. Bender
Oberstiftstraße 45
54338 Schweich
Germany
www.kunstschmiede-bender.de
Produkt — Product
Zaun-, Gitter- und Torelemente —
Fence-, grid- and gate elements
Design
Werksdesign / Inhouse

Unternehmen — Company
Küppersbusch Hausgeräte AG
Küppersbuschstraße 16
45883 Gelsenkirchen
Germany
www.kueppersbusch.de
Produkt — Product
Dampfgarer EKDG 6800.1M
Combined steam oven EKDG 6800.1M
Kücheneinbaugerät — Built-in appliance
Design
Werksdesign / Inhouse
Klaus Keichel

Unternehmen — Company
Küppersbusch Hausgeräte AG
Küppersbuschstraße 16
45883 Gelsenkirchen
Germany
www.kueppersbusch.de
Produkt — Product
EEH 6500.5MX
Einbauherd — Built-in electric cooker
Design
Werksdesign / Inhouse
Klaus Keichel

Unternehmen — Company
Küppersbusch Hausgeräte AG
Küppersbuschstraße 16
45883 Gelsenkirchen
Germany
www.kueppersbusch.de
Produkt — Product
TKD 9880.0GE
Dunstabzugshaube — Cooker hood
Design
Werksdesign / Inhouse
Klaus Keichel

Unternehmen — Company
Kvadrat GmbH
Schöne Aussicht 20
61348 Bad Homburg
Germany
www.kvadrat.de
Produkt — Product
Tord Boontje & Kvadrat
Möbelstoffe und Fenstertextilien —
Upholstery and window covering
Design
Tord Boontje
www.tordboontje.com

Unternehmen — Company
Kverneland Geldrop BV
Neunenseweg 165
5660 AA Geldrop
Netherlands
www.deutz-fahr.de
Produkt — Product
DEUTZ-FAHR Mäher KM 4.27 FS – 4.29 FS
Trommelmäher — Drum mower
Design
Kooymans Industrieel | Ontwerp
Stuart W. Purdy
www.kooymans.nu

Unternehmen — Company
KW43 BRANDDESIGN
Gladbacher Straße
40219 Düsseldorf
Germany
www.kw43.de
Produkt — Product
KW43 BRANDDESIGN
Corporate Design
Design
KW43 BRANDDESIGN
www.kw43.de
//
Werksdesign / Inhouse
Kirsten Koehler, Prof. Dr. Rüdiger Goetz

l

Unternehmen — Company
L2M3
Kommunikationsdesign GmbH
Hölderlinstraße 57
70193 Stuttgart
Germany
www.L2M3.com
Produkt — Product
L2M3 Geschäftsausstattung —
L2M3 Corporate Design
Geschäftsausstattung — Office equipment
Design
Werksdesign / Inhouse
Sascha Lobe, Ina Bauer

Unternehmen — Company
Laboratory Tickner AG
Industriestrasse 47
6304 Zug
Switzerland
www.pronovamedical.com
Produkt — Product
Tickner
Zeckenentferner — Tick removal system
Design
NPK Industrial Design
www.npk.nl

Unternehmen — Company
Langebartels & Jürgens
Druckereigesellschaft mbH
Friedensallee 120
22763 Hamburg
www.team120.de
Produkt — Product
Selbstdarstellung L & J — Self-expression
Aufklappbare Selbstdarstellung —
Fold-out self-expression
Design
Format Design
Knut Ettling
www.format-hh.com

Unternehmen — Company
Lebast
Reiserdorf 13
92721 Störnstein
Germany
Produkt — Product
Lehm-Trockenbauplatte — Clay dryboard
Design
Werksdesign / Inhouse
Peter Gmeiner

Unternehmen — Company
Lehnen GmbH
Zum Rachtiger Wald 3
54516 Wittlich
Germany
www.lehnen-edelstahl.de
Produkt — Product
Lehnen Evolution
Serie Lehnen Evolution – Komfortable
Sicherheit im Bad — Serie Lehnen Evolution –
Comfortable Safety in your Bathroom
Design
brodbeck design
Stefan Brodbeck, Henriette Lipinski
www.brodbeckdesign.de
//
Werksdesign / Inhouse
Alois Lehnen

Unternehmen — Company
LEICHT KÜCHEN AG
Gmünder Straße 70
73550 Waldstetten
Germany
www.leicht.de
Produkt — Product
INS / INZ
Innenschubkasten INS und Innenauszug
INZ mit Glasfronten — Inner drawers (INS)
and Inner pullouts (INZ) as high quality glass
furnishing to INS and INZ
Design
Werksdesign / Inhouse
Manfred Deffner

Unternehmen — Company
LEISE/Designentwicklung und Vertrieb
Töngesgasse 40
60311 Frankfurt am Main
Germany
www.leise-leise.com
Produkt — Product
Studio Kollektion/May, August, Julies —
Studio Collection/May, August, Julies
Möbel — Furniture
Design
Werksdesign/Inhouse
Knut Völzke

Unternehmen — Company
Leitner GmbH
Düsseldorfer Straße 14
71332 Waiblingen
Germany
www.leitner.de
Produkt — Product
L_12/L_12DS
Großraum-Messesystem/Doppelstock-Messe-
system — Large-scale system/double-deck
Design
Werksdesign/Inhouse
Dieter Kassner, Alexander Recher

Unternehmen — Company
lemonfish GmbH
Im Rank 10
73655 Plüderhausen
Germany
www.lemonfish.de
Produkt — Product
Spencer
Reisetaschenset — Duffle bag with side pocket
Design
Werksdesign/Inhouse
Bettina Burchard, Alexandra Dittrich
249

Unternehmen — Company
Leni Hoffmann
Uhlandstraße 23a
40237 Düsseldorf
Germany
Produkt — Product
leni hoffmann
beautiful one day – perfect the next
Künstlerbuch — Artist book
Design
Werksdesign/Inhouse
Leni Hoffmann
343

Unternehmen — Company
Lensvelt office furniture
Minervum 7003
4817 ZL Breda
Netherlands
www.lensvelt.nl
Produkt — Product
This chair
Konferenz-Stuhl, stapelbar —
Contract chair, stackable and connectable
Design
Richard Hutten Studio bv
Richard Hutten
www.richardhutten.com
220

Unternehmen — Company
Liebling Verlag GmbH
Torstraße 129
10119 Berlin
Germany
www.liebling-zeitung.com
Produkt — Product
Liebling
Zeitung — Newspaper
Design
Floor 5 DEsign (GbR)
Stefan Guzy Jens Pieper, Marek Polewski
www.floor5.de

Unternehmen — Company
lif GmbH
SCHOTT AG
Etzwiesenstraße 33
72108 Rottenburg
Germany
www.lif-germany.de
Produkt — Product
LightPoints
LEDs im PVB-Flachglas-Folienverbund —
LEDs in flatglas laminate
Design
Werksdesign/Inhouse
Christoph Döppner, Rolf A. O. Schneider,
Bernd Albrecht, Andreas Nickut
179

Unternehmen — Company
Ligalux GmbH
Waterloohain 5
22769 Hamburg
Germany
www.ligalux.de
Produkt — Product
„Mutbekenntnisse" —
"Declarations of Courage"
Weihnachtsmailing — Christmas Mailing
Design
Ligalux GmbH
Claudia Fischer-Appelt, Hannes Mussbach,
Vicky Arndt
www.ligalux.de

Unternehmen — Company
Lightwave B. Flury – LED Products
Schlattweg 558
5704 Egliswil
Switzerland
www.lightwave-led.com
Produkt — Product
12.6003 – UFO
Hängeleuchte — Pendant
Design
Werksdesign/Inhouse
Beat Flury, Steven Knobel
176

Unternehmen — Company
LINDBERG
Bjarkesvej 30
8230 Aabyhoej
Denmark
www.lindberg.com
Produkt — Product
LINDBERG 2000 Spirit Plate Temples
Brillen — Spectacles
Design
Werksdesign / Inhouse
LINDBERG Designgruppe
247

Unternehmen — Company
Linde AG
Abraham-Lincoln-Straße 21
65189 Wiesbaden
Germany
www.linde.com
Produkt — Product
Linde „Was zählt" — Linde "what matters"
Geschäftsbericht 2004 — Annual report 2004
Design
KW43 BRANDDESIGN
Christian Vöttiner
Armin Hanke
www.kw43.de

Unternehmen — Company
Linde Material Handling GmbH & Co. KG
Schweinheimer Straße 34
63743 Aschaffenburg
Germany
www.linde-mh.de
Produkt — Product
H100 - H180
Gabelstapler — Fork lift truck
Design
Porsche Engineering Group GmbH
www.porscheengineering.com

Unternehmen — Company
Linn Deutschland GmbH
Hühnerposten 1d
20097 Hamburg
Germany
www.linn.co.uk
Produkt — Product
KLIMAX Chakra Twin
Stereo-Endverstärker — Stereo power amplifier
Design
Werksdesign / Inhouse
Alastair Brown

Unternehmen — Company
Linn Products Limited
Glasgow Road
G760EQ Waterfoot, Eaglesham, Glasgow
Great Britain
www.linn.co.uk
Produkt — Product
KLIMAX Kontrol
Stereo-Vorverstärker — Stereo pre amplifier
Design
Werksdesign / Inhouse
Alastair Brown

Unternehmen — Company
Linotype GmbH
Du-Pont-Straße 1
61352 Bad Homburg
Germany
www.linotype.com
Produkt — Product
TRYTYPE - EXPERIMENTS IN TYPOGRAPHY
Buch — Book
Design
Fachhochschule Wiesbaden,
Fachbereich Gestaltung
Guido Ludes, Stefanie Weber
www.fh-wiesbaden.de

Unternehmen — Company
Linotype GmbH
Du-Pont-Straße 1
61352 Bad Homburg
Germany
www.linotype.com
Produkt — Product
www.linotype.com/fontcomparer
Online-Tool Font Comparer —
Online-Tool Font Comparer
Design
SCHOLZ & VOLKMER Intermediales
Design GmbH
www.s-v.de

Unternehmen — Company
Loewe AG
Industriestraße 11
96317 Kronach
Germany
www.loewe.de
Produkt — Product
Individual 32
LCD-TV — LCD-TV
Design
Phoenix Design
www.phoenixdesign.de
60 Silber

Unternehmen — Company
Loewe AG
Industriestraße 11
96317 Kronach
Germany
www.loewe.de
Produkt — Product
Spheros R 37
LCD-TV
Design
Phoenix Design
www.phoenixdesign.de

Unternehmen — Company
Loewe AG
Industriestraße 11
96317 Kronach
Germany
www.loewe.de
Produkt — Product
Loewe IFA 2005
Messestand — Fair stand
Design
Braun Wagner
www.braunwagner.de
//
Interbrand, Zintzmeyer & Lux
www.interbrand.de
64 Silber

Unternehmen — Company
Lonza AG
Muenchensteinerstrasse 138
4002 Basel
Switzerland
www.lonza.com
Produkt — Product
Lonza
Messestand — Fair stand
Design
Tribolet arch
Bernardo Tribolet
www.tribolet-arch.ch
//
Ueberholz GmbH
Nico Ueberholz
www.ueberholz.de
327

Unternehmen — Company
Ludwig Leuchten KG
Frühlingstraße 15
86415 Mering
Germany
www.ludwig-leuchten.de
Produkt — Product
421–388 FC10
Deckeneinbauleuchte —
Recessed ceiling luminaire
Design
Bartenbach Lichtlabor GmbH
www.bartenbach.com
177

Unternehmen — Company
Ludwig Schröder GmbH & Co KG
Tornescher Weg 105
25436 Uetersen
Germany
www.ludwigschroeder.de
Produkt — Product
Gürtelkollektion
Ludwig Schröder
Gürtel — Belts
Design
Werksdesign / Inhouse
Ludwig Schröder

Unternehmen — Company
Lumex Lichttechnik
Bonholzstraße 6
71111 Waldenbuch
Germany
www.leuchtrahmen.de
Produkt — Product
Line Light
LED-Leuchte — LED light
Design
Werksdesign / Inhouse
Gerhard Müller
181

m

Unternehmen — Company
MAB Betschart AG
Hauptstrasse 178
6436 Muotathal
Switzerland
www.mab-moebel.ch
//
die Collection
dC-Marketing GmbH
Schützenstraße 7
74722 Buchen/Odenwald
Germany
Produkt — Product
STRETCH
Kastenmöbel — Wall system (modular cabinets and shelving units)
Design
Thomas Althaus

Unternehmen — Company
Maetch
Naanhofsweg 76
6361 DP Nuth
Netherlands
Produkt — Product
Jef Wishaupt
Buch — Book
Design
Polka Design
Joep Pohlen, Dennis Schmitz
www.polka.nl

Unternehmen — Company
MAGIS SPA
Via Magnadola 15
31045 Motta di Livenza (TV)
Italy
www.magisdesign.com
Produkt — Product
TRIOLI
Kinderstuhl — Childrens' chair
Design
Aarnio Design Ltd.
Eero Aarnio
www.aarnio-design.com

Unternehmen — Company
Mammut Sports Group AG
Industriestrasse
5703 Seon
Switzerland
www.mammut.ch
Produkt — Product
Pulse Barryvox
Lawinenverschüttetensuchgerät — Avalanche beacon
Design
Rigami Design
Richard Amiel
www.rigami-design.ch

Unternehmen — Company
Mapa
NUK
Industriestraße 21–25
27404 Zeven
Germany
Produkt — Product
NUK Oral Care System
Kombi-Beißringe und Zahnpflege-Lernset — Massaging, training, and caring system for oral care
Design
yellow design yellow circle
Alexander Schlag
www.yellowdesign.com

Unternehmen — Company
markilux®
Schmitz-Werke GmbH + Co. KG
Hansestraße 87
48282 Emsdetten
Germany
www.markilux.com
Produkt — Product
markilux® 6000
Gelenkarm-Markise — Folding-arm awning
Design
kramerDesign
Andreas Kramer
www.kramer-produkt-design.de
//
Werksdesign/Inhouse
Sven Kröner, Luise Schniggendiller, Martin Dieckmann, Holger Honrath
166

Unternehmen — Company
Martin et Karczinski
Mauerkircherstraße 8
81679 München
Germany
www.martinetkarczinski.de
Produkt — Product
„Branding Interface, Gespräche über Markenkommunikation von morgen"
Buch — Book
Design
Werksdesign/Inhouse
Peter Martin, Daniel Karczinski, Daniela Scharf. Nicola Sacher, Andrea Schwabl, Frank Widemann, Antje Stumpe, Beate Mini

Unternehmen — Company
MARTOR KG
Heider Hof 60
42653 Solingen
Germany
www.martor.de
Produkt — Product
MEGASAFE
Sicherheitsmesser — Safety cutter
Design
Martin Rohrbach
www.rohrbach-design.de

Unternehmen — Company
Maschinen & Formenbau Leinetal MFL GmbH
Basserstraße 2a
31535 Neustadt
Germany
www.mfl-leinetal.de
Produkt — Product
Ei-Profi
Eiöffner — Eggcracker
Design
Werksdesign / Inhouse
Manfred Scholte

Unternehmen — Company
Max Weishaupt
Max Weishaupt Straße 14
88475 Schwendi
Germany
Produkt — Product
WM-G10
Weishaupt Gasbrenner —
Weishaupt gas burner
Design
TEAMS Design GmbH
www.teamsdesign.com

Unternehmen — Company
Maxfield GmbH
Kaistraße 2
40221 Düsseldorf
Germany
www.maxfield.de
Produkt — Product
MAX P-X Photo
MP3-Player
Design
Werksdesign / Inhouse
Hans Nopper

Unternehmen — Company
McDonald's Deutschland Inc.
Drygalsky Allee 51
81477 München
Germany
www.mcdonalds.de
Produkt — Product
Outtakes
McDonald's Restaurant - Fast Food
Design
Heye & Partner GmbH
Claudia Siegert, Andreas Hoffmann,
Thorsten Meier, Norbert Herold
www.heye.de

Unternehmen — Company
McDonald's Deutschland Inc.
Drygalski Allee 51
81477 München
Germany
www.mcdonalds.de
Produkt — Product
Catering McDonald's
TV Spot
Design
Heye & Partner GmbH
Thorsten Meier, Norbert Herold,
Claudia Siegert, Nico Wackerbarth
www.heye.de

Unternehmen — Company
McDonald's Deutschland Inc.
Drygalski Allee 51
81477 München
Germany
www.mcdonalds.de
Produkt — Product
Koma McDonald's — Coma McDonald's
TV Spot
Design
Heye & Partner GmbH
Alexander Bartel, Andreas Forberger,
Martin Kießling, Markus Lange
www.heye.de

Unternehmen — Company
MDF Italia
Via Morimondo 5/7
20143 Milano
Italy
www.mdfitalia.it
Produkt — Product
Eleen
Bett — Bed
Design
Soda Designers
Nada Nasrallah, Christian Horner
www.soda.at
210

Unternehmen — Company
MDF Italia
Via Morimondom 5/7
20143 Milano
Italy
www.mdfitalia.it
Produkt — Product
INSIDE ART — INSIDE LINE
Regalsystem — Shelving system
Design
Piergiorgio Cazzaniga
214

Unternehmen — Company
MEA Bausysteme GmbH
Sudetenstraße 1
86551 Aichach
Germany
Produkt — Product
MEAMAX Lichtschachtsystem —
MEAMAX lightwell system
Kellerlichtschacht — Basement lightwell
Design
Werksdesign / Inhouse

Unternehmen — Company
Medikom GmbH
Qvest Magazin
Ursulaplatz 1
50668 Köln
Germany
www.qvest.de
Produkt — Product
Architektur der Dichte
Architektur Bildreportage —
Architecture picture reportage
Design
Werksdesign / Inhouse
Constantin Rothenburg, Michael Wolf

Unternehmen — Company
medico international e.V.
Burgstraße 106
60389 Frankfurt
Germany
//
MeisterWerke Schulte GmbH
Zum Walde 16
59602 Rüthen
Germany
www.meister-leisten.de
Produkt — Product
Das virtuelle Minenfeld — The Virtual Minefield
Bodeninstallation — Floor installation
Design
Heine / Lenz / Zizka Projekt GmbH
www.virtuelles-minenfeld.de / www.hlz.de

Unternehmen — Company
Merida & Centurion Germany GmbH
Blumenstraße 49–51
71106 Magstadt
Germany
www.centurion.de
Produkt — Product
Backfire LRS HFS
Mountainbike
Design
Slogdesign
Jürgen Hinderhofer
www.slogdesign.de

Unternehmen — Company
Merten GmbH & Co. KG
Fritz-Kotz-Straße 8
51674 Wiehl
Germany
www.merten.de
Produkt — Product
AQUADESIGN
Unterputz-Spezialprogramm – spritzwasser-geschützt, IP44 — Special flush-mounted range – splash-proof, IP44
Design
Werksdesign / Inhouse
182

Unternehmen — Company
Merten GmbH & Co. KG
Fritz-Kotz-Straße 8
51674 Wiehl
Germany
www.merten.de
Produkt — Product
Lösungen für intelligente Gebäude —
Solutions for intelligent buildings
CD-ROM / Digital media animation —
CD-ROM / Digital media animation
Design
Zink & Krämer
Markus Junker, Hendrik Junker
www.zuk.de
330

Unternehmen — Company
Merten GmbH & Co. KG
Fritz-Kotz-Straße 8
51674 Wiehl
Germany
www.merten.de
Produkt — Product
M-SMART Aktivweiß — M-SMART Active White
Antibakterieller Schalter —
Anti-bacterial switch
Design
Werksdesign / Inhouse
183

Unternehmen — Company
Merten GmbH & Co. KG
Fritz-Kotz-Straße 8
51674 Wiehl
Germany
www.merten.de
Produkt — Product
Panzer — Anti-Vandalism
Unterputzspezialprogramm —
Special flush-mounted switch range
Design
Werksdesign / Inhouse
184

Unternehmen — Company
Merten GmbH & Co. KG
Fritz-Kotz-Straße 8
51674 Wiehl
Germany
www.merten.de
www.bedsidepanel.de
Produkt — Product
Merten Bed Side Panel
Zentrale Steuereinheit für Hotelzimmer —
Central control unit for hotel rooms
Design
Werksdesign / Inhouse
288

Unternehmen — Company
MEYRA Vertriebsgesellschaft mbH
Meyraring 2
32689 Kalletal
Germany
www.meyra.de
Produkt — Product
FX ONE
Starrrahmen-Aluminium-Aktivrollstuhl —
Aluminium rigid frame everydayuse wheelchair
Design
REALO Design
Christof Meier
www.realo-design.de
155

Unternehmen — Company
Midgard-Licht GmbH
Am Stadtpark 13
07955 Auma
www.midgard-licht.de
Produkt — Product
MIDGARD
Lenklampen 113 + 114 —
Task Light Types 113 + 114
Design
Curt Fischer

Unternehmen — Company
Miele & Cie. KG
Carl-Miele-Straße 29
33332 Gütersloh
Germany
www.miele.de
Produkt — Product
Baureihe T9000 — Series T 9000
Wäschetrockner — Tumble dryer
Design
Werksdesign / Inhouse

Unternehmen — Company
Miele & Cie. KG
Carl-Miele-Straße 29
33332 Gütersloh
Germany
www.miele.de
Produkt — Product
Baureihe W 4000 — Series W 4000
Waschvollautomat — Washing machine
Design
Werksdesign / Inhouse

Unternehmen — Company
Miele & Cie. KG
Carl-Miele-Straße 29
33332 Gütersloh
Germany
www.miele.de
Produkt — Product
Gewerbe-Waschmaschinengeneration PW 6101, 6131, 6161, 6201 — Commercial washer-extractor generation PW 6101, 6131, 6161, 6201
Gewerbe-Waschmaschine —
Commercial washer-extractor
Design
Werksdesign / Inhouse

Unternehmen — Company
Miele & Cie. KG
Carl-Miele-Straße 29
33332 Gütersloh
Germany
www.miele.de
Produkt — Product
S4
Kompakt-Bodenstaubsauger —
Compact vacuum cleaner
Design
Werksdesign / Inhouse

Unternehmen — Company
Mikron Machining Technology
Casella Postale 115
6903 Lugano
Switzerland
www.mikron.com
Produkt — Product
Mikron NRG-50
Rundtakt-Transfer-Maschine —
Rotary transfer machine
Design
process product design
Peter Wirz, Andreas Sickert
www.process.ch

Unternehmen — Company
Ministry of Sound Recordings
Pfuelstraße 5
10997 Berlin
Germany
www.ministryofsound.de
Produkt — Product
Plattenküche — Vinyl kitchen
Messestand — Fair stand
Design
,simple GmbH
Felix Hansen, Tobias Miethke
www.simple.de
326

Unternehmen — Company
Minox GmbH
Walter-Zapp-Straße 4
35580 Wetzlar
Germany
www.minox.com
Produkt — Product
BD8x44BP / BD10x44BP
Fernglas — Binoculars
Design
Volkswagen
Jens Kohlhase, Albrecht Stein
www.volkswagen.de

Unternehmen — Company
MMK Museum für Moderne Kunst
Domstraße 10
60311 Frankfurt am Main
Germany
www.mmk-frankfurt.de
Produkt — Product
„What´s New, Pussycat?"
Erscheinungsbild für eine Ausstellung zeitgenössischer Kunst — Design for an exhibition of contemporary art
Design
Wolff Kommunikation
Carsten Wolff, Thomas Rott, Christiane Wolff, Wim Schafspelz
www.wolff-kommunikation.de
352

Unternehmen — Company
mobilcom Communicationstechnik GmbH
Hollerstraße 126
24782 Budelsdorf
Germany
www.mobilcom.de
Produkt — Product
Messestand von mobilcom auf der CeBIT 2005 — Fair stand mobilcom CeBIT 2005
Messestand — Fair stand
Design
Triad Berlin Projektgesellschaft mbH
Daniel Strauß, Ulf Eberspächer
www.triad.de
315

Unternehmen — Company
Moralt Tischlerplatten GmbH & Co. KG
Lenggriesser Straße 52
83646 Bad Tölz
Germany
www.moralt-tischlerplatten.de
Produkt — Product
Moralt Lightwood Balsa
Moralt Lightwood Balsa als Anwendungsbeispiel in einem Couchtisch — Moralt Lightwood Balsa as example of use in a table
Design
Alfons Venjakob GmbH & Co. KG
Antonius Rehkemper
www.venjakob-moebel.de
//
Werksdesign / Inhouse
Ralf Pinnekamp, Christian Stöhr

Unternehmen — Company
Moravske sklarny Kvetna
nam. E. Zahna 329
687 66 Strani-Kvetna
Czech Republic
www.moravskesklarny.cz
Produkt — Product
Mr. Egg
Trinkbecher — Drinking glasses
Design
Studio Olgoj Chorchoj
www.olgojchorchoj.cz

Unternehmen — Company
M-real Zanders GmbH
An der Gohrsmühle
51465 Bergisch Gladbach
Germany
www.m-real.com
Produkt — Product
Chromolux Packaging
Verpackungsserie — Packaging-series
Design
zinnobergruen
Tobias Schwarzer, Bärbel Muhlack
www.zinnobergruen.de

Unternehmen — Company
Münchener Rückversicherungs-Gesellschaft
Königinstraße 107
80802 München
Germany
www.munichre.com
Produkt — Product
Innovationen bewegen – Geschäftsbericht der Münchener-Rück-Gruppe 2004 — Advancing innovation Munich Re Group 2004
Geschäftsbericht — Annual report
Design
höfelinger + wagner design gmbh
Vit Steinberger
www.hwdesign.de

Unternehmen — Company
munksjö Paper GmbH
Waldhäuser Straße 41
40211 Aalen
Germany
Produkt — Product
Messestand / premium performance by Munksjö — Stand / premium performance by Munksjö
Messestand — Fair stand
Design
Werksdesign / Inhouse
Frank Wagner, Thomas Häussler, Thomas Tscherter

Unternehmen — Company
Museum der Arbeit Hamburg
Wiesendamm 3
22305 Hamburg
Germany
www.museum-der-arbeit.de
Produkt — Product
neontigers
Photographs of Asian Megacites
Buch — Book
Design
Jung und Pfeffer: Visuelle Kommunikation
Peter Bialobrzeski, Florian Pfeffer, Till F. Teenck
www.jungundpfeffer.de

Unternehmen — Company
MuseumsQuartier Errichtungs- und
Betriebsgesmbh
Museumsplatz 1
1070 Wien
Austria
www.mqw.at
Produkt — Product
Hofmöblierung Enzi —
Yard furniture named Enzi
Multifunktionales Möbelelement —
Mulitfunctional furniture
Design
PPAG POPELKA PODUSCHKA Architekten
Anna Popelka, Georg Poduschka
www.ppag.at

n

Unternehmen — Company
Nationaltheater Mannheim
Mozartstraße 9
68161 Mannheim
Germany
Produkt — Product
„226"
Spielzeitbroschüre der 226. Spielzeit am
Nationaltheater Mannheim 2004/2005 —
Season brochure for the 226th season of the
National Theatre Mannheim 2004/2005
Programm-Broschüre/Druckerzeugnis —
Programme brochure/printed matter
Design
SEIDLDESIGN
Wolfgang Seidl
www.seidldesign.com
365

Unternehmen — Company
Naviflash Sp.z.o.o.
ul. Wojska Polskiego 4
39-300 Mielec
Poland
www.naviflash.de
Produkt — Product
Naviflash 1020
Navigationssystem — Navigation system
Design
Werksdesign/Inhouse
Andre Bury
287

Unternehmen — Company
Netras Mobile Systems
Schootense Dreef 21
5708 HZ Helmond
Netherlands
www.netras.nl
Produkt — Product
Netras City-L 3.5
LKW unter 3,500 kg —
Van for light weight distribution
Design
Motio Development
Arjan Steketee
www.motio.nl

Unternehmen — Company
Neue Gestaltung GmbH
Rudolf-Ditzen-Weg 9
13156 Berlin
Germany
www.neuegestaltung.de
Produkt — Product
Lookcookbook
Buch — Book
Design
Werksdesign/Inhouse

Unternehmen — Company
Neue Pinakothek
Barer Straße 29
80799 München
Germany
www.pinakothek.de
Produkt — Product
MANET MANET | Zwei Bilder, ein Raum —
MANET MANET | Two Paintings, one Room
Ausstellungskommunikation —
Communication concept
Design
KMS Team GmbH
Michael Keller, Christine Burkard,
Axel Sanjosé, Silke Streppelhoff
www.kms-team.de

Unternehmen — Company
NIBRA Dachkeramik GmbH & Co. KG
Magdeburger Straße 42
39326 Ammensleben
Germany
//
Teichgräber GmbH
Schwabener Weg 3a
85630 Grasbrunn/Neukeferloh
Germany
Produkt — Product
MS 5 „Dino"
Dachziegel aus gebranntem Ton —
Roof tile made of fired clay
Design
Werksdesign/Inhouse

Unternehmen — Company
Norbert Woll GmbH
Heinrich-Barth-Straße 7-11
66115 Saarbrücken
Germany
www.woll-cookware.com
Produkt — Product
Logic
Kochgeschirr — Cookware
Design
Design Mattis
Eberhard Friemel
www.design-mattis.de

Unternehmen — Company
Normann Copenhagen
Oesterbrogade 70
2100 Copenhagen East
Denmark
www.normann-copenhagen.com
Produkt — Product
Magma
Thermokanne — Vacuum jug
Design
Werksdesign / Inhouse
Francis Cayouette

Unternehmen — Company
Normann Copenhagen
Oesterbrogade 70
2100 Copenhagen East
Denmark
www.normann-copenhagen.com
Produkt — Product
Strainer
Sieb — Strainer
Design
Werksdesign / Inhouse
Boje Estermann

Unternehmen — Company
nya nordiska textiles GmbH
An den Ratswiesen
29451 Dannenberg
Germany
www.nya.de
Produkt — Product
Andorra
Dekoration — Decoration fabric
Design
Werksdesign / Inhouse
nya nordiska textiles GmbH
Design-Team nya nordiska

Unternehmen — Company
nya nordiska textiles GmbH
An den Ratswiesen
29451 Dannenberg
Germany
www.nya.de
Produkt — Product
Cassate
Dekorationsstoff — Decoration fabric
Design
Werksdesign / Inhouse
nya nordiska textiles GmbH
Design-Team nya nordiska

Unternehmen — Company
nya nordiska textiles GmbH
An den Ratswiesen
29451 Dannenberg
Germany
www.nya.de
Produkt — Product
Lucetta CS
Dekorationsstoff — Descoration fabric
Design
Werksdesign / Inhouse
nya nordiska textiles GmbH
Design-Team nya nordiska

Unternehmen — Company
nya nordiska textiles GmbH
An den Ratswiesen
29451 Dannenberg
Germany
www.nya.de
Produkt — Product
Oki
Dekorationsstoff — Decoration fabric
Design
Werksdesign / Inhouse
nya nordiska textiles GmbH
Design-Team nya nordiska

Unternehmen — Company
nya nordiska textiles GmbH
An den Ratswiesen
29451 Dannenberg
Germany
www.nya.de
Produkt — Product
STRATOS
Decorationsstoff — Decoration fabric
Design
Werksdesign / Inhouse
nya nordiska textiles GmbH
Design-Team nya nordiska

Unternehmen — Company
nya nordiska textiles GmbH
An den Ratswiesen
29451 Dannenberg
Germany
www.nya.de
Produkt — Product
Tiburon
Dekorationsstoff — Decoration fabric
Design
Werksdesign / Inhouse
nya nordiska textiles GmbH
Design-Team nya nordiska

O

Unternehmen — Company
o2 Germany GmbH & Co.OHG
Georg-Brauchle-Ring 23–25
80992 München
Germany
www.de.o2.com
Produkt — Product
Medienwolke — Media Cloud
Messeauftritt CeBIT 2005 —
Exhibition stand CeBIT 2005
Design
KMS Team GmbH / Schmidhuber+Partner
Michael Keller, Susanne Schmidhuber,
Christoph Rohrer, Birgit Vogel,
Wahan Mechitarian, Kerstin Arleth,
Julia Romeiß, Andreas Koch
www.kms-team.de

Unternehmen — Company
Object Carpet GmbH
Rechbergstraße 19
73770 Denkendorf
Germany
www.object-carpet.com
Produkt — Product
Pearl 1300
Teppichboden, Schlinge —
Wall to wall carpet, loop
Design
Werksdesign / Inhouse
Object Carpet GmbH

Unternehmen — Company
Océ Technologies
St. Urbanusweg 43
5900 MA Venlo
Netherlands
www.oce.com
Produkt — Product
Océ TCS4
Großformat-Farbscanner —
Wideformat colour scanner
Design
Werksdesign / Inhouse
Océ Design Team

Unternehmen — Company
OLYMP GmbH & Co. KG
Hohenheimer Straße 91–97
70184 Stuttgart
Germany
www.olymp.de
Produkt — Product
WaveLight
Lichtsegel — Light sail
Design
pan it's design
Oliver Panitz
www.pan-its-design.de
180

Unternehmen — Company
Olympus + Jbc GmbH
Kuehnstraße 61
22045 Hamburg
Germany
www.olympus-owi.com
Produkt — Product
OES Pro2
Resektoskop — Resektoscope
Design
Held + Team
Thomas Märzke
www.heldundteam.de

Unternehmen — Company
Olympus Winter & Jbc GmbH
Kuehnstraße 61
22045 Hamburg
Germany
www.olympus-owi.com
Produkt — Product
Endoeye HD 0°
Videolaparoskop — Videolaparoscope
Design
Held + Team
www.heldundteam.de

Unternehmen — Company
Olympus Winter + Jbc GmbH
Kuehnstraße 61
22045 Hamburg
Germany
www.olympus-owi.com
Produkt — Product
HiQ+
Nadelhalter – minimalinvasives chirurgisches
Instrument — Needleholder – minimalinvasive
surgical instrument
Design
Held + Team
Burkhard Peters
www.heldundteam.de

Unternehmen — Company
Orange Retail B.V.
t.a.v. Orange online shop
Groenhovenstraat 2
2596 HT Den Haag
Netherlands
http://shop.orange.nl
Produkt — Product
http://shop.orange.nl
Internetseite — Website
Design
Qua Asociates B.V.
www.qua.nl

Unternehmen — Company
Oregon Scientific (Deutschland) GmbH
Siemensstraße 1
63263 Neu-Isenburg
Germany
www.oregonscientific.de
Produkt — Product
MR 238
Wetterstation — Weather station
Design
Werksdesign / Inhouse

Unternehmen — Company
ORGENTEC Diagnostika GmbH
Carl-Zeiss-Straße 49
55129 Mainz
Germany
www.orgentec.com
//
BIT Analytical Instruments GmbH
Am Kronberger Hang 3
65824 Schwalbach
Germany
www.bit-instruments.com
Produkt — Product
ALEGRIA
Diagnosegerät — Diagnostic equipment
Design
Teleca System Design AB
www.teleca.se
//
Werksdesign / Inhouse
Dr. Wigbert Berg, Dr. Stephan Becker
303

Unternehmen — Company
ORION Versand GmbH & Co. KG
Schäferweg 14
24941 Flensburg
Germany
www.orion.de
Produkt — Product
Handtuch — Towel
Werbespot — Commercial
Design
Heye & Partner
Jan-Florian Ege, Nicole Oberberger,
Christiane Seifert
www.heye.de
328

Unternehmen — Company
ORTLIEB Sportartikel GmbH
Rainstraße 6
91560 Heilsbronn
Germany
www.ortlieb.com
Produkt — Product
Office-Bag2
Fahrradaktenkoffer — Briefcase for bicycle
Design
Werksdesign / Inhouse
Marcus Müller

Unternehmen — Company
OSCHWALD Boden aus Natur GmbH
Rudolf-Blassingstraße 7
79183 Waldkirch
Germany
www.oschwaldkirch.de
Produkt — Product
Geysira
Teppichboden-Kollektion — Floor carpet collection
Design
Werksdesign / Inhouse
189

Unternehmen — Company
Oticon A/S
Kongebakken 9
2654 Smørum
Denmark
www.oticon.com
Produkt — Product
Oticon Delta
Hörsystem — Hearing device
Design
lockenwitz i/d
Christian Lockenwitz
www.lockenwitz.dk
//
Werksdesign / Inhouse
Michael Frank Petersen, Claus Würfel,
Søren Skjærbæk, Lars Tuborg Jensen,
Søren Colding
38 Gold

Unternehmen — Company
Otto Bock HealthCare GmbH
Max-Näder-Straße 15
37115 Duderstadt
Germany
www.ottobock.de
Produkt — Product
Avantgarde VR Light
Manueller Rollstuhl — Manual wheelchair
Design
Werksdesign / Inhouse
Achim Eiser
154

Unternehmen — Company
Otto Graf GmbH
Kunststofferzeugnisse
Carl-Zeiss-Straße 2–6
79331 Teningen
Germany
www.graf-online.de
Produkt — Product
GRAF Sicker-Tunnel 300
Sicker-Tunnel — Sicker tunnel
Design
Werksdesign / Inhouse
Fabien Berhard

Unternehmen — Company
Ottweiler Druckerei
Johannes-Gutenbergstraße 14
66564 Ottweiler
Germany
www.mediengruppepaul.de
//
viasit Bürositzmöbel GmbH
Boxbergweg 4
66538 Neunkirchen
Germany
www.viasit.com
Produkt — Product
key
Elektronische Bedinungsanleitung — Electronical operational instruction
Design
pixellab digitale Werkstatt
Thomas Schneck
//
Werksdesign / Inhouse
Günter Weber

Unternehmen — Company
outermedia GmbH, Berlin, im Auftrag von FWU Institut für Film und bild, Grünwald
Zionskirchstraße 73
10119 Berlin
Germany
www.fwu.de
//
FWU Institut für Film und Bild in Wissenschaft und Unterricht gemeinnützige GmbH
Bavariafilmplatz 3
82031 Grünwald
Germany
www.fwu.de
Produkt — Product
Geschichte digital: Die Französische Revolution — History digital: The French Revolution
CD-ROM
Design
Werksdesign / Inhouse
Eva Duvenkamp, Elke Hanisch

P

Unternehmen — Company
PDR
Western Avenue
CF52YB Cardiff
Great Britain
www.pdr-online.co.uk
Produkt — Product
Smart Sharps
Abfallbehälter für scharfe / spitze Instrumente — Reusable bin for sharp medical waste
Design
Waste Solutions / Stericycle
Jason Hester

Unternehmen — Company
pearls acoustic identity GmbH
Hanauer Landstraße 11–13
60314 Frankfurt
Germany
www.pearls-music.com
Produkt — Product
Interaktives Showreel, Website
www.pearls-music.com
Intenetseite — Website
Design
Werksdesign / Inhouse

Unternehmen — Company
Philips Lighting
Mathildelaan 1
5611 bd Eindhoven
Netherlands
www.lighting.philips.com
Produkt — Product
www.dynamiclighting.philips.com
Internetseite — Website
Design
affe mit blumenkranz – internet
Daniel Salas, Boris Müller
www.affemitblumenkranz.de
//
jung und pfeffer: visuelle kommunikation
Ole Kaleschke, Florian Pfeffer
www.jungundpfeffer.de

Unternehmen — Company
Philips Lighting – UB der Philips GmbH
Postfach 104929
20034 Hamburg
Germany
www.philips.com
Produkt — Product
Philips Lighting
„Lighthouse Experiences"
Messeauftritt auf der Euroshop 2005, Düsseldorf — Trade fair appearance at the Euroshop 2005, Düsseldorf
Design
D'ART DESIGN GRUPPE
www.d-art-design.de
313

Unternehmen — Company
Phi-ton B.V.
Zijpendaalseweg 51A
6814 CD Arnhem
Netherlands
www.phi-ton.com
Produkt — Product
3D-Matratze — 3D-mattress
Schlaf- und Komfortprodukt — Sleeping and comfort product
Design
Werksdesign / Inhouse
Fried-Jan Unger
208

Unternehmen — Company
Phi-ton BV
Zypendaalseweg 51A
6014 Arnhem
Netherlands
www.phi-ton.com
Produkt — Product
Phi-ton
Bett — Bed
Design
Werksdesign / Inhouse
Fried-Jan Unger
209

Unternehmen — Company
PHOENIX CONTACT GmbH & Co. KG
Flachsmarktstraße 8
32825 Blomberg
Germany
www.phoenixcontact.com
Produkt — Product
DUPLICON
Steckverbinder für die Energieverteilung —
Plug connector for power distribution
Design
Tecform
Gerwin Müller
306

Unternehmen — Company
Photon Energy GmbH
Bräunleinsberg 10
91242 Ottensoos
Germany
www.photon-energy.de
Produkt — Product
NAUTILUS
Laserbeschriftungsgerät —
Laser marking system
Design
Werksdesign / Inhouse

Unternehmen — Company
PIIROINEN
Arvo Piiroinen Oy
Tehdaskatu 28
24100 Salo
Finland
www.piiroinen.com
Produkt — Product
Flakes
Stuhl — Chair
Design
schmidt hammer lassen
www.shl.dk

Unternehmen — Company
Piquadro S.p.A.
Località Sassuriano 246
40047 Silla di Gaggio Montano (BO)
Italy
www.piquadro.com
Produkt — Product
Voyager Collection
Reisegepäckkollektion — Travel collection
Design
Werksdesign / Inhouse
Cristian Righetti, Daniel Valsesia

Unternehmen — Company
PLANK
Plank Collezioni Srl
Via Nazionale 35
39040 Ora
Italy
www.plank.it
Produkt — Product
MIURA
Barhocker — Barstool
Design
Konstantin Grcic Industrial Design
Konstantin Grcic
www.konstantin-grcic.com
76 Silber

Unternehmen — Company
Plantronics GmbH
Gildenweg 7
50354 Hürth
Germany
www.plantronics.com
Produkt — Product
Discovery™ 640
Bluetooth-Headset — Bluetooth headset
Design
Werksdesign / Inhouse
Soohyun Ham, Darrin Caddes
286

Unternehmen — Company
PLAZAMEDIA GmbH
Münchener Straße 101
85737 Ismaning
Germany
www.plazamedia.de
Produkt — Product
DSF: Kinderlied Formel 1 —
DSF: Song for children formula 1
Programmtrailer — Promotion trailer
Design
Werksdesign / Inhouse
Daniele Cosentino, Volker Lachmann,
Harald König, Andreas Wein

Unternehmen — Company
PLEX GmbH
Köpenicker Straße 48 / 49
10179 Berlin
Germany
www.plexgroup.com
Produkt — Product
SunFuel Lab
Ausstellungsattraktion für die Autostadt —
Exhibition / Attraction at the Autostadt
Design
Werksdesign / Inhouse
Dietmar Mühr, Oliver Mühr

Unternehmen — Company
POC Sweden AB
Skogsövägen 22
133 33 Saltsjöbaden
Sweden
www.pocski.com
Produkt — Product
POC Skull Comp Helmet
Skihelm — Ski helmet
Design
Werksdesign / Inhouse
POC Design Team
50 Gold

Unternehmen — Company
Poem for Fashion and Message
Hauptmarkt 15
08056 Zwickau
Germany
www.po-m.de
Produkt — Product
Metropolitan - haute couture for the streets
Modekollektion — Fashion collection
Design
Elena Biedermann, Marko Stephan

Unternehmen — Company
Poggenpohl Möbelwerke GmbH
Poggenpohlstraße 1
32051 Herford
Germany
www.poggenpohl.de
Produkt — Product
PLUSMODO
Küchenarchitektur — Kitchen architecture
Design
Jorge Pensi Diseño S.L.
Jorge Pensi
//
Werksdesign / Inhouse
Poggenpohl Möbelwerke GmbH
204

Unternehmen — Company
Polar Electro
Polar Electro Oy
Professorintie 5
90440 Kempele
Finland
www.polar.fi
Produkt — Product
POLAR RS200
Laufcomputer — Running computer
Design
Werksdesign / Inhouse
Olli Komulainen, Lauri Lumme,
Heli Haapakoski

Unternehmen — Company
POOL-X design-kompetenz
Veldershofer Weg 16
91207 Lauf a. d. Pegnitz
Germany
www.pool-x.de
Produkt — Product
Buch im Sandwich-Druck —
Book printed in Sandwich print
Künstlerisches Unikat — Unique piece of art
Design
POOL-X design-kompetenz
Oliver J. Stumpf, Christian Oberlander

Unternehmen — Company
Porsche Lizenz- und
Handelsgesellschaft mbH & Co. KG
Porschestraße 1
74321 Bietigheim-Bissingen
Germany
www.porsche-design.com
Produkt — Product
Erscheinungsbild für Porsche Design —
Corporate Design for Porsche Design
Corporate Design
Design
KMS Team GmbH
Knut Maierhofer, Patrick Märki, Dirk Koy,
Christina Baur, Sandra Ehm
www.kms-team.de
120 Silber

Unternehmen — Company
Porsche Lizenz- und
Handelsgesellschaft mbH & Co. KG
Porschestraße 1
74321 Bietigheim-Bissingen
Germany
www.porsche-design.com
Produkt — Product
Porsche Design Corporate Design
Kampagne, Broschüren —
Campaign, brochures
Design
KW43 BRANDDESIGN
Margit Tabert
www.kw43.de
//
KMS Team GmbH
Knut Maierhofer, Patrick Märki
www.kms-team.de
369

Unternehmen — Company
Porzellanfabrik Schönwald
Rehauer Straße 44–54
95173 Schönwald
Germany
www.schoenwald.com
Produkt — Product
Signature Basic
Geschirrserie — China series
Design
Product Design - Interior Design
Carsten Gollnick
www.gollnick-design.de
269

Unternehmen — Company
PPGHJWT Artists
Rietlandpark 301
1019 DW Amsterdam
Netherlands
www.ppghjwt-artists.nl
Produkt — Product
VEA RE:set Reclame
Poster und Faltblätter — Posters and leaflets
Design
Werksdesign / Inhouse
John de Vries
362

Unternehmen — Company
Pressol Schmiergeräte GmbH
Bartholomäusstraße 26 d
90489 Nürnberg
Germany
www.pressol.com
Produkt — Product
Kraftstoffkanister — Petrol jerrycan
Design
Werksdesign / Inhouse

Unternehmen — Company
Procter & Gamble International Operations SA
PO Box 111, 47 route de Saint Georges
1213 Petit-Lancy 1
Switzerland
www.wella.com
Produkt — Product
Climazon2
IR Wärmebehandlungsgerät —
IT Heat Treatment Appliance
Design
Patzak Design
Peter Patzak
www.patzak-design.com
//
designafairs GmbH
Peter Riedl-Bob
www.designafairs.com

Unternehmen — Company
Prof. Marian Nestmann / Prof. Gregor Krizstian
Ober-Ramstädter Straße 98 M
64367 Mühltal
Germany
//
Evangelische Kirche in Hessen und Nassau
Paulusplatz 1
64285 Darmstadt
Germany
Produkt — Product
FÜNF JAHRESBERICHTE 2000–2005 —
5 annual reports 2000–2005
Jahresbericht — Annual report
Design
Werksdesign / Inhouse
Gregor Krizstian

Unternehmen — Company
Puma GmbH IP Solingen
An den Eichen 20–22
42699 Solingen
Germany
Produkt — Product
Nordmeer
Messer — Knife
Design
Heinz-Peter Knoop
www.knoop-art.de

q

Unternehmen — Company
Qvest Magazin
Medikom GmbH
Münzstraße 14
10178 Berlin
Germany
www.qvest.de
Produkt — Product
DANIEL RICHTER –
ACH DAS REDEN ÜBER DIE KUNST
Fotoreportage — Picture reportage
Design
Werksdesign / Inhouse
Constantin Rothenburg, Will McBride,
Lars Jensen, Kerstin Fritsche

r

Unternehmen — Company
Rado Watch Co. Ltd.
Bielstrasse 45
2543 Lengnau
Switzerland
www.rado.com
Produkt — Product
RADO Original Classic XL —
RADO Original Chronograph
Armbanduhren — Wristwatches
Design
Rado Produkte Lab

Unternehmen — Company
Rado Watch Co. Ltd.
Bielstrasse 45
2543 Lengnau
Switzerland
www.rado.com
Produkt — Product
V10K
Armbanduhr — Wrist watch
Design
Rado Produkte Lab

Unternehmen — Company
Rainer Euskirchen Büro Design GmbH
Jöllenbecker Straße 314
33613 Bielefeld
Germany
www.myponte.com /
www.euskirchen-moebel.de
Produkt — Product
System: MYponte
Sideboard
Design
Bic Design, Dirk Togtmeyer
www.bic-design.de
//
Werksdesign / Inhouse
213

Unternehmen — Company
RASTAL GMBH & CO. KG
Lindenstraße
56203 Höhr-Grenzhausen
Germany
www.rastal.com
Produkt — Product
CLUB
Seidel — Tankard
Design
Werksdesign / Inhouse
Marion Beyer

Unternehmen — Company
Raymund Henrich
Ammerthal 3a
85622 Weißenfeld
Germany
www.buchbinderei-henrich.de
Produkt — Product
Imagebroschüre — Image brochure
Design
Fa-Ro Marketing GmbH
Sabrina Ruchti
www.fa-ro.de
//
Werksdesign / Inhouse
Raymund Henrich
351

Unternehmen — Company
RDD industrial design network
Schneegasse 19
4551 Ried / Traunkreis
Austria
www.rdd.at
Produkt — Product
Vakumat-Pumpe Pumpenaggregat — Vakumat pump
Selbstsaugende Kreiselpumpe — Self-priming centrifugal pump unit
Design
RDD industrial design network
Rainer Atzlinger
www.rdd.at

Unternehmen — Company
Regent Beleuchtungskörper AG
Dornacherstrasse 390, Postfach 246
4018 Basel
Switzerland
www.regent.ch
Produkt — Product
NoLimit
Pendel-, Wand- oder Systemleuchte für Fluoreszenzlampe T5 — Pendant-, wall-mounted or system luminaire for T5-fluorescent tube
Design
Felice Dittli

Unternehmen — Company
REHAU AG + Co.
Otto-Hahn-Straße 2
95111 Rehau
Germany
www.REHAU.com
Produkt — Product
RAUVOLET acoustic-line
Schrankrollladensystem — Roller shutter system
Design
Werksdesign / Inhouse
Stefan Bräuer, Klaus Grießhammer, Klaus Schoberth, Carsten Neufert
228

Unternehmen — Company
Renault Deutschland
Renault Nissan Straße 6–10
50321 Brühl
Germany
www.renault.de
Produkt — Product
Renault Modus
Pkw — Passenger car
Design
Renault SAS
www.renault.com
146

Unternehmen — Company
Renault Nissan AG
Renault Nissan Straße 6–10
50321 Brühl
Germany
www.renault.de
Produkt — Product
Renault Kampagne — Renault Campaign
TV-Spot Kampagne — TV commercial campaign
Design
539090
Simona Daniel, Musse Hassevall, Jo Ekentorp
www.539090.com
//
Werksdesign / Inhouse
Claudia Willvonseder, Thomas Wiegand, Thorsten Albrecht, Holger Gaubatz

Unternehmen — Company
Reset Grafische Medien GmbH
Virchowstraße 8
22767 Hamburg
Germany
www.reset.de
Produkt — Product
Der Reset Kalender 2005 / Hamburg Architecture — The Reset Calendar 2005 / Hamburg Architecture
Kalender — Calendar
Design
Format Design
Knut Ettling, Oliver Heissner
www.formatdesign.net
//
Werksdesign / Inhouse
Achim Pettera

Unternehmen — Company
Rhombus Rollen GmbH & Co.
Albert-Einstein-Straße 15
42929 Wermelskirchen
Germany
www.rhombus-rollen.com
Produkt — Product
Serie 399
Krankenbettrollen — Hospital-bed castor
Design
Die Haptiker
Jörg Ostenrieder, Marcus Wiedemann
www.die-haptiker.de
//
Werksdesign / Inhouse
Roland Kausemann

Unternehmen — Company
Ricoh Europe B.V.
Oberrather Straße 6
40472 Düsseldorf
Germany
www.ricoh-europe.com
Produkt — Product
GR Digital
Digitalkamera — Digital camera
Design
Werksdesign / Inhouse
Masahiko Kurita, Tatsuo Okuda, Mikane Tago

Unternehmen — Company
Rieber GmbH & Co. KG
Hoffmannstraße 44
72770 Reutlingen
Germany
www.rieber.de
Produkt — Product
Waterstation cubic 600
Spüle — Sink
Design
Pearl Creative
Tim Storti, Christian Rummel
www.pearlcreative.com
//
Werksdesign / Inhouse
Max Maier

Unternehmen — Company
Rieber GmbH & Co. KG
Hoffmannstraße 44
72770 Reutlingen
Germany
www.rieber.de
Produkt — Product
Varithekbuffet mit Cooktainer —
Varithekbuffet with cooktainer
Speisenausgabe — Buffet
Design
Pearl Creative
Tim Storti, Christian Rummel
www.pearlcreative.com
//
Werksdesign / Inhouse
Max Maier

Unternehmen — Company
Rieber GmbH & Co. KG
Hoffmannstraße 44
72770 Reutlingen
Germany
www.rieber.de
Produkt — Product
Waterstation cubic 980
Spüle — Sink
Design
Pearl Creative
Tim Storti, Christian Rummel
www.pearlcreative.com
//
Werksdesign / Inhouse
Max Maier

Unternehmen — Company
Rieder Faserbeton Elemente GmbH
Am Glasberg 1
83059 Kolbermoor
Germany
www.rieder.cc
Produkt — Product
fibreStone
Mehrschichtplatten — Multilayer boards
Design
Werksdesign / Inhouse
161

Unternehmen — Company
riese und müller GmbH
Haasstraße 6
64293 Darmstadt
Germany
www.r-m.de
Produkt — Product
Avenue
Stadtfahrrad — City bicycle
Design
Werksdesign / Inhouse
Stijn Deferm, Markus Riese, Heiko Müller

Unternehmen — Company
RINCO ULTRASONICS AG
Industriestrasse 4
8590 Romanshorn
Switzerland
www.rincoultrasonics.com
Produkt — Product
AGM (Advanced Generator Module)
Ultraschall-Schweißgenerator —
Ultrasonic welding generator
Design
Werksdesign / Inhouse

Unternehmen — Company
RIVA
RSN METAIS LTDA
Abel Postalli Street, 445
95112-255 Caxias do Sul
Brasil
www.riva.com.br
Produkt — Product
MURAZZO
Viereckiges Salz- und Pfefferstreuer-Set —
Salt and pepper shaker set - Collection
Emporio Riva
Design
RIVA
RUBENS SIMÕES

Unternehmen — Company
Robert Bosch Hausgeräte GmbH
Carl-Wery-Straße 34
51739 München
Germany
www.bosch-hausgeraete.de
Produkt — Product
SGS09 S35 EU
Bosch - Vollautomatik Geschirrspüler Edelstahl — Bosch - fully automatic dishwasher stainless steel
Design
Werksdesign / Inhouse
Thomas Ott

Unternehmen — Company
Robert Bosch Hausgeräte GmbH
Carl-Wery-Straße 34
81739 München
Germany
www.bosch-hausgeraete.de
Produkt — Product
„piano á filo" PCL 795 FEU
Bosch Gas-Kochstelle — Bosch gas-fired hob
Design
Werksdesign / Inhouse
Alexander Marsch

Unternehmen — Company
Robert Bosch Hausgeräte GmbH
Carl-Wery-Straße 34
51739 München
Germany
www.bosch-hausgeraete.de
Produkt — Product
SGS55 M15 EU
Bosch Edelstahl-Standspüler — Bosch stainless steel freestanding dishwasher
Design
Werksdesign / Inhouse
Thomas Ott

Unternehmen — Company
Robert Bosch Hausgeräte GmbH
Carl-Wery-Straße 34
51739 München
Germany
www.bosch-hausgeraete.de
Produkt — Product
SGS55 T13 EU
Bosch Stand-Geschirrspüler Volledelstahl — Bosch freestanding dishwasher in solid stainless steel
Design
Werksdesign / Inhouse
Thomas Ott

Unternehmen — Company
Robert Bosch Hausgeräte GmbH
Carl-Wery-Straße 34
51739 München
Germany
www.bosch-hausgeraete.de
Produkt — Product
SGV67 T23 EU
Bosch Geschirrspüler Edelstahl — Bosch dishwasher stainless steel
Design
Werksdesign / Inhouse
Thomas Ott

Unternehmen — Company
Robert Bosch Hausgeräte GmbH
Carl-Wery-Straße 34
51739 München
Germany
www.bosch-hausgeraete.de
Produkt — Product
Worlddryer WTMC 632 SCN silber
Wäschetrockner — Dryer
Design
Werksdesign / Inhcuse
Jacques Helot, Ralph Staud, Thomas Tischer

Unternehmen — Company
Robert Bosch Hausgeräte GmbH
Carl-Wery-Straße 34
81739 München
Germany
www.bosch-hausgeraete.de
Produkt — Product
MUM 8200
Küchenmaschine — Kitchen machine
Design
Werksdesign / Inhouse
Tobias Krüger, Rolf Feil

Unternehmen — Company
Robert Bosch Hausgeräte GmbH
Carl-Wery-Straße 34
81739 München
Germany
www.bosch-hausgeraete.de
Produkt — Product
DIE 995 F
Insel-Esse Edelstahl — Island chimney hood, stainless steel
Design
Eisele-Kuberg-Design
www.eiselekubergdesign.de
//
Werksdesign / Inhouse
Roland Vetter

Unternehmen — Company
Robert Bosch Hausgeräte GmbH
Carl-Wery-Straße 34
81739 München
Germany
www.bosch-hausgeraete.de
Produkt — Product
HLN 45 44 50
Electro-Standherd — Electric freestanding cooker
Design
Eisele-Kuberg-Design
www.eiselekubergdesign.de
//
Werksdesign / Inhouse
Roland Vetter

Unternehmen — Company
Robert Bosch Hausgeräte GmbH
Carl-Wery-Straße 34
81739 München
Germany
www.bosch-hausgeraete.de
Produkt — Product
HSB 7450 55
Gas-Range-Cooker — Gas range cooker
Design
Eisele-Kuberg-Design
www.eiselekubergdesign.de
//
Werksdesign / Inhouse
Roland Vetter

Unternehmen — Company
Robert Bosch Hausgeräte GmbH
Carl-Wery-Straße 34
81739 München
Germany
www.bosch-hausgeraete.de
Produkt — Product
HSC 140 670
Einbau-Wärmeschublade Aluminium —
Built-in warming drawer aluminium
Design
Werksdesign / Inhouse
Roland Vetter

Unternehmen — Company
Robert Bosch Hausgeräte GmbH
Carl-Wery-Straße 34
81739 München
Germany
www.bosch-hausgeraete.de
Produkt — Product
side by side KAN 58 P90
Kühl-Gefrier Kombination —
Fridge / Freezer combination
Design
Werksdesign / Inhouse
Thomas Tischer

Unternehmen — Company
Robert Bosch Hausgeräte GmbH
Carl-Wery-Straße 34
81739 München
Germany
www.bosch-hausgeraete.de
Produkt — Product
LiftMatic HBN 77 P750
Wandbackofen — Wall oven
Design
Werksdesign / Inhouse
Roland Vetter

Unternehmen — Company
Robert Bosch Hausgeräte GmbH
Carl-Wery-Straße 34
81739 München
Germany
www.bosch-hausgeraete.de
Produkt — Product
PCL755 T NL
Gas-Kochstelle Edelstahl —
Gas-fired hole stainless steel
Design
Eisele Kuberg-Design
www.eiselekubergdesign.de
//
Werksdesign / Inhouse
Roland Vetter

Unternehmen — Company
Robert Bosch Hausgeräte GmbH
Carl-Wery-Straße 34
81739 München
Germany
www.bosch-hausgeraete.de
Produkt — Product
PKY 875 P02
Glaskeramik-Inselkochstelle in Edelstahl —
Glass ceramic island hob in stainless steel
Design
Werksdesign / Inhouse
Roland Vetter

Unternehmen — Company
Robert Bosch Hausgeräte GmbH
Carl-Wery-Straße 34
81739 München
Germany
www.bosch-hausgeraete.de
Produkt — Product
WFMC 640 SUC silber
Bosch Wäschetrockner silber —
Bosch tumble dryer silver
Design
Werksdesign / Inhouse
Carsten Weber, Ralph Staud, Thomas Fischer

Unternehmen — Company
ROBOT Visual Systems GmbH
Opladener Straße 202
40789 Monheim am Rhein
Germany
www.robot.de
Produkt — Product
Traffi Tower
Verkehrsüberwachungsanlage —
Traffic monotoring unit
Design
Büro für Gestaltung
Ralf Jakubowski

Unternehmen — Company
Robotoolz Inc.
404 Villa Street
CA 94041 Mountain View
USA
www.robotoolz.com
Produkt — Product
CrossLiner
Selbstnivellierender Linienlaser mit Justage-adapter — Automatic leveling line laser with attachment accessory
Design
Oxxid
Stefan Reichl, Andreas Rößner,
Sabrina Hofmann, Verena Kehrer
www.oxxid.de

Unternehmen — Company
Rodenstock GmbH
Isartalstraße 43
80469 München
Germany
www.rodenstock.de
Produkt — Product
R 3148 / R 3149
Sonnenbrillen — Sunglasses
Design
design schoenmueller
Harald Schönmüller
www.schoenmueller.de

Unternehmen — Company
rohi stoffe GmbH
Schönlinderstraße 1
82538 Geretsried
Germany
www.rohi.com
//
ANKER Teppichboden
Gebr. Schoeller GmbH & Co. KG
Zollhausstraße 112
52353 Düren
Germany
www.anker-dueren.de
Produkt — Product
rohi / ANKER Aircraft Collection 2006
Flugzeugtextilien (Sitzbezugstoffe, Vorhänge, Kopflätzchen, Teppichböden) — Aircraft textiles (seat-cover fabrics, curtains, headrests, carpets)
Design
Werksdesign / Inhouse
rohi designteam / ANKER Designteam

Unternehmen — Company
rose plastic GmbH
Rupolzer Straße 53
88138 Hergensweiler
Germany
www.rose-plastic.com
Produkt — Product
PremiumClip
Bohrereinzelaufhänger — Hanging clip for single drill
Design
Werksdesign / Inhouse
Markus Mayer

Unternehmen — Company
Rosenbauer International AG
Paschinger Straße 90
4060 Leonding
Austria
www.rosenbauer.com
Produkt — Product
HEROS-xtreme
Feuerwehrhelm — Fire fighting helmet
Design
Formquadrat GmbH
Stefan Degn, Mario Zeppetzauer
www.formquadrat.com

Unternehmen — Company
Rosenbauer International AG
Paschinger Straße 90
4060 Leonding
Austria
www.rosenbauer.com
Produkt — Product
Rosenbauer Panther 8x8
Flughafen-Feuerwehrfahrzeug — Airport fire engine
Design
Spirit Design / Innovation and Branding
Daniel Huber, Tomas Maly
www.spiritdesign.at
88 Silber

Unternehmen — Company
Rosenthal AG
Philip-Rosenthal-Platz 1
95100 Selb
Germany
www.rosenthal.de
Produkt — Product
Blu Canela
Vasenkollektion, Titanium bedampft — Vase series with titanium décor
Design
Harry & Camila
//
Werksdesign / Inhouse
Rosenthal Creative Center
272

Unternehmen — Company
Rosenthal AG
Philip-Rosenthal-Platz 1
95100 Selb
Germany
www.rosenthal.de
Produkt — Product
Tectonics
Schalenserie aus Porzellan, Titanium bedampft — Porcelain dishes with titanium décor
Design
Werksdesign / Inhouse
Rosenthal Creative Center
Cairn Young
273

Unternehmen — Company
Rosenthal AG
Philip-Rosenthal-Platz 1
95100 Selb
Germany
www.rosenthal.de
Produkt — Product
Vase of Phases
Vasenserie — Vase series
Design
Studio Dror
Dror Benshetrit
//
Werksdesign / Inhouse
Rosenthal Creative Center
274

Unternehmen — Company
ROSET S.A.
B.P.9
01470 Briord
France
www.ligne-roset.com
Produkt — Product
DOWNTOWN
Polstermöbel — Upholstery
Design
Pascal Mourgue Design
Pascal Mourgue
www.pascalmourgue.com

Unternehmen — Company
RÖSLE GmbH & Co. KG
Johann-Georg-Fendt-Straße 38
87616 Marktoberdorf
Germany
www.roesle.de
Produkt — Product
Clevere Löffel
Intelligente Utensilien — Intelligent utensils
Design
forma+
Margherita Croci, Sascha Klingholz
www.formaplus-design.de

Unternehmen — Company
Rotho Kunststoff AG
Industriestrasse Althau 11
5303 Würenlingen
Switzerland
www.rotho.com
Produkt — Product
COOL & FRESH
Salat Butler — Salad butler
Design
Werksdesign / Inhouse
Rotho Kunststoff AG

Unternehmen — Company
Royal Mosa
Meerssenerweg 358
6201 BA Maastricht
Netherlands
www.mosa.nl
Produkt — Product
Terra XXL
Fliesen-Kollektion — Collection of tiles
Design
Werksdesign / Inhouse
197

Unternehmen — Company
RTI Sports GmbH
Rudolf-Diesel-Straße 21
56220 Urmitz
Germany
www.rtisports.de
Produkt — Product
Ergon Race Grip R2
Ergonomischer Fahrradgriff mit Barend —
Ergonomic bicycle grip with bar end
Design
Ergonomidesign AB
Peter Ejvinsson, Anna Carell, Thomas Nilsson
www.ergonomidesign.com
//
Werksdesign / Inhouse
Franc Arnold, Kim Tofaute

Unternehmen — Company
RTI Sports
Rudolf-Diesel-Straße 21
56220 Urmitz
Germany
www.rtisport.de
Produkt — Product
Micro Rocket Alu & Carbon
Fahrrad-Miniluftpumpe — Bicycle minipump
Design
Werksdesign / Inhouse
Topeak Design Team

Unternehmen — Company
RTI Sports
Rudolf-Diesel-Straße 21
56220 Urmitz
Germany
www.rtisport.de
Produkt — Product
PropShock
Fahrrad-Dämpferpumpe — Suspension-pump
Design
Werksdesign / Inhouse
Topeak Design Team

Unternehmen — Company
Ruckstuhl AG
St. Urbanstrasse 21
4901 Langenthal
Switzerland
www.ruckstuhl.com
Produkt — Product
PLUS, Edition 2005
Teppichkollektion aus Wollfilz —
Carpet range of woolfelt
Design
Design Development
Alfredo Häberli, Theo Gschwind
www.alfredo-haeberli.com

Unternehmen — Company
Rudolf Thurner GmbH
Thurner wo_menswear
Hauptstraße 54
7000 Eisenstadt
Austria
www.thurner-wm.at
Produkt — Product
Thurner Wo_Menswear
Geschäftsgestaltung — Showroom design
Design
Martin Mostböck Design
Martin Mostböck d.i. arch.
www.martin-mostboeck.com

Unternehmen — Company
RUF-Bett International GmbH & Co. KG
Im Wöhr 1
76437 Rastatt
Germany
www.ruf-betten.de
Produkt — Product
Corner
Polsterbett — Upholstery bed
Design
Designbüro Heiliger
Prof. Stefan Heiliger
www.heiliger-design.de

Unternehmen — Company
Rundfunk Berlin Brandenburg (RBB)
Marlene-Dietrich-Allee 20
14482 Potsdam
Germany
www.rbb-online.de
Produkt — Product
TypOmat http:/audiotain.exozet.com/Typomat/
Internetseite — Website
Design
interactive tools, Agentur für digitale Medien

Unternehmen — Company
rutan GmbH
Pallaswiesenstraße 28
64293 Darmstadt
Germany
www.rutan.de, www.pitgreen.com
Produkt — Product
PIT GREEN / MICRO GOLF
Golf Accessoire — Golf accessory
Design
Werksdesign / Inhouse
Matthias Lenz, Oliver Johannson,
Axel Hildenbrand, Jörg Meister

Unternehmen — Company
RWE AG
Opernplatz 1
42128 Essen
Germany
Produkt — Product
agenda edition
Unternehmenspublikation — Corporate book
Design
HOFFMANN UND CAMPE Corporate Publishing
Dirk Linke, Sabine Keller, Jessica Winter
www.hoca-cp.de
354

Unternehmen — Company
RWE AG
Opernplatz 1
45128 Essen
Germany
www.rwe.de
Produkt — Product
RWE Digital Brand Management System
Online-Corporate-Design-Portal —
Online Corporate Design portal
Design
MetaDesign AG
www.metadesign.de

S

Unternehmen — Company
S. Siedle & Söhne
Telefon- und Telegrafenwerke OHG
Bregstraße 1
78120 Furtwangen
Germany
www.siedle.de
Produkt — Product
Siedle-Steel
Orientierungssystem — Orientation system
Design
Werksdesign / Inhouse
S. Siedle & Söhne
Telefon-und Telegrafenwerke OHG
Eberhard Meurer
186

Unternehmen — Company
S. Weisz Uurwerken BV
Schubertstraat 21
1077 GP Amsterdam
Netherlands
www.jacob-jensen.com
//
Filius & Haake GmbH & Co. KG
Kuckucksweg 55
33607 Bielefeld
Germany
www.filius-haake.de
Produkt — Product
Jacob Jensen Chronograph 600
Titan Armbanduhr — Titanium watch
Design
Jacob Jensen Design
Timothy Jacob Jensen
www.jacob-jensen.com
245

Unternehmen — Company
Sabine Samuel
Bockumer Straße 357
40489 Düsseldorf
Germany
Produkt — Product
Solo
Haarspange — Hair slide
Design
Werksdesign / Inhouse

Unternehmen — Company
Sahm GmbH & Co. KG
Westerwaldstraße 13
56203 Höhr-Grenzhausen
Germany
www.sahm.de
Produkt — Product
Tokyo
Bierbecher — Beer tumbler
Design
Michael Boehm

Unternehmen — Company
Sahm GmbH & Co. KG
Westerwaldstraße 13
56203 Höhr-Grenzhausen
Germany
www.sahm.de
Produkt — Product
Nova
Bier-Pokal — Beer goblet
Design
Design Werkstatt
Jutta Hohngarten-Feistel

Unternehmen — Company
Sal. Oppenheim jr. & Cie. KGaA
Unter Sachsenhausen 4
50667 Köln
Germany
Produkt — Product
215plus / 216plus / 217plus –
Ein Magazin für die besondere Perspektive
Kundenmagazin — Customer magazine
Design
Simon & Goetz Design GmbH & Co. KG
Bernd Vollmöller
www.simongoetz.de

Unternehmen — Company
SALEWA Sportgeräte GmbH
Bereich DYNAFIT
Saturnstraße 63
85609 Aschheim
Germany
www.dynafit.de
Produkt — Product
DYNAFIT TLT Vertical FT
Skitouren-Bindung — Skitouring binding
Design
Target Design
www.target-design.com

Unternehmen — Company
Samsung Electronics GmbH
Am Kronberger Hang 6
65824 Schwalbach
Germany
www.samsung.com
Produkt — Product
„Beyond The Games"
Messeauftritt von Samsung Electronics bei den World Cyber Games – Samsung European Championship 2005 während der CeBIT 2006 — Trade fair stand of Samsung Electronics at the World Cyber Games – Samsung European Championship 2005 at CeBIT 2006
Design
ideea, Messe-und Dekorationsbau GmbH
www.ideea.de
//
Werksdesign / Inhouse
Jürgen Schmidt-André
314

Unternehmen — Company
Sartorius Hamburg GmbH
Meiendorfer Straße 205
22145 Hamburg
Germany
www.sartorius.com
Produkt — Product
PR 6202
Hygienic Load Cell — Hygienic load cell
Design
Design 3 Produktdesign
Wolfgang Wagner
www.design3.de
//
Werksdesign / Inhouse
Klaus Vayhinger

Unternehmen — Company
Satelliten Media Design
Hohe Bleichen 5
20354 Hamburg
Germany
www.satelliten.de
Produkt — Product
www.slogans.de – Datenbank der Werbe-slogans — www.slogans.de – Database of advertising slogans
Internetseite — Website
Design
Werksdesign / Inhouse
Alexander Hahn, Inga Wermuth
332

Unternehmen — Company
SCHAEFER GmbH
Winterlinger Straße 4
72488 Sigmaringen
Germany
www.ws-schaefer.de
Produkt — Product
IceCube
Glastableau in 3D — 3D Glass Fixture
Design
piu products Integrated Design
Torsten Gratzki
www.piuproducts.com

Unternehmen — Company
SCHLARAFFIA
RECTICEL-SCHLAFKOMFORT GmbH
Schlaraffiastraße 1–10
44867 Bochum
Germany
www.schlaraffia.de
Produkt — Product
Flexmont
Motorrahmen — Motor frame
Design
Werksdesign / Inhouse
Giacomo Lanza, Andreas Schrade, Pierre Bourgeois, Markus Veutgen
207

Unternehmen — Company
SCHLARAFFIA
RECTICEL-SCHLAFKOMFORT GmbH
Schlaraffiastraße 1–10
44867 Bochum
Germany
www.schlaraffia.de
Produkt — Product
Media®
Lattenrost — High performance frame
Design
Werksdesign / Inhouse
Giacomo Lanza, Andreas Schrade,
Pierre Bourgeois, Markus Veutgen
206

Unternehmen — Company
Schmitz u. Söhne GmbH & Co. KG
Zum Ostenfeld 29
58739 Wickede (Ruhr)
Germany
www.schmitz-soehne.de
Produkt — Product
STL 285
Patienten-Transporter für Notfall und Ambulanz — Patient transporter for emergencies, Outpatient Departments
Design
Rainer Schindhelm Industrie-Design
Rainer Schindhelm

Unternehmen — Company
Schmitz-Leuchten GmbH & Co. KG
Niedereinerfeld 29
59823 Arnsberg
Germany
www.schmitz-leuchten.de
Produkt — Product
15868.25
Control-Strahler — Control spotlight
Design
Werksdesign / Inhouse

Unternehmen — Company
Schmitz-Leuchten GmbH & Co. KG
Niedereinerfeld 29
59823 Arnsberg
Germany
www.schmitz-leuchten.de
Produkt — Product
TURN
Pendelleuchte — Pendant light
Design
Klaus Baulmann

Unternehmen — Company
Schmitz-Leuchten GmbH & Co.KG
Niedereimerfeld 29
59823 Arnsberg
Germany
www.schmitz-leuchten.de
Produkt — Product
INSIDE
Wandleuchte — Wall light
Design
Eva Louis

Unternehmen — Company
Schneider „Der Ideenschreiner" GmbH
Porschestraße 6
70736 Fellbach
Germany
www.der-ideenschreiner.de
Produkt — Product
BIG APPLE
Esstisch — Dining table
Design
Werksdesign / Inhouse
217

Unternehmen — Company
Schneider GmbH
Martin-Heinrich-Klaproth-Straße 28
38855 Wernigerode
Germany
www.dachslschneider.de
Produkt — Product
Base Kid
Schreiblernfüller — Fountain pen for beginners
Design
Atelier für Gestaltung
Andreas Papenfuss, Christina Albert,
Josef Landmann, John Mitsching,
Ulrike Papenfuss
www.atelierpapenfuss.de
233

Unternehmen — Company
schneiderschram GmbH
Architektenmöbel
Linienstraße 215
10119 Berlin
Germany
www.schneiderschram.de
Produkt — Product
schneller raum — instant space
Raumteiler, Projektionswand —
Room divider, projection screen
Design
Léon Wohlhage Wernik Architekten GmbH
Konrad Wohlhage
www.leonwohlhagewernik.de

Unternehmen — Company
schneiderschram GmbH
Architektenmöbel
Linienstraße 215
10119 Berlin
Germany
www.schneiderschram.de
Produkt — Product
Tafelschrank — Wide board
Schrank — Cupboard
Design
Eike Becker Architekten GmbH
Eike Becker
www.eikebeckerarchitekten.com

Unternehmen — Company
Schöck Bauteile GmbH
Vimbucher Straße 2
76534 Baden-Baden
Germany
www.schoeck.de
//
Fiberline Composites A/S
Nr. Bjertvej 88
6000 Kolding
Denmark
www.fiberline.com
Produkt — Product
ComBAR
GFK Bewehrung — Glassfibre reinforcement
Design
Werksdesign/Inhouse

Unternehmen — Company
Schock GmbH
Gmünder Straße 65
73614 Schorndorf
Germany
www.schock.de
Produkt — Product
Cristalite granite
Wand- und Bodenbelag —
Flooring and wall cladding
Design
Werksdesign/Inhouse
Dr. Gudrun Janke, Martin Kirschbaum
195

Unternehmen — Company
Schönbuch Collection
Möbelmarketing GmbH
Industriestraße 11
97631 Bad Königshofen
Germany
www.schoenbuch.com
Produkt — Product
UPON
Dielenmöbel, Garderobenständer und
Wandgarderobe — Entrance hall furniture,
coat rack and wardrobe
Design
Stefan Diez
www.stefan-diez.com

Unternehmen — Company
Schreinerei Anderegg AG
Olensbachstrasse 7
9631 Ulisbaach
Switzerland
www.anderegg-ideen.ch
Produkt — Product
Holzteppich — Wooden carpet
Bodenbelag — Carpet
Design
Werksdesign/Inhouse
Christian Anderegg

Unternehmen — Company
Schreinerei Schleiter/Holzgestaltung
Hopfenmühle 1
95233 Helmbrechts
Germany
www.andreas-schleiter.com
Produkt — Product
Orplid
Stuhl — Chair
Design
Werksdesign/Inhouse
Andreas Schleiter
218

Unternehmen — Company
Schüco International KG
Karolinenstraße 1–15
33609 Bielefeld
Germany
www.schueco.de
Produkt — Product
Schüco Griff-Designlinie —
The new Schüco design handles
Griffe — Handles
Design
Teams Design GmbH
www.teamsdesign.com
//
Werksdesign/Inhouse
Antje Bönisch, Konrad Kaiser
168

Unternehmen — Company
Schüco International KG
Karolinenstraße 1–15
33609 Bielefeld
Germany
www.schueco.de
Produkt — Product
Schüco TipTronic
Mechatronische Beschlagsgeneration —
The mechatronic generation of fittings
Design
Teams Design GmbH
www.teamsdesign.com
//
Werksdesign/Inhouse
Konrad Kaiser, Antje Bönisch
169

Unternehmen — Company
SCM Microsystems GmbH
Oskar-Messter-Straße 13
85737 Ismaning
Germany
www.scmmicro.com
Produkt — Product
e20
Mobiler Chipkartenleser —
Mobile smart card reader
Design
Werksdesign/Inhouse
Remco Willemse

Unternehmen — Company
SCOPETIME GmbH
Krausestraße 102
22049 Hamburg
Germany
www.neolog.cc
Produkt — Product
NEOLOG A-24 II
Armbanduhr — Wristwatch
Design
Werksdesign/Inhouse
Arman Emami
244

Unternehmen — Company
sd & m AG
software design & management
Carl-Wery-Straße 42
81739 München
Germany
www.sdm.de
Produkt — Product
Magazin m & it
Magazin zur Kundenansprache —
Magazine for customers
Design
Horst Metzner GmbH
Horst Metzner, Thomas Scheuplein
www.horstmetzner.de
//
Werksdesign / Inhouse
Edmund Küpper, Rüdiger Azone,
Dr. Uwe Dunslaff
358

Unternehmen — Company
sdr+
Schaevenstraße 7
50676 Köln
Germany
www.sdr-plus.com
Produkt — Product
606
Regalsystem — Shelving System
Design
Dieter Rams

Unternehmen — Company
Sector 7G
Textilgatan 43
120 30 Stockholm
Sweden
www.drosselmeyer.se
Produkt — Product
Drosselmeyer
Nussknacker — Nut cracker
Design
Drosselmeyer Deutschland
Andreas Illing, Erik von Schoultz, Joakim Norin
www.drosselmeyer.se

Unternehmen — Company
SELECTED SOUND
MUSIKVERLAG GMBH & Co. KG
Alsterufer 1
20354 Hamburg
Germany
www.selected-sound.de
Produkt — Product
Selected Sound
Corporate Design
Design
form one - visuelle kommunikation
Niels Bünemann
www.form-one.de
364

Unternehmen — Company
SENATOR Schreibgeräte GmbH & Co. KGaA
Bahnhofstraße 57
64401 Groß-Bieberau
Germany
www.senator-pen
Produkt — Product
@tract
Kugelschreiber — Ballpoint pen
Design
Rommel und Schoen Design
Volkmar Rommel
www.rommelundschoendesign.de
239

Unternehmen — Company
SENATOR Schreibgeräte GmbH & Co. KGaA
Bahnhofstraße 57
64401 Groß-Bieberau
Germany
www.senator-pen.com
Produkt — Product
New Spring Chrome
Kugelschreiber — Ballpoint pen
Design
Rommel und Schoen Design
Volkmar Rommel
www.rommelundschoendesign.de
236

Unternehmen — Company
SENATOR Schreibgeräte GmbH & Co. KGaA
Bahnhofstraße 57
64401 Groß-Bieberau
Germany
www.senator-pen.com
Produkt — Product
Aqua-Lighter®
Textmarker — Highlighter
Design
Werksdesign / Inhouse
Georg Keil
234

Unternehmen — Company
SENATOR Schreibgeräte GmbH & Co. KGaA
Bahnhofstraße 57
64401 Groß-Bieberau
Germany
www.senator-pen.com
Produkt — Product
Upside Down
Drehkugelschreiber — Ballpoint pen
Design
Werksdesign / Inhouse
Hans-Jürgen Beyer
238

Unternehmen — Company
SENATOR Schreibgeräte GmbH & Co. KGaA
Bahnhofstraße 57
64401 Groß-Bieberau
Germany
www.senator-pen.com
Produkt — Product
AQUA-WONDER®
Fasermaler — Fibre pen
Design
Werksdesign / Inhouse
Georg Keil
235

Unternehmen — Company
SENATOR Schreibgeräte GmbH & Co. KGaA
Bahnhofstraße 57
64401 Groß-Bieberau
Germany
www.senator-pen.com
Produkt — Product
SOLARIS
Metall-Schreibgeräte —
Metall writing instruments
Design
Hariolf Briggl
237

Unternehmen — Company
Sennheiser electronic GmbH & Co. KG
Am Labor 1
30900 Wedemark
Germany
www.sennheiser.com
Produkt — Product
MX 75 Sport
Stereo-Ohrhörer — Sport earphones series
Design
berlinlondon
Susanne Ewert, Markus Hohl, Lars Milde
www.berlinlondon.com

Unternehmen — Company
Sennheiser electronic GmbH & Co. KG
Am Labor 1
30900 Wedemark
Germany
www.sennheiser.com
Produkt — Product
MXL 70 VC
Stereo-Ohrhörer mit Lanyard und Lautstärkeregler — Stereo earphone with lanyard and volume control
Design
londonberlin
Susanne Ewert, Markus Hohl
www.berlinlondon.com

Unternehmen — Company
Sennheiser electronic GmbH & Co. KG
Am Labor 1
30900 Wedemark
Germany
www.sennheiser.com
Produkt — Product
OMX 90 VC Style
Stereo-Ohrbügelhörer mit Lautstärkeregler — Stereo clip-on earphone with volume control
Design
londonberlin
Susanne Ewert, Markus Hohl
www.berlinlondon.com

Unternehmen — Company
Sennheiser electronic GmbH & Co. KG
Am Labor 1
30167 Wedemark
Germany
www.sennheiser.com
Produkt — Product
HD 205
Dynamischer Kopfhörer — Dynamic headphone
Design
Brennwald Design
Jörg Brennwald
www.brennwald-design.de

Unternehmen — Company
serien Raumleuchten GmbH
Hainhäuser Straße 3–7
63110 Rodgau
Germany
www.serien.com
Produkt — Product
Zoom
Pendelleuchte — Suspension lamp
Design
Werksdesign / Inhouse
Floyd Paxton

Unternehmen — Company
Serioplast GmbH
Güterstraße 14
29549 Bad Bevensen
Germany
Produkt — Product
Monoverpackung
Verschlussprinzip für Kunststoffverpackungen — Closure principle for plastic bottles
Design
iondesign
Anja Götz, Christoph Fleckenstein
www.iondesign.de

Unternehmen — Company
SGM Technology for Lighitng SPA
Via Pio La Torre, 1
61010 TAVULLIA
Italy
www.sgm.it
Produkt — Product
PALCO LED
LED-Farbwechsler — LED colour changer
Design
PROMAU R.P.D. srl
www.promau.com

Unternehmen — Company
sieger design GmbH & Co. KG
Schloss Harkotten
48336 Sassenberg
Germany
www.sieger-design.com
Produkt — Product
Chaise Couture No.1
Polstermöbel — Seating programme
Design
Werksdesign / Inhouse
Michael Sieger

Unternehmen — Company
sieger design GmbH & Co. KG
Schloss Harkotten
48336 Sassenberg
Germany
www.sieger-design.com
Produkt — Product
sieger insight No. 3
Kundenmagazin / Unternehmensbroschüre — Customer magazine
Design
Werksdesign / Inhouse
Michael Sieger

Unternehmen — Company
sieger design GmbH & Co. KG
Schloss Harkotten
48336 Sassenberg
Germany
www.sieger-design.com
Produkt — Product
Best of Mickey, Best of Snoopy
Internetseite — Website
Design
avency
Roland Albers
www.avency.de
//
Werksdesign / Inhouse
Christian Sieger, Michael Sieger

Unternehmen — Company
Siemens AG
Hofmannstraße 51
81359 München
Germany
www.siemens.de
Produkt — Product
Gigaset M2 professional
Schnurlostelefon — Cordless telefone
Design
designafairs GmbH
Bianca Fleischer
www.designafairs.com

Unternehmen — Company
Siemens AG
Industrial Solutions and Services, Postal Automation
Bücklestraße 1–5
78467 Konstanz
Germany
www.siemens.com / postalautomation
Produkt — Product
Open Mail Handling System (OMS)
Großpostsortieranlage — Flats sorting machine
Design
die:haptiker GmbH
Jörg Ostenrieder
www.die-haptiker.de
//
Werksdesign / Inhouse
Andreas Preussner, Ernst Haerle, Josef Mok, Christian Gude

Unternehmen — Company
Siemens AG, Bereich Automation and Drives
Gleiwitzerstraße 555
90475 Nürnberg
Germany
www.siemens.com / automation
Produkt — Product
SIMATIC MOBILE PANEL 277
Mobiles Handbediengerät — Mobile hand control unit
Design
at-design
Jan Andersson, Christoph Tomczack
www.atdesign.de
//
Werksdesign / Inhouse
Siemens AG A & D Design Management
Gunter Ott, Nina Wendt
305

Unternehmen — Company
Siemens AG, Bereich Automation and Drives
Gleiwitzerstraße 555
90475 Nürnberg
Germany
www.siemens.com / automation
Produkt — Product
SINUMERIK HT8
Mobiles Handbediengerät — Mobile hand control unit
Design
at-design
Jan Andersson, Christoph Tomczack
www.atdesign.de
//
Werksdesign / Inhouse
Siemens AG A & D Design Management
Gunter Ott, Nina Wendt
304

Unternehmen — Company
Siemens AG, Medical Solutions Special Systems
Hartmannstraße 48
91052 Erlangen
Germany
www.siemens.com / medical
Produkt — Product
MODULARIS Variostar
Mobiler Lithotripter — Mobile lithotripter
Design
designafairs GmbH
Dirk Lamprecht, Gerd Helmreich, Thomas Maurer
www.designafairs.com
//
Werksdesign / Inhouse
Jens Fehre, Matthias Mahler

Unternehmen — Company
Siemens Home and Office
Communication Devices GmbH & Co. KG
St.-Martin-Straße 47
81541 München
Germany
www.siemens.com/gigaset
Produkt — Product
Gigaset S455
Schnurlostelefon — Cordless telefone
Design
designafairs GmbH
Christian Rogge, Julia Steiger
www.designafairs.com

Unternehmen — Company
Sigrid Reeckmann Weberei
Koberger Straße 49
90409 Nürnberg
Germany
Produkt — Product
Webobjekte — Woven objects
Bügeltaschen, Kissen, Seidenschals — Handbags, cushions, silk scarfs
Design
Werksdesign/Inhouse

Unternehmen — Company
SILHOUETTE International Schmied AG
Ellbognerstraße 24
4021 Linz
Austria
www.silhouette.com
Produkt — Product
Model 7522/10 6090 53/16
Optische Brille — Optical frame
Design
Werksdesign/Inhouse
Gerhard Fuchs

Unternehmen — Company
Silit-Werke GmbH & Co. KG
Neufraer Straße 6
88499 Riedlingen
Germany
www.silit.de
Produkt — Product
Studio
Antihaft-Pfannen — Non-stick frying pans
Design
Werksdesign/Inhouse

Unternehmen — Company
Silit-Werke GmbH & Co. KG
Neufraer Straße 6
88499 Riedlingen
Germany
www.silit.de
Produkt — Product
Profi
Reibe — Grater
Design
Teams Design
www.teamsdesign.com

Unternehmen — Company
SINNSALON
Jarrestraße 20
22303 Hamburg
Germany
www.sinnsalon.de
Produkt — Product
Ticket zum Glück 2005
Neujahrskarte — New Years greeting card
Design
Werksdesign/Inhouse
336

Unternehmen — Company
Sireo – Real Estate Asset Management
Seizstraße 16
80538 München
Germany
www.sireo.de
Produkt — Product
Residenzpost München
Buch — Book
Design
UNID – One Voice Branding
Ralph Missy, Martina Donat, Stanislav Erman
www.unid.de
347

Unternehmen — Company
Skantherm GmbH & Co. KG
Lümernweg 188 a
33378 Rheda-Wiedenbrück
Germany
www.skantherm.de
Produkt — Product
ZIO
Kaminofen drehbar mit Wasserverdunster — Turnable stove with water evaporator
Design
HELIX-Akantus Formgestaltung GmbH & Co. KG
www.akantus.de

Unternehmen — Company
Skantherm GmbH & Co. KG
Lümernweg 188 a
33378 Rheda-Wiedenbrück
Germany
www.skantherm.de
Produkt — Product
emotion
Kaminofen für Holzbefeuerung — Wood-burning stove
Design
Prof. Wulf Schneider und Partner
Wulf Schneider
www.profwulfschneider.de

Unternehmen — Company
Skantherm GmbH & Co. KG
Lümernweg 188 a
33378 Rheda-Wiedenbrück
Germany
www.skantherm.de
Produkt — Product
SHAKER
Kaminofen für Holzbefeuerung —
Wood-burning stove
Design
Antonio Citterio Architetto
Antonio Citterio, Toan Nguyen
www.antoniocitterioandpartners.it

Unternehmen — Company
SKF GmbH
Gunnar-Wester-Straße 12
97421 Schweinfurt
Germany
www.skf.de
Produkt — Product
SKF
Tretlagereinheit für Fahrräder —
Bottom bracket unit for bicycles
Design
I.D.M. Müsing
Bernd Müsing
//
Werksdesign / Inhouse
SKF GmbH
153

Unternehmen — Company
SKS Metaplast Scheffer-Klute GmbH
Zur Hubertushalle 4
59846 Sundern
Germany
Produkt — Product
X-Blade / Shockblade 5168
Radschutz-Set für Mountainbiker —
Dirtboard-set for mountainbiking
Design
Flöz Industriedesign
Oliver Wahl
www.floez.de

Unternehmen — Company
SolarWorld AG
Kurt-Schumacher-Straße 12–14
53113 Bonn
Germany
www.solarworld.de
Produkt — Product
Vögel — Birds
Imageanzeige —
Advertisement for an image campaign
Design
POLARWERK E.K.
Thomas Thesseling, Sebastian Kühnel,
Bastian Nemitz
www.polarwerk.de

Unternehmen — Company
sorø ole bjerregaard
Dantes Plads 6
1556 Kopenhagen
Denmark
www.soroe.com
Produkt — Product
H-line
Konferenz-, Arbeits- und Esstisch —
Conference-, work- and dining table
Design
knud holscher design
Knud Holscher, Per Kristian Dahl
www.knudholscher.dk

Unternehmen — Company
spectral Gesellschaft für Lichttechnik mbH
Bötzinger Straße 31
79111 Freiburg
Germany
www.spectral-online.de
Produkt — Product
LED-Farblichtleuchten —
LED luminaires with chromatic light
Leuchten — Luminaires
Design
Werksdesign / Inhouse
Paul Heinrich Neuhorst

Unternehmen — Company
spectral Gesellschaft für Lichttechnik mbH
Bötzinger Straße 31
79111 Freiburg
Germany
www.spectral-online.de
Produkt — Product
Astral.ed
Leuchtenfamilie — Luminaire family
Design
Werksdesign / Inhouse
Paul Heinrich Neuhorst

Unternehmen — Company
Spemot AG
Industriestrasse 70
4657 Dulliken
Switzerland
www.soulution-audio.com
Produkt — Product
soulution 710
Audio-Verstärker — Audio amplifier
Design
greutmann bolzern designstudio
Urs Greutmann, Carmen Greutmann
www.gbdesign.ch
282

Unternehmen — Company
Spemot AG
Industriestrasse 70
4657 Dulliken
Switzerland
www.soulution-audio.com
Produkt — Product
soulution 720
Audio-Vorverstärker — Audio preamplifier
Design
greutmann bolzern designstudio
Urs Greutmann, Carmen Greutmann
www.gbdesign.ch
283

Unternehmen — Company
Spennare
KG Spennare AB
Hastholmsvagen 32
131 30 Nacka
Sweden
www.spennare.com
Produkt — Product
Banner Stand S1
Portables Präsentations System —
Portable display system
Design
Werksdesign / Inhouse
Jerk Thomsgard

Unternehmen — Company
Spennare
KG Spennare AB
Hastholmsvagen 32
131 30 Nacka
Sweden
www.spennare.com
Produkt — Product
Roll Up S1
Portables Präsentations-System —
Portable display system
Design
Werksdesign / Inhouse
Jerk Thomsgard

Unternehmen — Company
Spielkartenfabrik Altenburg GmbH
Leipziger Straße 7
04600 Altenburg
Germany
www.spielkarten.com
Produkt — Product
T-Spielkarten — T-Playing Cards
Werbemittel — Advertising tool
Design
Heine / Lenz / Zizka Frankfurt / Berlin
www.heine-lenz-zizka.com
//
Teunen Konzepte GmbH
www.teunen-konzepte.de

Unternehmen — Company
Spinlock Ltd
41 Birmingham Road
PO31 7BH Cowes, Isle of Wight
Great Bitain
www.spinlockdeckware.com
Produkt — Product
SPINLOCK Deckware Deck Pro
Schwimmweste — Lifejacket harness
Design
Werksdesign / Inhouse
Chris Hill

Unternehmen — Company
Spring Switzerland GmbH
Hörnlistrasse 14
8280 Eschlikon
Switzerland
www.spring.ch
Produkt — Product
Raclette2+
Modulares Raclette Elektrogerät —
Modular electrical raclette unit
Design
Fa. Dubach-Design
Fredi Dubach, Stefan Stauffacher
www.dubach-design.ch

Unternehmen — Company
SQ-lab GmbH
Ergonomisches Radsportzubehör
Frundsbergstraße 29
82064 Strasslach
Germany
www.sq-lab.com
Produkt — Product
SQ-Seat 602 Ergolux
Sattel — Saddle
Design
Designgestalten
Dirk Kuntz
www.designgestalten.de
//
Werksdesign / Inhouse
Dr. Stefan Staudte

Unternehmen — Company
SRAM
Romstraße 1
97424 Schweinfurt
Germany
www.sram.com
Produkt — Product
X.0 Trigger
Schalter, X.0 Gruppe, Mountainbike
Komponenten — Shifter, X.0 group,
Mountainbike components
Design
Werksdesign / Inhouse
Christian Buettner

Unternehmen — Company
Staatliches Baumanagement Osnabrück
Hakenstraße 15
49074 Osnabrück
Germany
Produkt — Product
Orientierungssystem Fachhochschule
Osnabrück — Signage system Osnabrück
University Of Applied Sciences
Orientierungssystem — Signage system
Design
Büro Uebele
Gerd Häußler, Prof. Andreas Uebele
www.uebele.com

Unternehmen — Company
Staatskanzlei Saarland
Ludwigsplatz 14
66117 Saarbrücken
Germany
www.staatskanzlei.saarland.de
Produkt — Product
www.buergerdienste-saar.de
Internetseite / Bürgerportal — Website / Portal
Design
Werksdesign / Inhouse
Christoph Damanik, Thomas Ultes
www.somneo.de

Unternehmen — Company
Stadt Augsburg, Kulturreferat
Bahnhofstraße 18 1/3 a
86150 Augsburg
Germany
Produkt — Product
Stadt Augsburg Buch „Bewerbung der Stadt Augsburg zur Kulturhauptstadt Europas 2010" — City of Augsburg book "Application of the city of Augsburg for cultural capital of Europe 2010"
Bewerbung der Stadt Augsburg — Application of the city of Augsburg
Design
Factor Design AG
Johannes Erler, Jindrich Novotny, Frizzi Kurkhaus, Christian Tönsmann
www.factordesign.com

Unternehmen — Company
STAGETEC Gesellschaft für professionelle Audiotechnik GmbH
Tabbertstraße 10
12459 Berlin
Germany
www.stagetec.com
Produkt — Product
aurus
Digitales Tonmischpultsystem — Digital sound mixer system
Design
Schünemann DesignSolutions
Arno Schünemann
www.schuenemann-design.de

Unternehmen — Company
Starmed GmbH
Im Lehrer Feld 30
89081 Ulm
Germany
www.starmed.eu
Produkt — Product
Silentsafe
Leichentransport- und Bergesystem — System to transport and salvage dead bodies
Design
Reform Design
Matthias Wagner
www.refom-design.de
//
Werksdesign / Inhouse
René Stern
84 Silber

Unternehmen — Company
Stash
Wyckerbrugstraat 45
6121 EB Maastricht
Netherlands
www.s-t-a-s-h.com
Produkt — Product
Stash, retail design
Inneneinrichtung für ein Taschengeschäft — Interiordesign for a handbagshop
Design
Maurice Mentjens Design
Maurice Mentjens
www.mauricementjens.com
68 Silber

Unternehmen — Company
STEINEL
Vertrieb GmbH
Dieselstraße 80–84
33442 Herzebrock-Clarholz
Germany
www.steinel.de
Produkt — Product
sensIQ
Infrarot-Bewegungssensor — Infrared motion sensor
Design
Kurz Kurz Design
Dorian Kurz
www.kurz-kurz-design.de
173

Unternehmen — Company
steute Schaltgeräte GmbH & Co. KG
Brückenstraße 19
32584 Löhne
Germany
www.steute.de
Produkt — Product
TGM
Türgriffschalter — Door handle switch
Design
Margit Herzog Design
Margit Herzog
www.margitherzog-design.de
185

Unternehmen — Company
Stiftung Brandenburgische Gedenkstätten
Heinrich-Grüber-Platz 3
16515 Oranienburg
Germany
Produkt — Product
Gedenkstätte Sachsenhausen „Station Z" — Sachsenhausen Memorial "Station Z"
Planung des Gedenkortes der Vernichtungsstation — Planning of the extermination place memorial
Design
hg merz architekten museumsgestalter
www.hgmerz.com
322

Unternehmen — Company
STILL GmbH
Berzeliusstraße 10
22113 Hamburg
Germany
www.still.de
Produkt — Product
STILL EXU
Niederhubwagen, deichselgeführt —
Pedestrian low lift pallet truck
Design
TEAMS Design GmbH
Reinhard Renner, Ulrich Warth
www.teamsdesign.com

Unternehmen — Company
stoffel design
Inhaber Thomas Stoffel
Hauptstraße 53
55758 Stipshausen
Germany
www.stoffel-design.com
Produkt — Product
Ring Eggs
Edelstein Ring — Gems ring
Design
Werksdesign / Inhouse
Thomas Stoffel

Unternehmen — Company
Storck Bicycle GmbH
Carl-Zeiss-Straße 4
65520 Bad Camberg
Germany
www.storck-bicycle.de
Produkt — Product
Adrenalin Carbon mit Carbon Hinterbau —
Adrenalin carbon with carbon rear frame
Voll gefedertes Mountainbike —
Full suspension mountainbike
Design
Werksdesign / Inhouse
Markus Storck

Unternehmen — Company
Storck Bicycle GmbH
Carl-Zeiss-Straße 4
65520 Bad Camberg
Germany
www.storck-bicycle.de
Produkt — Product
Multitask Comfort
Fahrrad — Bicycle
Design
Werksdesign / Inhouse
Markus Storck

Unternehmen — Company
Storck Bicycle GmbH
Carl-Zeiss-Straße 4
65520 Bad Camberg
Germany
www.storck-bicycle.de
Produkt — Product
Scenario CD 1.0 10th Anniversary
Carbon-Rennradrahmen und -Gabel Set —
Carbon-fibre racing bike frame and fork set
Design
Gregor Hans
//
Werksdesign / Inhouse
Markus Storck

Unternehmen — Company
Storyboard Furniture GbR
Steingrubenweg 11 b
91639 Wolframs-Eschenbach
Germany
www.storyboard-furniture.de
Produkt — Product
CLAP
Regalsystem — Shelving system
Design
CarlosKönigDesigners
Carlos König
www.carloskoenigdesigners.de

Unternehmen — Company
Strange Attractors Design
Paramaribostraat 18–I
2585GN The Hague
Netherlands
www.strangeattractors.com
Produkt — Product
Little Yellow Writing Hood
Animation / Typografisches Märchen —
Animation / Typographic fairytale
Design
Strange Attractors Design
Ryan Pescatore Frisk
www.strangeattractors.com

Unternehmen — Company
Streetlife
Oude Singel 144
2312 RG Leiden
Netherlands
www.streetlife.nl
Produkt — Product
Rough & Ready 6
Bank, R & R-6-2340 aus der R & R-5 / 6 Serie —
Bench, R & R-6-2340 from the R & R-5 / 6 series
Design
Werksdesign / Inhouse
Peter Krouwel, Arjan van der Vegte
216

Unternehmen — Company
Ströer Out-Of-Home Media AG
media for mobile people
Ströer Allee 1
50999 Köln
Germany
www.stroeer.de
Produkt — Product
City Light Poster – Solitaire
Outdoor-Werbeträger — Outdoor media
Design
Wittenberg Hübener
Prof. Torsten Wittenberg
www.whcdesign.de

Unternehmen — Company
Ströer Out-Of-Home Media AG
media for mobile people
Ströer Allee 1
50999 Köln
Germany
www.stroeer.de
Produkt — Product
Ströer City Light Poster & Dynamische Fahrgast-Information — Ströer City Light Poster with passenger information system
Werbeträger / Verkehrsinformationssystem — Advertisement produkt / Public information system
Design
Werksdesign / Inhouse
Andreas Tipmann

Unternehmen — Company
Ströer Out-Of-Home Media AG
media for mobile people
Ströer Allee 1
50999 Köln
Germany
www.stroeer.de
Produkt — Product
City Guide
Informationsterminal im Außenbereich — Public information terminal
Design
Wittenberg Hübener
Prof. Torsten Wittenberg
www.whcdesign.de

Unternehmen — Company
Studio rUZ
Herengracht 62
1015BP Amsterdam
Netherlands
www.studioruz.com
Produkt — Product
Kampagne Festival für frühe Musik 2005 — Campaign Festival for early music 2005
Kampagne Bild, Tasche Heft, Konzertprogramm, Festival-Programm, Fahne, Plakat — Campaign image, pocket folder, concert programme, festival programme, banner, posters
Design
Studio rUZ
Daan Noppen
www.studioruz.com

Unternehmen — Company
Süddeutsche Zeitung GmbH
Sendlinger Straße 8
80331 München
Germany
www.sueddeutsche.de
Produkt — Product
Klavier Kaiser
Piano Editon — Audio booklets
Design
Werksdesign / Inhouse
Eberhard Wolf
342

Unternehmen — Company
Südtirol Marketing Gesellschaft (SMG)
Pfarrplatz Nr. 11
39100 Bozen
Italy
www.suedtirol.info
Produkt — Product
Südtirol — South Tyrol
Corporate Design Dachmarke Südtirol — Corporate Design umbrella brand South Tyrol
Design
MetaDesign AG
www.metadesign.de

Unternehmen — Company
Sunbeam Corporation
Lord
2019 Botany
Australia
www.sunbeam.com.au
Produkt — Product
Cafe Aromatic Machine EM5800
Espressomaschine — Espresso machine
Design
Werksdesign / Inhouse
Sunbeam Design Team

Unternehmen — Company
SUPERNOVA LIGHTING SYSTEMS
Industriestraße 11
79194 Freiburg
Germany
www.supernova-lights.com
Produkt — Product
SUPERNOVA P99 – D SUPERNOVA TAIL LIGHT
Xenon Fahrradbeleuchtungssystem mit LED-Rücklicht — H.I.D. Xenon bicylce lighting system with LED tail light
Design
SUPERNOVA DESIGN
Marcus Wallmeyer
www.supernova-lights.com
152

Unternehmen — Company
Sutter Medizintechnik GmbH
Tullastraße 87
79108 Freiburg
Germany
www.sutter-med.de
Produkt — Product
CURIS
Radiofrequenz-Chirurgiegerät — Radiofrequency generator
Design
item product design
Frank Detering
www.itempd.de

Unternehmen — Company
Suunto
Valimotie 7
01510 Vantaa
Finland
www.suunto.com
Produkt — Product
Suunto D9
Tauchcomputer — Diving computer
Design
Werksdesign / Inhouse
Mikko Ahlström

Unternehmen — Company
Svensk Form
Holmamiralens väg 2
11149 Stockholm
Sweden
Produkt — Product
„Made in Sweden"
Plakat — Poster
Design
Gabor Palotai Design
Gabor Palotai
Sweden
www.gaborpalotai.com
361

Unternehmen — Company
Swany Europe GmbH
Tösstalstrasse 212
8405 Winterthur
Switzerland
www.swanyeurope.com
Produkt — Product
Renegade AX-16L
Skihandschuh — Glove for snowsports
Design
VISTAPARK GmbH
Joachim Rao, Matthias Ocklenburg
www.vistapark.de
279

Unternehmen — Company
SWANY Europe GmbH
Tösstalstrasse 212
8405 Winterthur
Switzerland
www.swanyeurope.com
Produkt — Product
NFX-8 Lunar
Technische Schneesporthandschuhe — Technical glove for snowsports
Design
Werksdesign / Inhouse
Markus Kellermüller, Masahiko Kato, Ichiro Kuwahara
278

t

Unternehmen — Company
t + t netcom
Elektrastraße 6
81925 München
Germany
www.tt-netcom.com
Produkt — Product
tt7000
Industrie-PDA — Industrial PDA
Design
SquareOne GmbH
Gert Trauernicht, Eckhard Ruoff, Stefan Mostert
www.squareonedesign.de
293

Unternehmen — Company
TAG Systems AS
Søndre gate 11
8607 Mo i Rana
Norway
www.tag.no
//
IKANO Finans AS
Box 295
1372 ASKER
Norway
www.ikano.no
Produkt — Product
Credit card kash® Visa
Kreditkarten-Kollektion — Credit card collection
Design
SUPERLOW / TBWA\OSLO
Halvor Bodin
www.superlow.com

Unternehmen — Company
Tangram Design
Holbergs gt 21
0130 Oslo
Norway
www.salmonbrands.no
Produkt — Product
SALMA
Lachs — Fish loin form salmon
Design
Salmon Brands
Bjørn Rybakken
www.salma.no

Unternehmen — Company
Tanja Schickert
Konkordiastraße 52
40219 Düsseldorf
Germany
www.sinnvolleschokolade.de
Produkt — Product
Sinnvolle Schokolade
Schokolade — Chocolate
Design
KW43 BRANDDESIGN
Tanja Schickert, Margit Tabert, Katja Hopp, Mirjam Zastrau
www.kw43.de

Unternehmen — Company
TaylorMade-adidas Golf Company, Inc.
5545 Fermi Court
92008 Carlsbad, CA
USA
www.tmag.com
Produkt — Product
TAYLORMADE r7 — TaylorMade r7 TP
Golf Produkt — Golf product
Design
Priority Designs
Paul Kolada, Robert Nunez, Chad Haas
www.prioritydesigns.com
//
Werksdesign / Inhouse
Craig Barez

Unternehmen — Company
TEAM 7 Natürlich Wohnen GmbH
Braunauer Straße 26
4910 Ried im Innkreis
Austria
www.team7.at
Produkt — Product
MAGNUM Freischwinger —
MAGNUM Cantilever Chair
Stuhl, Freischwinger — Chair, cantilever chair
Design
Design Ballendat
Martin Ballendat
www.ballendat.de

Unternehmen — Company
Tenuta Kornell-Staves Weingut
Bozner Straße 28
39018 Siebeneich
Italy
www.staves.it
Produkt — Product
STAVES Weingut — Staves winery
Corporate Design
Design
Heye & Partner GmbH
Florian Drahorad, Dominik Neubauer
www.heye.de

Unternehmen — Company
tesa AG
Quickbornstraße 24
20253 Hamburg
Germany
www.tesa.de
Produkt — Product
tesa Korrektur- und Kleberoller „Roll & Relax" — tesa Correction and Adhesive Rollers "Roll & Relax"
Korrektur- und Kleberoller —
Correction and adhesive rollers
Design
Koop Industrial Design
Norbert Koop
www.koopdesign.de
232

Unternehmen — Company
Tetra GmbH
Herrenteich 78
49324 Melle
Germany
www.tetra.net
Produkt — Product
Tetra AquaArt
Aquarium-Komplett-Set 60 l
Aquarium-Komplett-Set 30 l
Aquarium — Aquarium
Design
Delta Form productdesign
Detlef Belter
www.deltaform.de

Unternehmen — Company
Thalia Theater Halle
Thaliapassage
06108 Halle
Germany
www.thaliatheaterhalle.de
Produkt — Product
Hotel Neustadt
Kampagne — Campaign
Design
Neue Gestaltung GmbH
Eva Wendel
www.neuegestaltung.de

Unternehmen — Company
Thalia Theater Halle
Thaliapassage
06108 Halle
Germany
www.thaliatheaterhalle.de
Produkt — Product
Hotel Neustadt Buch — Hotel Neustadt Book
Buch — Book
Design
Matthias Rick
www.raumlabor-berlin.de
//
Neue Gestaltung GmbH
Eva Wendel, David Voss
www.neuegestaltung.de

Unternehmen — Company
Thebingbuß GmbH
Voßkamp 24
46414 Rhede
Germany
www.thebingbuss.com
Produkt — Product
„Das Chorpodest für Ihren Auftritt" —
„The choir landing for your appearance"
Mobile Chorpodestanlage —
Mobile choir landing plant
Design
Werksdesign / Inhouse
Josef Thebingbuß

Unternehmen — Company
thema communications ag
Hanauer Landstraße 190
60314 Frankfurt
Germany
www.thema-communications.de
Produkt — Product
Stimmen zum Geschäftsbericht – Der Geschäftsbericht zwischen Image und Information — Opinions on the Annual Report – The Annual Report between Image and Information
Buch — Book
Design
Werksdesign / Inhouse
344

Unternehmen — Company
Thermoplastik Erich Müller GmbH
Gewerbestraße 1
64807 Dieburg
Germany
www.thermoplastik.de
//
CarbonFunctions Vertriebs GmbH
Hachelallee 7
75179 Pforzheim
Germany
www.carbonfunctions.de
Produkt — Product
MANTIS HE
Elektro Golfcaddy — Caddy with electric drive
Design
Studio Wagner: Design
Wolf Udo Wagner
www.wolf-udo-wagner.com

Unternehmen — Company
thinknewgroup
Lindwurmstraße 71
80337 München
Germany
www.thinknewgroup.com
Produkt — Product
thinknewgroup Corporate Design
Firmenimage — Corporate identity
Design
Werksdesign / Inhouse
Gabi Lück, Florian Fischer
368

Unternehmen — Company
Thomas Leuchten
Höpflerg 17
1230 Wien
Austria
www.thomasleuchten.at
Produkt — Product
LA LINEA
Wand- und Deckenleuchte —
Wall and ceiling light
Design
Werksdesign / Inhouse
Thomas Gruber

Unternehmen — Company
ThyssenKrupp AG
August-Thyssen-Straße 1
40211 Düsseldorf
Germany
www.thyssenkrupp.com
Produkt — Product
Discover Steel #1 Duisburg
Buch-Serie — Book series
Design
Werksdesign / Inhouse
Annette Häfelinger, Kerstin Weidemeyer

Unternehmen — Company
ThyssenKrupp AG
August-Thyssen-Straße 1
40211 Düsseldorf
Germany
www.thyssenkrupp.com
Produkt — Product
ThyssenKrupp Discover
Internetseite — Website
Design
Scholz & Volkmer
www.s-v.de

Unternehmen — Company
Töpferei Trommler
Peesten 19
95359 Kasendorf
Germany
www.toepferei-trommler.de
Produkt — Product
Gestellofen
Heizofen — Heater
Design
Werksdesign / Inhouse
Gerhard Trommler, Barbara Kühn

Unternehmen — Company
Tombow Pencil and Pencil GmbH
Am Südpark 12
65451 Kelsterbach
Germany
www.tomboweurope.com
Produkt — Product
ZOOM 414
Multifunktions-Schreibgerät —
Multifunctional writing instrument
Design
Werksdesign / Inhouse
Kazunori Katami

Unternehmen — Company
Totems Communication GmbH
Ludwigstraße 59
70176 Stuttgart
Germany
www.totems.com
Produkt — Product
Yura Yura
Messestand Köln 2004 —
Fair stand, Cologne Trade fair 2004
Design
Werksdesign / Inhouse
Yukiko Nezu

Unternehmen — Company
Totems Communication GmbH
Ludwigstraße 59
70176 Stuttgart
Germany
www.totems.com
Produkt — Product
„Die Villa des Jägers – my mercedes" —
"The villa of the hunter – my mercedes"
Mercedes-Benz Wanderausstellung —
Mercedes-Benz touring exhibition
Design
Werksdesign / Inhouse

Unternehmen — Company
Totems Communication GmbH
Ludwigstraße 59
70176 Stuttgart
Germany
www.totems.com
Produkt — Product
Neues Begreifen - DaimlerChrysler Automechanika 2004 — New grasp - DaimlerChrysler Automechanika 2004
Kommunikationskonzept — Communication concept
Design
Werksdesign / Inhouse
Frank C. Ulrich, Ariane Raab, Jean-Louis Vidière

Unternehmen — Company
Tourismus+Congress GmbH Frankfurt am Main
Kaiserstraße 56
60329 Frankfurt am Main
Germany
www.tcf.frankfurt.de
Produkt — Product
KLANG|PASSAGEN - Museumsuferfest Frankfurt 2004 — SOUND|PASSAGES - 2004 Frankfurt Museum Riverbank Festival
Licht-und Medieninszenierung — Light and media installation
Design
Atelier Markgraph GmbH
Roland Lambrette, Andreas Lorenschat, Jürgen Zeller
www.markgraph.de
//
Meso
www.meso.net
//
Parviz Mir-Ali
www.sl15.com
//
Procon MuliMedia AG
www.procon-online.de
//
XL Video GmbH
//
Procédés Chénel Werbegesellschaft
www.procedes.de
310

Unternehmen — Company
Transsolar Energietechnik GmbH
Curiestraße 2
70563 Stuttgart
Germany
www.transsolar.com
Produkt — Product
Transsolar Klima Engineering
Buch — Book
Design
büro uebele
visuelle kommunikation
Prof. Andreas Uebele, Christine Voshage, Katrin Häfner
www.uebele.com

Unternehmen — Company
Trendbüro
Beratungsunternehmen für gesellschaftlichen Wandel
JK. PW. GmbH
Hohe Brücke 1
20459 Hamburg
Germany
www.trendbüro.de
Produkt — Product
www.designklicks.de
Internetseite — Website
Design
Deutschlandrock Design & Software Büro
Stefan Landrock
www.deutschlandrock.com
136 Silber

Unternehmen — Company
Trerè s.r.l.
Via Modena, 18
46041 Asola-Mantova
Italy
www.x-bionic.com
//
X-Technology Swiss
Samstagernstrasse 45
8832 Wollerau
Switzerland
www.x-technology.com
Produkt — Product
X-Bionic®
Bekleidung / Unterwäsche — Clothing / Underwear
Design
X-Technology Swiss
Prof. Bodo W. Lambertz
www.x-technology.com
277

Unternehmen — Company
TRILUX-Lenze GmbH + Co KG
Heidestraße 4
59759 Arnsberg
Germany
www.trilux.de
Produkt — Product
Athenik LED
Leuchte — Luminaire
Design
Werksdesign / Inhouse
172

Unternehmen — Company
Troika domovari GmbH
Nisterfeld 11
57629 Müschenbach
www.troika.de
Produkt — Product
Abrollding
Design-Klebebandabroller aus Metallguss matt oder glänzend — Design tape dispenser, metal casting, matt or shiny
Design
ding3000 GbR
Ralf Webermann
www.ding3000.com

Unternehmen — Company
Troll Lighting
Passeig de la Ribera 109
08420 Canovelles
Spain
www.troll.es
Produkt — Product
CustomLight
Beleuchtungssystem — Lighting system
Design
Werksdesign / Inhouse
Troll Lighting Team
178

Unternehmen — Company
Truma Gerätetechnik GmbH & Co. KG
Wernher-von-Braun-Straße 12
85640 Putzbrunn
Germany
www.truma.com
Produkt — Product
Mover SE
Rangiersystem —
Manoeurvring systems for caravans
Design
Werksdesign / Inhouse
Peter Wickelmaier

Unternehmen — Company
TRUMPF Medizin Systeme GmbH
Carl-Zeiss-Straße 5
07318 Saalfeld
Germany
www.trumpf-med.com
Produkt — Product
SHUTTLE
Transportvorrichtung für Operationstisch-
platten und Patienten — Transport device for
operating table tops and patients
Design
Gotha Design
Klaus Blechschmidt
//
Werksdesign / Inhouse
Arnd Kuchenbecker, André Griesel

Unternehmen — Company
TUI AG
Karl-Wiechert-Allee 4
30625 Hannover
Germany
www.tui.de
Produkt — Product
Die Entdeckung des Lächelns —
Discovering the smile
Imagefilm
Design
Klaus E. Küster Werbeagentur GmbH
Klaus Erich Küster

Unternehmen — Company
Tupperware Belgium N.V.
Pierre Corneliskaai 35
9300 Aalst
Belgium
www.tupperware.com
Produkt — Product
Käse Max
Käsebehälter — Cheese servers
Design
Tupperware General Services N.V.
Victor Cautereels
www.tupperware.com
256

Unternehmen — Company
Tupperware Belgium N.V.
Pierre Corneliskaai 35
9300 Aalst
Belgium
www.tupperware.com
Produkt — Product
Quick-Chef 2 & Happy Chef —
Quick-Chef 2 & Happy Chopper
Küchenmaschine und Zerkleinerer —
Ergonomic food processor and chopper
Design
Tupperware General Services N.V.
Jan-Hendrik de Groote, Vincent Jalet
www.tupperware.com
258

Unternehmen — Company
Tupperware France S.A.
Route de Monts
37300 Joué-Les-Tours
France
www.tupperware.com
Produkt — Product
Mini-Max™
Zusammenfaltbare Lebensmittel-Behälter —
Collapsible food containers
Design
Tupperware Corporation
Tupperware Global Design
www.tupperware.com
257

Unternehmen — Company
TVS Spa
Via G. Galilei, 2
61033 Fermignano (PU)
Italy
www.tvs-spa.it
Produkt — Product
Archetun
Kochgeschirr — Non-stick aluminium cookware
Design
STUDIO MATTEO THUN
Matteo Thun
www.matteothun.com
42 Gold

Unternehmen — Company
twh
Täschnerwerkstatt Hermann
H.-Abendrothstraße 06
99423 Weimar
Germany
www.twh-weimar.de
Produkt — Product
Speed
Taschenserie — Bag series
Design
Werksdesign / Inhouse
Petra Hermann

Unternehmen — Company
type directors club of new york, german liaison committee
Robert-Koch-Straße 8
55129 Mainz
Germany
www.type-director.de
Produkt — Product
type directors club new york „ausstellungs-plakat tdc show" — type directors club new york" tdc show exhibition poster"
Plakat — Poster
Design
büro uebele
visuelle kommunikation
Michaela Burger, Prof. Andreas Uebele,
Sandra Zellmer
www.uebele.com

Unternehmen — Company
Typotron AG
Scheidwegstrasse 18
9016 St. Gallen
Switzerland
www.typotron.ch
Produkt — Product
Mördermuscheln, Tritonshörner und Perlboote
Broschüre — Brochure
Design
Designgruppe Merkwürdig
Kai Staudacher, Nadine Bendzulla,
Jennifer Staudacher
www.merkwuerdig.com

u

Unternehmen — Company
Ueberholz GmbH
Warndtstraße 7
42285 Wuppertal
Germany
www.ueberholz.de
Produkt — Product
Eigendarstellung auf der EuroShop: Ueberholz. GmbH, Büro für temporäre Architektur — Self-Presentation at Euroshop 2005: Ueberholz. GmbH, Office for temporary architecture
Messestand — Fair stand
Design
Werksdesign / Inhouse
Nico Ueberholz, André Füsser
320

Unternehmen — Company
Ulrich Alber GmbH
Vor dem Weißen Stein 21
72461 Albstadt
Germany
www.alber.de
Produkt — Product
e-fix E25
Zusatzantrieb für manuelle Rollstühle — Power add-on for manual wheelchairs
Design
einmaleins Büro für Gestaltung
Marc-Gregor Weidt, Rupert Bläsi
www.einmaleins.net

Unternehmen — Company
Ulrike Krasemann Eventmanagement
Papenhuder Straße 45
22087 Hamburg
Germany
www.ulrikekrasemann.de
Produkt — Product
Ulrike Krasemann Visitenkarten-Kampagne „Unforgettable Events" — Ulrike Krasemann Calling Cards Campaign "Unforgettable Events"
Visitenkarten-Serie — Calling card series
Design
Jochens.Hilbert KG
Oliver Jochens, Alexander Hilbert,
Nico Döbereiner
www.jochenshilbert.de

Unternehmen — Company
Universal Music Domestic Division – A Division of Universal Music GmbH
Stralauer Allee 1
10245 Berlin
Germany
www.umusik.com
Produkt — Product
Paul van Dyk / Peter Heppner „Wir Sind Wir"
Video
Design
Katapult Filmproduktion GmbH
Jörn Heitmann, Ingo Georgi
www.katapult-film.com
//
Werksdesign / Inhouse
Robert Eysholdt, Martin Rutter

Unternehmen — Company
USED GmbH
Am Bahnhof 3
49610 Quakenbrück
Germany
www.used-HQ.com
Produkt — Product
ScooterTrike Roadster 1.5
Dreirad — Three wheeled bicycle for special needs
Design
Velvet Systems
Klaus Schröder
www.velvet-systems.com

Unternehmen — Company
usus.berlin gmbh
Schlesische Straße 27
10997 Berlin
Germany
www.usus-berlin.de
Produkt — Product
usus.isis ivory
Schreibgerät, Kugelschreiber —
Writing instrument, ballpoint pen
Design
Werksdesign / Inhouse
Hilmar Nicolay, Lennart Kleiber

Unternehmen — Company
Ute Soelch
architektur.innenarchitektur.design
Hopfenstraße 2
95652 Waldsassen
Germany
Produkt — Product
keep well 163
Bindesystem — Binder system
Design
Werksdesign / Inhouse
Ute Soelch
338

Unternehmen — Company
UVEX ARBEITSSCHUTZ GmbH
Würzburger Straße 181–189
90766 Fürth
Germany
www.uvex-safety.de
Produkt — Product
athletic workwear
Berufsbekleidung: Latzhose und Bundjacke —
Workwear: dungarees and jacket
Design
Birmelin Design Studio
Elisabetta Birmelin

V

Unternehmen — Company
V & B Fliesen GmbH
Rotensteinerweg
66663 Merzig
Germany
www.villeroy-boch.com
Produkt — Product
Pro Architectura New - Sinus
Fliesenserie mit Formfliesensystemen —
Tile collection with pre-shaped tile systems
Design
Architekturbüro Frank Huster
Frank Huster
www.frankhuster.de
196

Unternehmen — Company
Vaillant GmbH
Berghauser Straße 40
42859 Remscheid
Germany
www.vaillant.de
Produkt — Product
atmoTEC / turboTEC
Gas-Wandheizgerät — Gas-Wall hung boiler
Design
Vistapark
www.vistapark.de

Unternehmen — Company
Vaillant GmbH
Berghauser Straße 40
42859 Remscheid
Germany
www.vaillant.de
Produkt — Product
ecoCRAFT
Bodenstehender Gas-Brennwertkessel —
Floor standing condensing gas boiler
Design
Vistapark
www.vistapark.de

Unternehmen — Company
Vaillant GmbH
Berghauser Straße 40
42859 Remscheid
Germany
www.vaillant.de
Produkt — Product
ecoLEVEL
Kondensatpumpe — Condensate pump
Design
Vistapark
www.vistapark.de

Unternehmen — Company
Vaillant GmbH
Berghauser Straße 40
42859 Remscheid
Germany
www.vaillant.de
Produkt — Product
geoTHERM
Luft / Wasser-Wärmepumpe mit Nutzung
der Wärmequelle Außenluft —
Air / water heat pump using external air
as the heat source
Design
Vistapark
www.vistapark.de

Unternehmen — Company
Vaillant GmbH
Berghauser Straße 40
42859 Remscheid
Germany
www.vaillant.de
Produkt — Product
VRT 50
Raumtemperaturregler —
Room temperature control
Design
IDA Design
www.idadesign.fr

Unternehmen — Company
VARTA Microbattery GmbH
Daimlerstraße 1
73479 Ellwangen
Germany
www.powerone-batteries.de
Produkt — Product
power one pen charger
Mobiles Ladegerät für Hörgerätebatterien — Mobile charger for hearing-aid batteries
Design
Werksdesign / Inhouse
Herbert Schein, Michael Kron, Eduard Pytlik, Eugen Rechtenbacher
296

Unternehmen — Company
Verkehrsverbund Bremen / Niedersachsen (VBN)
Willy-Brandt-Platz 7
28357 Bremen
Germany
www.vbn.de
Produkt — Product
VBN Verbundbericht 2004 — VBN Annual Report 2004
Geschäftsbericht — Annual report
Design
construktiv GmbH
Miriam Edelmann, Kai Tietjen, Svenja Miller, Stefan Dziallas
www.construktiv.de
356

Unternehmen — Company
Verlag Hermann Schmidt Mainz
Robert-Koch-Straße 8
55129 Mainz
Germany
www.typografie.de
Produkt — Product
Kaleidotype
Computerprogramm / Software — Computer programme / software
Design
Buttgereit und Heidenreich GmbH
Michael Buttgereit, Wolfram Heidenreich, Martin Winter, Jörg Fassmann
www.b-und-h.de, www.kaleidotype.de
294

Unternehmen — Company
vibracoustic GmbH & Co. KG
Höhnerweg 2–4
69465 Weinheim
Germany
www.vibracoustic.com
Produkt — Product
Vibracoustic Corporate Website
Internetseite — Website
Design
designverign GmbH
Christoph Scheufeld, Christian Hohn, Stefan Tomski
www.designverign.de
331

Unternehmen — Company
VIEGA
Ennester Weg 9
57439 Attendorn
Germany
www.viega.de
Produkt — Product
VISIGN M3
Überlaufgarnitur — Overflow system
Design
ARTEFAKT industriekultur
Achim Pohl, Thomas Fiegl
www.artefakt.de
201

Unternehmen — Company
Viessmann Werke GmbH & Co. KG
Viessmannstraße 1
35107 Allendorf / Eder
Germany
www.viessmann.com
Produkt — Product
Vitocell-H 300
Speicher-Wassererwärmer — DHW cylinder
Design
Phoenix Design
Andreas Haug, Fabian Kollmann
www.phoenixdesign.de

Unternehmen — Company
Viessmann Werke GmbH & Co. KG
Viessmannstraße 1
35107 Allendorf / Eder
Germany
www.viessmann.com
Produkt — Product
Vitocell-W 100 CUG
Speicher-Wassererwärmer — DHW cylinder
Design
Phoenix Design
Andreas Haug, Fabian Kollmann
www.phoenixdesign.de

Unternehmen — Company
Viessmann Werke GmbH & Co. KG
Viessmannstraße 1
35107 Allendorf / Eder
Germany
www.viessmann.com
Produkt — Product
Vitodens 343
Gas-Brennwert-Kompaktgerät — Compact condensing tower
Design
Phoenix Design
Andreas Haug, Fabian Kollmann
www.phoenixdesign.de

Unternehmen — Company
Viking Fottoy AS, Oslo
Ole Deviks vei 46
0614 Oslo
Norway
www.vikingfootwear.com
Produkt — Product
PU
Kinderstiefel — Kid boots
Design
scherfdesign
Rudolf Alfred Scherf
www.scherfdesign.de

Unternehmen — Company
Villeroy & Boch AG
Saaruferstraße
66693 Mettlach
Germany
www.villeroy-boch.com
Produkt — Product
Bellevue
Badkollektion — Bathroom collection
Design
.molldesign Reiner Moll & Partner
Daniel Mader
www.molldesign.de
199

Unternehmen — Company
visiomatic GmbH
Weissacher Straße 1
70499 Stuttgart
Germany
www.visiomatic.de
Produkt — Product
visiomatic Home System / visiPad®SL 18
Display
Design
wiege Entwicklungsgesellschaft mbH
Michael Englisch, Torsten Wittenberg,
Britta Clausen
291

Unternehmen — Company
Visiomed AG
Universitätsstraße 160
44801 Bonn
Germany
www.visiomed.de
Produkt — Product
microDERM
Hautkrebsdiagnosesystem —
Skin cancer detection system
Design
Piu Products
Torsten Gratzki
www.piuproducts.com

Unternehmen — Company
VITA Zahnfabrik H. Rauter GmbH & Co. KG
Spitalgasse 3
79713 Bad Säckingen
Germany
www.vita-zahnfabrik.com
Produkt — Product
VITA VM 13 – Auf die Gerüste! Fertig! Los! —
VITA VM 13 – On your frameworks! Set! Go!
Anzeige — Advertisement
Design
Studio Laeis, Design & Werbung
www.laeis.com
341

Unternehmen — Company
VITA Zahnfabrik H. Rauter GmbH & Co. KG
Spitalgasse 3
79713 Bad Säckingen
Germany
www.vita-zahnfabrik.com
Produkt — Product
VITA Zähne — VITA Teeth
Anzeigenkampagne — Advertising campaign
Design
Studio Laeis, Design & Werbung
www.laeis.com
363

Unternehmen — Company
Vitra AG
Klünenfeldstrasse 22
4127 Birsfelden
Switzerland
www.vitra.com
Produkt — Product
Vitra Broschüre „Growing a Chair" —
Vitra Brochure "Growing a Chair"
Buch — Book
Design
Wolfgang Scheppe
www.wolfgangscheppe.com

Unternehmen — Company
VitrA Tiles
E-5 Karayolu Uzeri Sifa Mah. Ataturk Cad.
34941 Istanbul
Turkey
www.vitrakaro.com
Produkt — Product
Defne Koz Iznik Serie
Keramik Fliesen — Ceramic tile
Design
Defne Koz

Unternehmen — Company
Vivanco Gruppe AG
Ewige Weide 15
22926 Ahrensburg
Germany
www.vivanco.de
Produkt — Product
VIVANCO CHH 100
Digitales Funkkopfhörersystem mit
USB-Dongle — Digital RF headphones system
with USB-dongle
Design
Werksdesign / Inhouse

Unternehmen — Company
Volkswagen AG
Berliner Ring
38440 Wolfsburg
Germany
www.volkswagen.de
Produkt — Product
Passat / Passat Variant
Pkw — Passenger car
Design
Werksdesign / Inhouse

Unternehmen — Company
Volkswagen AG
Postfach 19060
38436 Wolfsburg
Germany
www.volkswagen.de
Produkt — Product
Volkswagen Corporate Design Essentials
Corporate Book — Corporate book
Design
MetaDesign AG
www.metadesign.de
124 Silber

Unternehmen — Company
VOX Film-und Fernseh GmbH & Co. KG
Richard-Byrd-Straße 6
50829 Köln
Germany
www.vox.de
Produkt — Product
SOS für Kinder - Die VOX-Hilfsaktion 2004 und 2005 — SOS for children - The VOX-Charity 2004 and 2005
TV-Spots mit Spendenaufruf — TV fund-raising campaign
Design
Werksdesign / Inhouse

W

Unternehmen — Company
W.L. Gore & Associates GmbH
Aiblinger Straße 60
83620 Feldkirchen-Westerham
Germany
www.gorebikewear.com / de
GORE BIKE WEAR. XENON Shirt
Produkt — Product
Rennrad-Trikot — Race biking shirt
Design
Werksdesign / Inhouse
Clemens Deilmann

Unternehmen — Company
W.L. Gore & Associates GmbH
GORE BIKE WEAR
Aiblinger Straße 60
83620 Feldkirchen-Westerham
Germany
www.gorebikewear.com / de
Produkt — Product
XENON Lady Jacke
Damen Rennrad Jacke — Ladies race cycling jacket
Design
Werksdesign / Inhouse
Clemens Deilmann

Unternehmen — Company
WACOM Europe GmbH
Europark Fichtenhain A9
47807 Krefeld
Germany
www.wacom.com
Produkt — Product
CINTIQ 21UX
Interaktives Display — Interactive pen display
Design
Werksdesign / Inhouse

Unternehmen — Company
Waidmann / Post
Büro für Konzeption und Gestaltung
Kaffeetwete 3
38100 Braunschweig
Germany
www.waidmannpost.de
Produkt — Product
ach anna, seufzerkalendarium — oh anna, sigh kalendarium
Buch — Book
Design
Werksdesign / Inhouse
Mechthild Waidmann, Stefan Waidmann

Unternehmen — Company
Wall AG
Friedrichstraße 118
10117 Berlin
Germany
www.wall.de
Produkt — Product
e-info Terminal — bluespot terminal
Multimedia-Terminal für den öffentlichen Raum — Street furniture
Design
ION industrial design berlin
Christoph Fleckenstein, Anja Götz
www.iondesign.de
292

Unternehmen — Company
Wall AG
Friedrichstraße 118
10117 Berlin
Germany
www.wall.de
Produkt — Product
Designlinie Landmark — Design landmark
Stadtmöbelfamilie — Street furniture
Design
Hadi Teherani AG
Hadi Teherani
www.haditeherani.de
160

Unternehmen — Company
Walter Knoll AG & Co. KG
Bahnhofstraße 25
71083 Herrenberg
Germany
www.walterknoll.de
Produkt — Product
FK Lounge
Schalensessel — Bucket chair
Design
Preben Fabricius, Jörgen Kastholm

Unternehmen — Company
Walter Knoll AG & Co. KG
Bahnhofstraße 25
71083 Herrenberg
Germany
www.walterknoll.de
Produkt — Product
Turtle
Schalenstuhl — Bucket chair
Design
PearsonLloyd
Luke Pearson, Tom Lloyd
www.pearsonlloyd.co.uk

Unternehmen — Company
Walter Knoll AG & Co. KG
Bahnhofstraße 25
71083 Herrenberg
Germany
www.walterknoll.de
Produkt — Product
Circle
Möbel — Furniture
Design
UNStudio
Ben van Berkel
www.unstudio.com
72 Silber

Unternehmen — Company
Walter Knoll AG & Co. KG
Bahnhofstraße 25
71083 Herrenberg
Germany
www.walterknoll.de
Produkt — Product
Together Sofa
Möbel — Furniture
Design
EOOS Design GmbH
Martin Bergmann, Gernot Bohmann, Harald Gründl
www.eoos.com

Unternehmen — Company
Waste Solutions: Smart Sharps
2 Mill Court, Goat Mill Rd
CF47 0AH Merthyr Tydfil
Great Britain
www.stericycle.com
Produkt — Product
Smartsharp
Mülleimer für medizinischen Abfall — Medical waste bin
Design
PDR
Stuart Cole, Robert Beveridge, Jason Hester
www.pdr-online.co.uk

Unternehmen — Company
WDR
Appellhofplatz 1
50667 Köln
Germany
www.wdr.de
Produkt — Product
WDR: DANKE! — WDR: Thank you!
Imagekampagne — Image campaign
Design
Werksdesign / Inhouse
Heinz Peter Brück, Mike Fröhlich, Andrea Linder

Unternehmen — Company
WDR
Appellhofplatz 1
50667 Köln
Germany
www.wdr.de
Produkt — Product
WDR: Fest versprochen — WDR: Happy Christmas – Promised!
On-Air-Promotion-Kampagne — On-air promotion campaign
Design
Werksdesign / Inhouse
Lars Eberhardt, Peter Jakobs, Karin Sarholz, Michael Worringen

Unternehmen — Company
WDR
Appellhofplatz 1
50667 Köln
Germany
www.wdr.de
Produkt — Product
WDR: Gott und die Welt — WDR: God and the World
On-Air-Promotion Spot — On-air promotion spot
Design
Werksdesign / Inhouse
Bernd Pick, Elena Rauen, Gunar Wardenbach

Unternehmen — Company
WDR
Appellhofplatz 1
50667 Köln
Germany
www.wdr.de
Produkt — Product
WDR: Jeck auf WDR — WDR: Crazy for WDR
Karneval Event-Promotion — Carnival event-promotion
Design
Werksdesign / Inhouse
Michael Freiwald, Karin Sarholz, Christina Sattmann, Mathias Wesselmann, Harald Schröder, Heinz-Peter Brück

Unternehmen — Company
WDR
Appellhofplatz 1
50667 Köln
Germany
www.wdr.de
Produkt — Product
WDR: Abenteuer Erde — WDR: Adventure Earth
On-Air-Promotion-Kampagne —
On-air promotion campagne
Design
ACT Videoproduktion GmbH
Georg Gilcher
www.actvideo.de
//
Werksdesign / Inhouse
Mike Fröhlich

Unternehmen — Company
Wehkamp
Meeuwenlaan 2
8011 BZ Zwolle
Netherlands
www.wehkamp.nl
Produkt — Product
www.wehkamp.nl
Hopping Site für den größten Versandhandel in den Niederlanden — eCommerce site
Design
Clockwork
Rolf Geerts, Martien ter Veen, Marnix de Vries
www.clockwork.nl

Unternehmen — Company
Weidmüller Interface GmbH & Co. KG
Klingenbergstraße 16
32758 Detmold
Germany
www.weidmueller.com
Produkt — Product
Buchsenleisten mit Zugfederanschlusstechnik —
Socket block with tension clamp connection
Steckverbinder — Omnimate range
Design
Werksdesign / Inhouse
308

Unternehmen — Company
Wella AG
Berliner Allee 65
64274 Darmstadt
Germany
www.wella.de
Produkt — Product
Xpert
Profi-Haarschneidemaschine (Akku-und Kabelbetrieb) — Professional clipper (rechargeable battery and plug-in option)
Design
designafairs GmbH
Bernd Eigenstetter, Markus Mayer
www.designafairs.com

Unternehmen — Company
Wepro AG
Huebstrasse 11
8633 Wolfhausen
Switzerland
www.wepro.ch
Produkt — Product
STABILO point 88 Mini pockets
Etui aus Weprotek —
Trendy pen wallet made of Weprotek
Design
pro in motion
Joachim Dreher
www.pro-in-motion.com
//
STABILO International GmbH
Katrin Hanke
www.stabilo.com

Unternehmen — Company
White Elephant Industrial Design
Kastaniengasse 33
2630 Ternitz
Austria
www.white-elephant.at
Produkt — Product
Akjinflate
Aufblasbarer Rettungsschlitten —
Inflateable alpine rescue device
Design
Werksdesign / Inhouse
Hermann Trebsche

Unternehmen — Company
Wienand Verlag & Medien GmbH
Weyertal 59
50937 Köln
Germany
Produkt — Product
Photokontakt. Benjamin Katz: Georg Baselitz
Katalog — Catalogue
Design
QWER
Annette Süß, Iris Utikal
www.qwer.de

Unternehmen — Company
Wiesmann Manufaktur der Individualisten
An der Lehmkuhle 87
48249 Dülmen
Germany
www.wiesmann.com
Produkt — Product
Wiesmann GT
Pkw — Passenger car
Design
Werksdesign / Inhouse
Martin Wiesmann
145

Unternehmen — Company
Wilde Metallbau GmbH & steeltec37.
Finsterwalder Straße 15
03238 Massen
www.wilde-metallbau.de
Produkt — Product
Modulhaus go-tic
Modulares Hausbausystem —
Modular system for house constructions
Design
eckedesign
Albrecht Ecke
www.eckedesign.de

Unternehmen — Company
Wilkhahn
Wilkening + Hahne GmbH+Co. KG
Fritz-Hahne-Straße 8
31848 Bad Münder
Germany
www.wilkhahn.com
Produkt — Product
Programm 230 Aline
Mehrzweck Stuhl — Multi-purpose chair
Design
Störiko product design GmbH
Andreas Störiko
224

Unternehmen — Company
Wilo AG
Nortkirchenstraße 100
44263 Dortmund
Germany
www.wilo-ag.de
Produkt — Product
WILO-Stratos PARA
Umwälzpumpe — Circulation pump
Design
mehnert corporate design gmbh & co. kg
www.mehnertdesign.de

Unternehmen — Company
Windmühlenmesser Manufactur Robert Herder GmbH & Co. KG
Ellerstraße 16
42697 Solingen
Germany
www.windmuehlenmesser.de
Produkt — Product
FÜR ZWEI
FOR TWO
Handwerklich gefertigtes Mehrzweckmesser mit Griff aus Pflaumen-, Oliven- oder Birnenholz — Hand crafted utility knife with handle made of plum-, olive- or pearwood
Design
Werksdesign / Inhouse
Giselheid Herder-Scholz, Tomoyuki Takada

Unternehmen — Company
Windmühlenmesser Manufactur Robert Herder GmbH & Co. KG
Ellerstraße 16
42697 Solingen
Germany
www.windmuehlenmesser.de
Produkt — Product
Parmoulin-Fromage
Käsemesser für den Tisch und für die Reise — Cheese knife for the table and out and about
Design
Werksdesign / Inhouse
Giselheid Herder-Scholz, Tomoyuki Takada

Unternehmen — Company
WISAP Gesellschaft für wissenschaftlichen Apparatebau mbH
Rudolf-Diesel-Ring 20
82054 Sauerlach
Germany
www.wisap.de
Produkt — Product
é-PUG 2 (personal users gadget)
Digitales Aufzeichnungsgerät — Digital recording device
Design
IMAGO DESIGN
Bernhard Schwanitz, Stefan Thalhammer
www.imago-design.de
295

Unternehmen — Company
WMF Württembergische Metallwarenfabrik AG
Eberhardstraße
73312 Geislingen
Germany
www.wmf.de
Produkt — Product
Lounge
Schalen und Vasen — Bowls and vases
Design
James Irvine
www.james-irvine.com

Unternehmen — Company
WMF Württembergische Metallwarenfabrik AG
Eberhardstraße
73312 Geislingen
Germany
www.wmf.de
Produkt — Product
Kult
Besteck — Cutlery
Design
freiTraum
Fritz Strauss
www.freitraum.at

Unternehmen — Company
WOLF-Garten GmbH & Co. KG
Industriestraße 83–85
57518 Betzdorf
Germany
www.WOLF-Garten.com
Produkt — Product
>> Comfort << RA-X
Vielzweckschere — Multi-purpose secateurs
Design
Cognito GmbH
Steffen Müller
www.cognito.de
//
Werksdesign / Inhouse
Andreas Heuel

Unternehmen — Company
WZB Werkstatt für Behinderte der Lebenshilfe GmbH
Am Beckerwald 31
66538 Spiesen-Elversberg
Germany
www.wzb.de
Produkt — Product
WZB-Science Wear
Arbeitsbekleidung für Reinräume — Wear for cleanrooms
Design
Werksdesign / Inhouse
Diane Petereit

X

Unternehmen — Company
X-Technology Swiss
Samstagern Straße 45
8832 Wollerau
Germany
www.x-socks.com
Produkt — Product
X-Socks® Poster
Grafik Design — Visual communication
Design
BBDO Berlin
Tom Fischer
www.bbdo-campaign.de

y

Unternehmen — Company
YXLON International X-Ray GmbH
Essener Bogen 15
22419 Hamburg
Germany
ww.yxlon.com
Produkt — Product
YXLON Systemdesign (Baureihen Y.Wheel/Y.Multiplex) — YXLON system design (series Y.Wheel/Y.Multiplex)
Strahlenschutzkabinen/Bedienterminals/Graphical User Interface — X-Ray cabinet/user terminal/graphical user interface
Design
bgp design
Knut Braake, Stefan Grobe
www.bgp-design.com

Z

Unternehmen — Company
ZDF
Zweites Deutsches Fernsehen
ZDF-Straße 1
55127 Mainz
Germany
www.zdf.de
Produkt — Product
www.bianca.zdf.de
Internetseite — Website
Design
gosub communications GmbH
Ralf Hoffmann, Reimar Servas, Florian Sprenger, Dominic Trautvetter
www.gosub.de
//
Werksdesign/Inhouse
Bernhard Fritsch, Isa Ostertag, Annika Kersting, Nanette Schmetz

Unternehmen — Company
ZDF
Zweites Deutsches Fernsehen
ZDF-Straße 1
55127 Mainz
Germany
www.zdf.de
Produkt — Product
ZDF: Zweisam im Zweiten — ZDF: Twosome on ZDF
Marketing Kampagne Fernsehen — TV marketing campaign
Design
Service Plan
Ewald Pusch
www.serviceplan.de
//
X-tension
Holger Gericke
www.x-tension.de
//
Werksdesign/Inhouse
Judith Herrmann, Kirstin Kröger, Robert Windisch, Britta Zentel

Unternehmen — Company
Zeitform-Design GmbH
Möbelbau, Design
Neuenreuth 13
95336 Mainleus
Germany
www.zeitform-design.de
Produkt — Product
Eckverbindung — Corner joint
Regalsystem — Shelving system
Design
Werksdesign/Inhouse
Günter Weber

Unternehmen — Company
Zeitform-Design GmbH
Möbelbau, Design
Neuenreuth 13
95336 Mainleus
Germany
www.zeitform-design.de
Produkt — Product
Typ D
Mobiles Fertighaus — Mobile prefab house
Design
Werksdesign/Inhouse
Günter Weber

Unternehmen — Company
Zisterzienserstift Zwettl, Pater Maximilian Krausgruber
Stift Zwettl 1
3910 Zwettl
Austria
www.stift-zwettl.at
Produkt — Product
Traum der Weisen/Wer's glaubt, wird selig? — Philosopher's Dream/Blessed are those who believe?
Ausstellungsdesign — Exhibition design
Design
Jäger & Jäger
Kommunikationsdesign
Regina Jäger, Olaf Jäger
www.jaegerundjaeger.de

Unternehmen — Company
Zweibrüder Optoelectronics GmbH
Kronenstraße 5-7
42699 Solingen
Germany
www.zweibrueder.com
Produkt — Product
LED-LENSER dot.spot triplex
LED-Spotleuchte, Wohnraumlampe —
LED spot light, living room lamp
Design
Werksdesign / Inhouse
Harald Opolka

Unternehmen — Company
Zweibrüder Optoelectronics GmbH
Kronenstraße 5-7
42699 Solingen
Germany
www.zweibrueder.com
Produkt — Product
als Serie: LED Lenser dot.spot Moon; dot.spot Power und dot.spot Nano Power —
production run: LED-Lenser dot.spot moon; dot.spot Power and dot.spot nano power
LED Wohnraumspots; Tischlampen —
LED spots; living room lamps
Design
Werksdesign / Inhouse
Harald Opolka

Unternehmen — Company
Zweibrüder Optoelectronics GmbH
Kronenstraße 5-7
42699 Solingen
Germany
www.zweibrueder.com
Produkt — Product
LED-LENSER Lily Marleen
LED-Tischlampe — LED living room lamp
Design
Werksdesign / Inhouse
Rainer Opolka

Unternehmen — Company
Zweibrüder Optoelectronics GmbH
Kronenstraße 5-7
42699 Solingen
Germany
www.zweibrueder.com
Produkt — Product
LED-LENSER David 15
LED-Taschenlampe — LED torch light
Design
Werksdesign / Inhouse
Rainer Opolka

Unternehmen — Company
Zweibrüder Optoelectronics GmbH
Kronenstraße 5-7
42699 Solingen
Germany
www.zweibrueder.com
Produkt — Product
LED-LENSER Head Fire Duplex
LED-Kopflampe — LED head lamp
Design
Werksdesign / Inhouse
Rainer Opolka

Unternehmen — Company
Zweibrüder Optoelectronics GmbH
Kronenstraße 5-7
42699 Solingen
Germany
www.zweibrueder.com
Produkt — Product
LED-Lenser® Frogman
LED-Taucherlampe — LED diver lamp
Design
Werksdesign / Inhouse
Harald Opolka

Unternehmen — Company
Zwiesel Kristallglas AG
Dr.-Schott-Straße 35
94227 Zwiesel
Germany
www.zwiesel-kristallglas.com
Produkt — Product
Okiu
Schalen — Bowls
Design
Prof. Michael Boehm
268

Unternehmen — Company
Zwiesel Kristallglas AG
Dr.-Schott-Straße 35
94227 Zwiesel
Germany
www.zwiesel-kristallglas.com
Produkt — Product
The First by Enrico Bernardo
Dekanter und Kelchgläser —
Decanters and stemware
Design
Werksdesign / Inhouse
Irmgard Braun-Ditzen
270

DIE STIFTERVERSAMMLUNG wählt aus ihrem Kreis das Präsidium des Rat für Formgebung, dem zwölf Personen und die Vertreter der öffentlichen Zuwendungsgeber angehören. Das Präsidium fällt als höchstes Gremium der Stiftung alle strategischen Entscheidungen für die Arbeit des Rat für Formgebung.

THE PATRONS MEETING selects from among its members the Presiding Committee of the German Design Council, which is made up of twelve persons and representatives of the public bodies that provide financing. As the foundation's uppermost board, the Presiding Committee takes all the strategic decisions for the work of the German Design Council.

Helmut Lübke, † 2006 (Präsident / President)
Dr. Michael Peters, Geschäftsführer / Managing Director, Messe Frankfurt GmbH (Vizepräsident / Vice-President)
Eckhard Tischer, Berater / Consultant (Schatzmeister / Treasurer)
Barbara Friedrich, Chefredakteurin / Editor-in-chief, Architektur & Wohnen
Prof. Achim Heine, Universität der Künste / University of Arts, Berlin
Burkhardt Leitner, Geschäftsführer / Managing Director, Burkhardt Leitner constructiv GmbH & Co. KG
Prof. Stefan Lengyel, Ehrenpräsident des VDID / Honorary President of VDID
Prof. Hansjerg Maier-Aichen, Hochschule für Gestaltung / Academy of Design Karlsruhe
Nils Holger Moormann, Geschäftsführer / Managing Director, Nils Holger Moormann GmbH
Prof. h.c. Peter Pfeiffer, Senior Vice President Design, DaimlerChrysler AG
Ruth Soénius, Leiterin / Head of Brand Strategy, Siemens AG
Prof. Dr. h.c. Erik Spiekermann, SpiekermannPartners, Berlin und HfK/Academy of Arts Bremen

Das Präsidium des Rat für Formgebung
The Presiding Committee of the German Design Council

Unternehmensmitglieder —
Corporate members

Aloys F. Dornbracht GmbH & Co. KG
www.dornbracht.com/de

anthologie quartett
www.anthologiequartett.de

Architektur & Wohnen
www.awmagazin.de

Atelier Markgraph GmbH
www.markgraph.de

ATOMIC Austria GmbH
www.atomicsnow.com

AUDI AG
www.audi.com

Barkow Leibinger Architekten
www.barkowleibinger.com

Berker GmbH & Co. KG
www.berker.de

Birkhäuser Verlag GmbH –
form Zeitschrift für Gestaltung
www.form.de

BMW AG
www.bmw.de

Braun GmbH
www.braun.de

Bree Collection GmbH & Co. KG
www.bree.de

brühl & sippold GmbH
www.bruehlsippold.de

büro uebele
visuelle kommunikation
www.uebele.com

C. Josef Lamy GmbH
www.lamy.de

Carl Mertens CMS-Grasoli
Besteckfabrik GmbH + Co. KG
www.carl-mertens.de

COR Sitzmöbel Helmut Lübke GmbH & Co. KG
www.cor.de

creative inneneinrichter GmbH
www.creative-inneneinrichter.de

DaimlerChrysler AG
www.daimler-chrysler.de

dan pearlman markenarchitektur gmbh
www.danpearlman.com

Demat
Direct Exposition And Managing Trading GmbH
www.demat.de

designafairs GmbH
www.designafairs.de

Deutsche Bank AG
www.deutsche-bank.de

e15 Design und Distributions GmbH
www.e15.com

EDAG Engineering + Design AG
www.edag.de

ERCO Leuchten GmbH
www.erco.com

Factor Design AG
www.factordesign.com

Ferdinand Holzmann Verlag GmbH
www.arcade-xxl.de

Festo AG & Co. KG
www.festo.com

Elmar Flötotto Service Holding GmbH
www.efh.de

FSB Franz Schneider Brakel GmbH + Co. KG
www.fsb.de

gantenhammer
Werbung & Design GmbH & Co. KG
www.gantenhammer.de

glaskoch B. Koch jr. GmbH & Co. KG
www.leonardo.de

Grohe Water Technology AG
GROHE Design Center
www.grohe.de

Hansgrohe AG
www.hansgrohe.de

HEWI Heinrich Wilke GmbH
www.hewi.de

Interbrand Zintzmeyer & Lux AG
www.interbrand.de

Hyundai Motor Europe Technical Center GmbH
www.hyundai-europe.com

interluebke Gebr. Lübke GmbH&Co. KG
www.interluebke.de

KEUCO GmbH & Co. KG
www.keuco.de

Koelnmesse GmbH
www.koelnmesse.de

Konradin Verlag Robert Kohlhammer GmbH
www.design-report.de

Koralle Sanitärprodukte GmbH
www.koralle.de

Leica Camera AG
www.leica-camera.de

Leitner GmbH
www.leitner.de

Loewe AG
www.loewe.de

LPK GmbH
www.lpk.com

Mauser Einrichtungssysteme GmbH & Co. KG
www.mauser-moebel.de

Messe Frankfurt GmbH
www.messefrankfurt.com

MetaDesign AG
www.metadesign.de

Miele & Cie. KG
www.miele.de

MINOX GmbH
www.minox.de

mode...information Heinz Kramer GmbH
www.modeinfo.de

Modular Lighting Instruments NV
www.supermodular.com

modulor GmbH
www.modulor.de

Neuhaus Einrichtungen GmbH & Co. KG
www.neuhausgmbh.de

nya nordiska textiles gmbH
www.nya.com

Pantone Europe GmbH
www.pantone.de

Porsche Design GmbH
www.porsche-design.com

Rodenstock GmbH
www.rodenstock.de

Rosenthal AG
www.rosenthal.de

Roset Möbel GmbH
www.ligne-roset.de

Seibel Designpartner GmbH
www.seibel-designpartner.de

S. Siedle & Söhne
Telefon- und Telegrafenwerke OHG
www.siedle.de

Siemens AG
www.siemens.de

SIGN Kommunikation GmbH
www.sign.de

Sinus Sociovision GmbH
www.sinus-sociovision.de

stilwerk Center-Management GmbH
www.stilwerk.de

Studio Peter Maly Design und Innenarchitektur
www.peter-maly.de

Stylepark AG
www.stylepark.com

Vieler International GmbH + Co. KG
www.vieler.net

Viessmann Werke GmbH & Co KG
www.viessmann.de

Vitra GmbH
www.vitra.com

wiege Entwicklungsgesellschaft mbH
www.wiege.com

WINI Büromöbel Georg Schmidt GmbH & Co. KG
www.wini.de

WMF Württembergische Metallwarenfabrik AG
www.wmf.de

Institutionelle Mitglieder —
Institutional members

Bayern Design Forum e.V.
www.bayern-design.de

BDG Bund Deutscher Grafik-Designer e.V.
www.bdg-deutschland.de

BDI Bundesverband der deutschen Industrie e.V
www.bdi.de

BDIA Bund Deutscher Innenarchitekten e.V
www.bdia.de

Bremer Design GmbH
www.designzentrumbremen.de

DDC Deutscher Designer Club e.V.
www.ddc.de

DDV Deutscher Designer Verband e.V.
www.germandesign.de

Design Center Stuttgart
www.design-center.de

Design Management Institute
www.dmi.org

Design Zentrum Nordrhein Westfalen e.V.
www.design-germany.de

Deutscher Werkbund e.V.
www.deutscher-werkbund.de

Deutscher Designertag e.V.
www.designertag.de

DIHK Deutscher Industrie- und
Handelskammertag
www.dihk.de

iF International Forum Design GmbH
www.ifdesign.de

Industrie- und Handelskammer
Frankfurt am Main
www.ihk-frankfurt.de

Institut für Neue Technische Form e.V.
www.intef.de

IDZ Internationales Design Zentrum Berlin e.V.
www.idz.de

Markenverband e.V.
www.markenverband.de

VDID Verband Deutscher Industrie Designer e.V.
www.vdid.de

VDM Verband der Deutschen Möbelindustrie e.V.
www.wohninformation.de/german

VKI Verband der Keramischen Industrie e.V.
www.keramverband.de

ZVEI Zentralverband Elektrotechnik-
und Elektronikindustrie e.V.
www.zvei.de

Personenmitglieder – Personal members

Marcus Botsch
Christoph Burkardt
Prof. Michael Burke
Kai Ehlert
Gerdum Enders
Fritz Hahne
Prof. Achim Heine
Jörg Heithoff
Albrecht Hotz
Armin Illion
K. Michael Kühne
Michael Knuf
Stephan Koziol
Burkhardt Leitner
Lothar S. Leonhard
Prof. Hansjerg Maier-Aichen
Reiner Moll
Nils Holger Moormann
Thomas Nagel
Thomas Rempen
Michael Rösing
Christian Sieger
Prof. Erik Spiekermann
Philipp Thonet
Roland Wagner
Prof. Othmar Wickenheiser

Ehrenmitglied – Honorary member

Prof. Dr. h.c. Dieter Rams

(Stand: Januar 2007/Status: January, 2007)

Designpreis der Bundesrepublik Deutschland 2007 — Design Award of the Federal Republic of Germany 2007

www.designpreis.de

Rat für Formgebung / German Design Council — Ludwig-Erhard-Anlage 1 — D-60327 Frankfurt am Main — E-Mail: info@german-design-council.de — www.german-design-council.de **Herausgeber / Editor** — Rat für Formgebung / German Design Council **Konzept / Concept** Lutz Dietzold, Andrej Kupetz, Stephan Ott **Gestaltung / Design** buebery park, Armin Illion, Selters im Taunus, www.bueberypark.com — Raasch & Partner GmbH, Dreieich **Fotografie / Photography** Ingmar Kurth, www.ingmarkurth.com, Lutz Sternstein, LutzSternstein@web.de **Produktfotos und -texte / Photos of the Products and Product Information Texts** Beteiligte Unternehmen / By the participating companies **Fotos / Photos** Stash, retail design (S. 68) Arjen Schmitz, Die Neue Sammlung (S. 319) Rainer Viertlböck **Redaktion und Korrektur / Editorial Department and Correction** Stephan Ott, Frankfurt am Main, ott@reality-fiction.com **Übersetzung / Translation** Jeremy Gaines **Lithos / Lithography** Raasch & Partner GmbH, Dreieich **Druck / Printing** Fgb Freiburger Grafische Betriebe, Freiburg **Papier / Paper** ProfiSilk 150 und / and 135 g / qm, ein Produkt von / a product of SAPPI Fine Paper **Projektmanagement / Project Management** Katalog, Ausstellung, Internetseite / Catalogue, Exhibition, Website — Rat für Formgebung / German Design Council – Antonia Roth, Florian Stryj — Birkhäuser Verlag AG (Katalog / Catalogue) – Nadine Haisch **Produktion / Production** Birkhäuser Verlag AG – Marion Plassmann **Ausstellung / Exhibition** Konzept und Gestaltung / Concept and Design — Johannes Fuchs Product Design, Frankfurt am Main, www.johannes-fuchs.de **Website** Redaktion / Editorial — Rat für Formgebung / German Design Council – Antonia Roth — Stephan Ott, Frankfurt am Main, ott@reality-fiction.com **Technische Umsetzung / Technical Realisation** OScomputing AG, Frankfurt am Main, enter@oscomputing.de — Gestaltung / Design — buebery park, Armin Illion, Selters im Taunus, www.bueberypark.com

Impressum
Imprint

Bibliographic information published by Die Deutsche Bibliothek
Die Deutsche Bibliothek lists this publication in the Deutsche Nationalbibliografie;
detailed bibliographic data is available in the Internet at http://dnb.ddb.de

Birkhäuser Verlag AG, P.O. Box 133
4010 Basel, Switzerland.
Member of Springer Science + Business Media

Printed in Germany

ISBN-13: 978-3-7643-8092-2
ISBN-10: 3-7643-8092-6

9 8 7 6 5 4 3 2 1

http://www.birkhauser.ch